Astronomers' Universe

For further volumes:
http://www.springer.com/series/6960

Duncan Lunan

Incoming Asteroid!

What Could We Do About It?

 Springer

Duncan Lunan
Troon, Ayrshire, UK

ISSN 1614-659X
ISBN 978-1-4614-8748-7   ISBN 978-1-4614-8749-4 (eBook)
DOI 10.1007/978-1-4614-8749-4
Springer New York Heidelberg Dordrecht London

Library of Congress Control Number: 2013947790

Printed on acid-free paper

Springer is part of Springer Science+Business Media (www.springer.com)

To recently absent friends: Iain Banks, John Braithwaite, Philip Coppens, Archie Roy
All greatly missed

Preface

The major concept that underlies my published work was thrashed out between myself and my late friend John Braithwaite over the winter of 1968–1969, with some input from Charles Muir and others. This concept has been the core of almost all my published work since—all the nonfiction and almost all the fiction has been compatible with it, if not generated or influenced by it.

The key concept, as outlined by John Braithwaite, was for a "Politics of Survival"—a program that would ensure that at least one viable segment of human life and civilization would survive any imaginable catastrophe—inevitably requiring the preservation of other forms of life as well. John and I disagreed pretty strongly from the outset about how that might be brought about. In particular, although the concept had sprung from a discussion about asteroid and cometary impacts, John came to believe that threats from off-planet should be ignored, to focus on what he considered to be more immediate ones of human origin. By contrast I felt that we should start with those, partly because they were cross-party issues and partly because they were more immediate. In one such argument, he challenged me: "Seriously, which is the more immediate issue—global starvation or asteroid impacts?" To which I replied, "Put it this way – if something a mile across hit the Pacific twenty minutes ago, we aren't going to finish these cups of coffee."

Completing my own formulation was highly liberating, coming at the end of a decade in which the destruction of humanity had come to seem increasingly inevitable. But I quickly learned that the very concept was a long way ahead of its time. At the end of the 1960s, everyone I spoke to about it was convinced either that doom was inevitable or that there were no problems at all. Either way, there was no point in talking about solutions. So I gave up on talking about it and simply got on with it, within ASTRA, the

Association in Scotland to Research into Astronautics. ASTRA is the Scottish counterpart of the British Interplanetary Society (BIS) and was a Scottish branch of the BIS for ten years before becoming independent as ASTRA in 1963.

In the early 1970s, I ran discussion projects within ASTRA that became parts of my book *Man and the Stars* [1], and a second series that became *New Worlds for Old* [2] and *Man and the Planets* [3]. The projects were run on Politics of Survival guidelines; I simply never mentioned it, and everyone supposed I had forgotten it, even as they were helping me think it through in more and more detail. But in 1977, the first of a new series of meetings was scheduled for October 1, which made "October the First Is Not Too Late" the obvious title, paraphrasing Sir Fred Hoyle. ASTRA council members Bill Ramsay and Gavin Roberts, backed up by the late Prof. Oscar Schwiglhofer, the founder of the society, demanded that I come clean about the Politics of Survival—and to their amazement, they found that they knew it all already.

After years of my keeping it low-key, in 1986 the late George Hay approached me with a challenge to write it up for the journal *Science and Public Policy*, edited by his colleague Dr. Morris Goldsmith, where it appeared in February 1987 (see Appendix 1) [4]. My Slovenian friend, the editor Samo Resnik, arranged for it to be published in translation in a student journal called *Katedra* and actually beat *Science and Public Policy* into print by a month. I was asked to launch the concept publicly, in advance of publication, at the Environmental Sciences Society of the University of Stirling, but even though *Science and Public Policy* very kindly mailed out 20 copies, at their invitation, to people who I thought should be interested, there was no feedback at all.

In 2001, ASTRA council member Andy Nimmo proposed that we devote a special issue of ASTRA's journal *Asgard* to the Politics of Survival. After the events of September 11, 2001, it was agreed that we would set up a new discussion project on it, dedicating four issues of the journal to the results, edited by Jamie McLean and myself with Andy as consultant.

The first issue in March 2002 introduced the Politics of Survival, and the rest of it was devoted to the impact threat and ways to counter it. Partly that was for historical reasons, but also it was because in 1998 ASTRA had become affiliated with Spaceguard

UK, run by Jay Tate and subsequently based at the Spaceguard Observatory in Powys. Jay was a guest speaker at ASTRA meetings in the Airdrie Arts Centre and in the Central Hotel, Glasgow, on March 20 and 21, 1998. We had promised to devote one of ASTRA's publications to Spaceguard, and we now ran an introduction to it and a discussion of the impact threat by Jay Tate [5], followed by an updated reprint of a paper by past president Gordon Ross and myself on using the "Solaris" solar sail as a comet or asteroid deflector (see Chap. 5) [6].

Bill Ramsay, also a past president of ASTRA, then proposed that we set up a new project, or a subproject, to address the question, "What would we do if we knew there was to be an impact in ten years' time?" It led to a fascinating series of meetings, including a conference at the Spaceguard Observatory in October 2003 and a seminar in the Bridie Library of Glasgow University Union in 2012, and the results make up the content of this book.

Troon, Ayrshire, UK   Duncan Lunan

About the Author

Duncan Lunan is a graduate of Glasgow University in Scotland, an M.A. with Honors in English and Philosophy plus Physics, Astronomy, and French, and has a postgraduate Diploma in Education. He is a full-time author and speaker with an emphasis on astronomy, spaceflight, and science fiction. He has contributed to 24 books and published over 950 articles and 33 short stories. As manager of the Glasgow Parks Department Astronomy Project, 1978–1979, he designed and built the first astronomically aligned stone circle in Britain for over 3,000 years, as described in his book *The Stones and the Stars: Building Scotland's Newest Megalith* (Springer, 2012).

Duncan was a curator of the Airdrie Public Observatory for 18 years, and in 2006–2009 he ran an educational outreach project for schools from the observatory, funded by the National Lottery. His other interests include ancient and medieval history, jazz, folk music, and hill walking. After 30 years in Glasgow, he recently returned to his hometown of Troon, Ayrshire, where he lives with his wife Linda.

Works By the Same Author

Man and the Stars: Contact and Communication with Other Intelligence
New Worlds for Old: The New Look of the Solar System
Man and the Planets: The Resources of the Solar System
Children from the Sky: A Speculative Investigation of a Mediaeval Mystery, the Green Children of Woolpit
The Stones and the Stars: Building Scotland's Newest Megalith
Starfield, Science Fiction by Scottish Writers (edited)
With Time Comes Concord and Other Stories

Acknowledgments

Many people have contributed to this book, in different ways. The initial idea came from Bill Ramsay and was a spin-off of a publishing project by myself and James McLean (see Preface). Among the major participants were Dr. David Asher, Craig Binns, the late John Braithwaite and Dr. Arthur Hodkin, Andy Nimmo, Chris O'Kane, Gordon Ross, and Jay Tate of Spaceguard UK. Others who took part in the Glasgow discussions included Dr. Gregory Beekman, Ed Buckley, Kirsty Campbell, Alan Cayless, Graham Dale, Vince Docherty, Martin Esslemont, Willie Fleming, Cameron Forrester, David Galloway, Bob Graham, Mark Hayward, William Laurie, Robert Law, Keith Llewellyn, Linda Lunan, Jamie McLean, Lesley Monro, John Morrison, and John Stark. Guest speakers at Glasgow meetings included Prof. Colin McInnes, Prof. Ted Nield, Lembit Öpik MP, Jay Tate, and Dr. Max Vasile, while Prof. Richard Crowther supplied very valuable input from Oxford. Thanks are due to the late Prof. Archie Roy for the orbital calculations relating to nuclear waste disposal from low Earth orbit (Chaps. 7 and 8); to Prof. Angus McAllister of Paisley University for participating as guest speaker in the discussion on extraterrestrial resources quoted in Chap. 8; in particular to Prof. Colin McInnes and Dr. Max Vasile for sharing the results of their work, often in advance of publication; to Ed Buckley, Tom Campbell, Sydney Jordan, Andy Paterson, Gavin Roberts, and Gordon Ross for the use of their artwork; and to Chris O'Kane and Robert Law for photographs.

Like most people, until the 1960s I greatly underestimated the hazard from incoming asteroids and comets and greatly overestimated the size that would be needed to cause serious harm. The first revelations came in a two-part article in *Analog* by Ralph A. Hall ("Secondary Meteorites," January 1964); J. E. Enever's "Giant Meteor Impact" in *Analog*, March 1966; and in Isaac Asimov's article "The Rocks of Damocles" in *Fantasy & Science Fiction* the same month (full references in chapters below). I gladly repeat here my gratitude to the late Dr. Asimov, Prof. Carl Sagan, and Sir Arthur C. Clarke for the generous and ongoing permissions to quote from their works, which all three gave me in the early 1970s, a welcome boost for a new writer in the field. Any errors or misquotations in the text are entirely my own.

I should add here my long overdue thanks to my friend and colleague Jamie Bentley, who for nigh on 40 years has kept me supplied with science news and related topics from Australia, where the press often carries different stories or has a different take from Down Under, as well as bombarding me with ideas of his own, many of which I have used and not always been able to credit. The number of Australian sources cited in my work is testimony to his efforts. Similar thanks are due to Fraser Cain and his contributors to Universe Today (www.universetoday.com), an invaluable daily online compilation of astronomy and space news, which I have followed as assiduously as time allowed since 2000.

A key step in the book's evolution came in 1982, when David Langford commissioned and accepted an article from me on the threat from Comet Swift-Tuttle, for the short-lived and lamented magazine *Extro*. It would have been in issue No. 4 had the magazine survived the distributors' failure with No. 3. A subsequent essay competition entry in 1986 prompted Gordon Ross to devise the first version of the Solaris "Comet-Chaser" (Chap. 5), inspired by the development of the adaptive optics flexible mirror at the University of Strathclyde. That project was headed by Dr. Peter Waddell with the late John Braithwaite as a consultant. The deflector concept was strongly backed by the late Chris Boyce of the Glasgow *Herald*, but did not achieve publication until 1992, when Frances Brown accepted two letters on it for *Space Policy*. Dr. Stanley Schmidt then accepted a new version of the article that appeared in *Analog* in 1994; ASTRA published expanded versions in *Asgard* in 1995 and 2002, the latter reprinted by Dr. Benny Peiser that year in *Cambridge Conference Net*.

As its name implies, the British Rocketry Oral History Program (BROHP) was founded to record the memories of participants in the UK's missile and rocket programs of the 1950s–1970s.

Linked closely as they were to an independent nuclear deterrent, much of that work was top secret at the time, and in BROHP's annual conferences at Charterhouse School it was an extraordinary privilege to meet the people responsible for Skylark, Black Knight, Blue Streak, Blue Steel, Black Arrow, Prospero, and the SR-53, as well as many programs even less well known. With each year the net grew wider. The use of German hydrogen peroxide technology in the SR-53 rocket aircraft led naturally to the Me-163 and to Britain's planned attempt on the sound barrier with the M52, to the redeployment of key personnel to the Avro Arrow program in Canada, to their mass recruitment by NASA when that program ended and so led to their key roles in manned spaceflight, up to and including the space shuttle. But "year after year, old men disappear," and as venerable figures dropped from sight, the organizers widened the brief further to include current British space projects, renaming the event as the UK Space Conference. As early as 2001, a major session was devoted to the impact threat and to British involvement, with such key participants as Colin Hicks, Nigel Holloway, Benny Peiser, Duncan Steel, and Jay Tate. For all of this, thanks are due to the dedicated BROHP team of Dave and Lesley Wright, Kate Pyne, Nicholas Hill, and Roy Dommett. The conference continues annually under the auspices of the new UK space agency, but to judge by the 2013 program, the historical element is no longer emphasized.

I must also thank Leslie Banks for the outstanding series of IBM Heathrow Conferences, 1982–1990; the late Victor Hirt for the View from Earth 1984 conference in Big Bear Lake, California; the organizers of the Space Development Conference, Washington, and the Space Manufacturing Conference, Princeton, both in 1985; and the many other organizations whose lectures and conferences I have attended over the years, many of which I have cited here for the first time.

For permissions to use individual quotations and illustrations, my thanks to the Geological Society of Denmark and Greenland, The Planetary Society, Dr. Philip Lubin, Daniel Machacek, Glyn Maxwell, Prof. Dana Newman and Robin Williamson.

In 2003, the UK Spaceguard Center in Powys hosted the Spaceguard ASTRA Asteroid Mitigation Conference (SAAM), courtesy of Jay and Ann Tate.

Donations to Spaceguard UK's Project Drax (see page 122) can be made online at www.spaceguarduk.com, or by post or bank transfer – details are on the website.

Thanks are due to the Airdrie Arts Centre; the Central Hotel, Glasgow; the Glasgow Council for the Voluntary Sector; and to Father Griffiths of the Ogilvie Centre of St. Aloysius' Church, for providing venues for the additional meetings, and to Gemma Higgins for the 2012 seminar in the Bridie Library of Glasgow University Union. In chronological order, representing key stages in the generation of the present book over the last 3 years, I have to thank my wife Linda, John Watson, Maury Solomon, and the editorial and production team at Springer, for the parts they have played in making it a reality.

Contents

Part I
Is There a Danger?

1. Comets

So the star, with the pale moon in its wake, marched across the Pacific, trailed the thunder-storms like the hem of a robe, and the growing tidal wave that toiled behind it, frothing and eager, poured over island after island and swept them clear of men: until that wave came at last – in a blinding light and with the breath of a furnace, swift and terrible it came – a wall of water, fifty feet high, roaring hungrily, upon the long coasts of Asia, and swept inland across the plains of China. For a space the star, hotter now and larger and brighter than the sun in its strength, showed with pitiless brilliance the wide and populous country; towns and villages with their pagodas and trees, roads, wide cultivated fields, millions of sleepless people staring in helpless terror at the incandescent sky; and then, low and growing, came the murmur of the flood. And thus it was with millions of men that night – a flight no whither, with limbs heavy with heat and breath fierce and scant, and the flood like a wall swift and white behind. And then death.

—H. G. Wells, "The Star"

What H. G. Wells has envisaged here is not a collision with Earth but a stray planet from interstellar space colliding with Neptune with enough force to hurl the combined incandescent mass sunwards [1]. When he published "The Star" in 1897, the scientific view was that the impact of a comet would do no physical harm. His novel *In the Days of the Comet* (1906) brings a comet upon us with no actual impact, the gases mixing peacefully with Earth's atmosphere [2]. There is a mysterious constituent, revealed before the encounter by a green line in the comet's spectrum, but that's just a device by which to change the character of the human race and bring in a socialist utopia. Earth passed through the tail of Halley's Comet in 1910 without harmful effects, not even creating a utopia, and in 1913 "The Poison Belt," one of Sir Arthur Conan

D. Lunan, *Incoming Asteroid!: What Could We Do About It?*,
Astronomers' Universe, DOI 10.1007/978-1-4614-8749-4_1,
© Springer Science+Business Media, LLC 2014

Doyle's 'Professor Challenger' stories, imagined simply a cloud of gas in space, without a comet to generate it [3]. But the effects of an actual impact by a large comet or asteroid would be analogous to what Wells described, as we shall see.

The nearest approach to a major comet scare came when Comet Swift-Tuttle (the Great Comet of 1862) was expected to return in late 1982. The possibility that it might hit Earth made waves in amateur astronomy, professional astronomy, the media, the political sphere, and the military-industrial one; and it was interesting that the various groups seemed hardly to be talking to one another much of the time. The whole debate about the need to protect Earth was reopened.

Swift and Tuttle were both American astronomers of the mid-nineteenth century, and both discovered several comets that were named after them. Comet Tuttle references are generally to the Great Comet of 1858 and not to the faint comet 1862 I, which he discovered in January that year but which never grew bright enough to be seen by the naked eye. On July 2, J. F. Julius Schmidt in Athens found a comet 1862 ll that was named after him.

Lewis Swift discovered 1862 III, the Great Comet of that year, on July 15 [4]. At first it was so like Schmidt's 1862 II that Swift didn't realize it was different, until Tuttle independently found it and announced it 3 days later. The astronomical world split the honors, hence Comet Swift-Tuttle, but older texts often confusingly call it Comet Tuttle.

In August 1862 the comet put on a spectacular display in northern hemisphere skies, traveling past Polaris from Camelopardis and developing a tail 25° long. In the telescope it was seen to be throwing out luminous jets that looped around the more normal tail in a most unusual way [5, 6] (Figs. 1.1 and 1.2a, b). When the orbit of the comet was calculated, it was found to coincide with the Perseid meteor shower through which Earth passes every August; the biggest display takes place between August 12 and 14 [7] (Fig. 1.3). The link between comets and meteor showers was big news in the 1860s, and the apparent link with the Perseids, the best-known meteor shower of all, was enough to keep the comet in the literature for the rest of the nineteenth century, although the first two comets of the year were forgotten and it was called simply 'the comet' or 'the great comet' of 1862.

Fig. 1.1 Comet Swift-Tuttle, also called 1862 III (Copyright © Sydney Jordan, 1994, after a nineteenth century lantern-slide loaned by John Braithwaite) [5]

The Swiss astronomer Plantamour lectured on the link with the Perseid meteors in the winter of 1871–1872, mentioning that Earth would next encounter the meteors on August 12, 1872. Newspapers reported that Earth would collide with the actual comet on that date, causing considerable public alarm [8]. Nineteenth-century astronomers thought such fears were groundless. Comets had passed close to Mercury, and between the moons of Jupiter, without producing any noticeable perturbations; so their masses had to be low. And Earth had passed through the tail of the great comet of 1861 without any noticeable effect, so the tails had to be tenuous gases; and when the head of one passed in front of the star Arcturus without dimming it significantly, that seemed to clinch the matter—comets were entirely gaseous. In the twentieth century it came to be accepted that there must be *something*

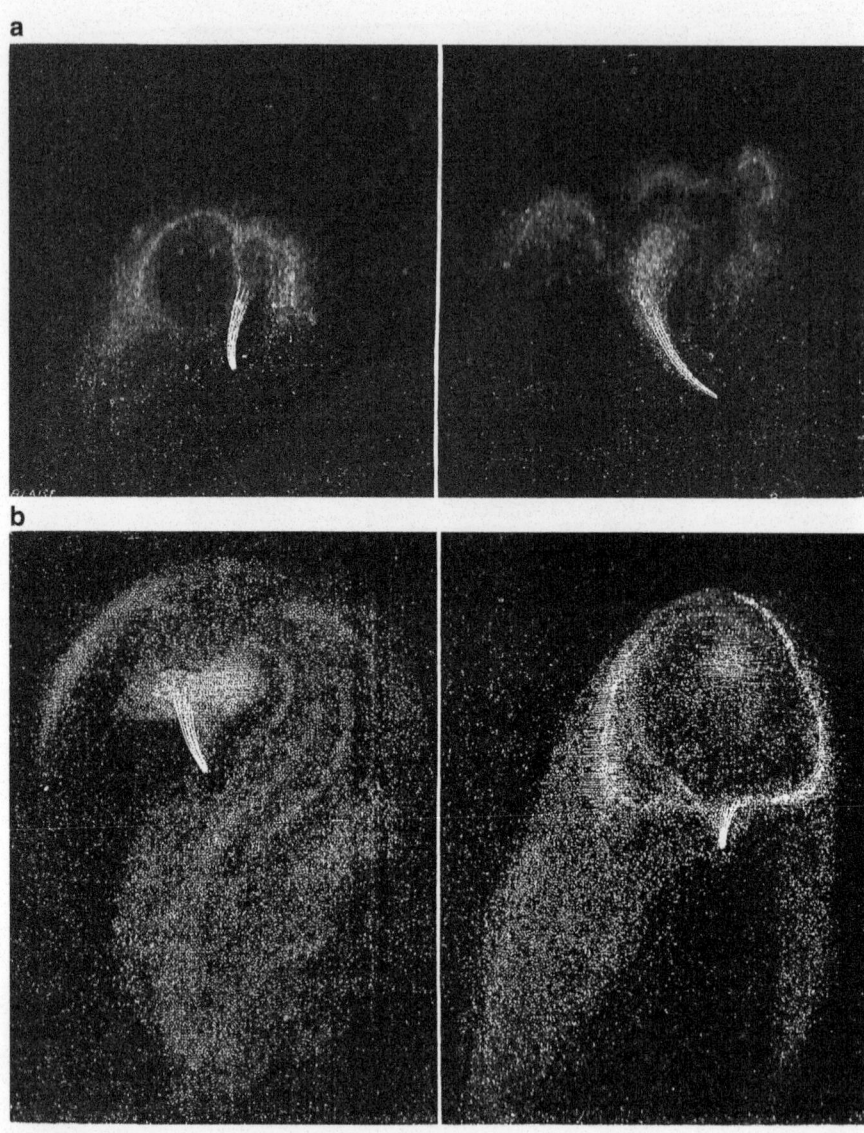

FIG. 1.2 (a, b) Observations of jets from Comet Swift-Tuttle (From Amédée Guillemin, *The Heavens*, 1871) [6]

inside the great heads of the comets, but whatever it was, no doubt it was too small and flimsy to be dangerous.

Science fiction writers, of course, ignore astronomers when it suits them (For example, many twentieth-century astronomers believed this was the only planetary system in our galaxy). In *Off on a Comet* and its sequel, Jules Verne described a comet made of

FIG. 1.3 Orbits of Earth and Perseid meteors (From F. Chambers, *A Handbook of Descriptive and Practical Astronomy*, 4th ed., 1889) [7]

solid gold telluride, so that it could knock lumps off Earth for the purposes of his story [9]. H. G. Wells stuck to prevailing theory for *In the Days of the Comet*, as above. In Arthur C. Clarke's short story "Into the Comet," first published in 1960, the spaceship *Challenger* (note the name) is able to penetrate the core of a comet because it's a loose cluster of dirty gray icebergs, giving off jets of methane and ammonia [10].

Clarke claimed to be quoting F. L. Whipple, but Whipple himself envisaged the 'dirty ice' as water and as a single mass typically 1–10 km in diameter [11]. This author's own story, based on his amateur observations of Comet Bennett in 1970, made the nucleus a single mass of ice and rock, surrounded by shoals of drifting ice, and Isaac Asimov said that story "pictures a Whipple-like comet with considerable accurate detail." [12] Within the head of Comet Bennett there was a bright blue patch surrounded by what appeared to be a diffraction pattern, and satellite observations showed the comet had a hydrogen halo 13 million km across, implying far more ice in the nucleus than the 'boulder' or 'sandbank' models would allow [13].

Lucifer's Hammer, the comet that hits Earth in the novel by Larry Niven and Jerry Pournelle, has a nucleus initially similar. Trying to envision the impact, the characters use the analogy of 'hot fudge sundae'—with the ice cream representing the 'foamy ice'—whose overall density was thought to be considerably less than water—with embedded crushed nuts representing the rocks. But there's so much vaporization as the comet rounds the Sun that what approaches Earth is a boulder field embedded in gas, much like Clarke's description [14]. Multiple impacts knock out civilization as we know it all over the world, and the Russians and the Chinese see it as the excuse for a nuclear war, so there's not much chance of a socialist utopia afterwards. All fiction writers use the models that best fits the needs of their plots.

At the end of March 1982, it was announced in the press that there was a chance Comet Swift-Tuttle would hit Earth in August that year [15]. The nucleus was thought then to be fairly small, and although the impact would be retrograde at 66°.26 inclination to the ecliptic, its blow could devastate a country if it fell on land. If it fell in the North Atlantic, the waves would break over the British Isles into the North Sea, rather than rolling across Europe to the Urals.

According to the *Telegraph*, there would be no danger unless the comet went through perihelion (its closest point to the Sun) on August 12, and the odds against that were given as two million to one. But as it says on the back cover of the novel, "The chances that LUCIFER'S HAMMER would hit Earth head-on were one in a million, then one in a thousand, then one in a hundred. And then…" [14].

The *Telegraph* gave as its source Dr. Brian Marsden, of the Harvard-Smithsonian Center for Astrophysics, whose preliminary work on the comet had been summarized by John Bortle of the W. R. Brooks Observatory in 1981. Marsden and Yeomans, of the Jet Propulsion Laboratory, had independently recalculated the orbit and predicted perihelion passage in June or September 1981 [16].

The problem was that the comet didn't seem to have any respectable antecedents. Giovanni Schiaparelli, better known for the first account of *canali* on Mars, calculated that the period of the orbit was 120 years and coincided with that of the meteors. Other nineteenth-century estimates were 121.5 years [17] and 142 years [18]. But on any of these, then even allowing for planetary perturbations, the historical record should contain bright August comets that were previous visits by Swift-Tuttle. Marsden didn't find such sightings. The best available correlation was with Kegler's Comet of 1737, which would be no threat to Earth in 1982. However, it would mean that the comet was being acted on by powerful nongravitational forces—most probably, that the jets coming off from the nucleus were generating thrust like rockets and altering its orbit. The comet could return to perihelion in November 1992, with a serious prospect of a collision one or two orbits later [19].

The Return of Swift-Tuttle

Look, where it comes again!

—*Hamlet*, Act I, Scene i

Bortle had never suggested a possible collision in 1982 [16], but although the comet didn't return that year, the amateur observers were now aroused. Meteor studies are one of the many areas of astronomy still dependent on amateurs for detailed, labor-intensive observations. From these are calculated the zenith hourly rates for each year's shower, and with sufficiently detailed records the interior structure of the shower can be mapped, sometimes identifying streams of meteors that have been emitted by the comet on previous passes around the Sun, but at least recognizing which parts of the overall stream contain more meteors than others. The Perseids' rates fell to 5 or 10 per hour in the 1920s, then rose slowly as the century went on. In the 1970s rates rose to 80 or more (the 1977 shower was well observed), and in 1980 there was a 50% increase. In 1981 and 1982 the rates appeared lower, but the Moon interfered with both. In 1983 there was no sharp rise, and if the 120-year period was roughly correct, the comet had probably passed on the far side of the Sun unobserved.

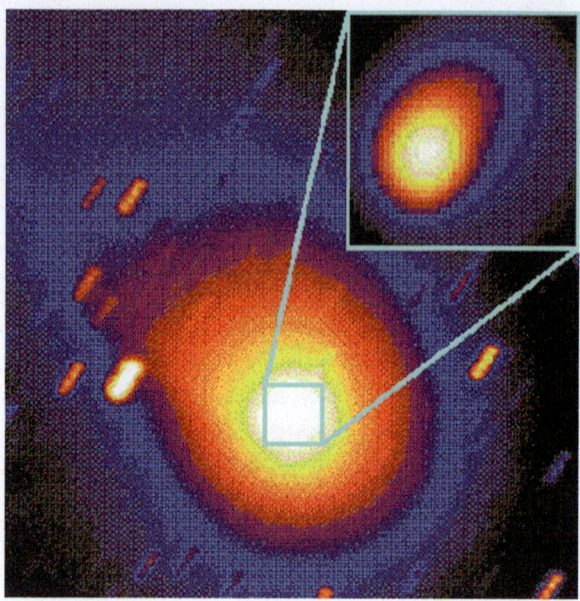

FIG. 1.4 Comet Swift-Tuttle, 1992 (NASA)

Marsden had made his 1992 prediction diffidently in 1973, and repeated it still more diffidently in July 1991. Just 12 days later, an unexpected peak in the Perseid shower suggested something was about to happen; it was repeated in 1992; and on September 26 the comet was recovered by Tsuruhiko Kiuchi of Japan (Fig. 1.4). Getting to grips with the inconsistencies in previous sightings of what *could* be the same comet, Marsden found himself facing the possibility of an impact in 2126. The comment appeared in the January 1993 *Sky & Telescope*, accompanied by a Don Davis cover showing the nucleus grazing Earth's atmosphere at the terminator, with a sea of fire below as the dust grains on parallel tracks met their end. Would the main body strike, or miss [19]?

If it proved necessary to try to deflect it from Earth, Marsden recommended intercepting the comet between the orbits of Saturn and Uranus in 2122. Because of Swift-Tuttle's high orbital inclination (113°), he assumed a rendezvous at the ascending node where the comet next crosses Earth's orbital plane [19]. Sending conventional spacecraft out of the ecliptic requires heavy fuel expenditure or gravitational slingshot. Low-thrust, continuous propulsion systems don't have that limitation, so the comet chasers discussed in Chap. 5 could do it, given time, or catch threatening comets on return to the Sun.

On November 3, 1992, Donald Yeomans ruled out non-gravitational effects on the comet, accounted for the sightings in 1737, 1862 and 1992, and ruled out a collision in 2126 [20]. Now that the orbit had been charted with such accuracy, it seemed that the net effect of the jets was nil. That's not to say that it will stay that way. Comet Encke's orbit has undergone multiple changes due to varying jet thrust since its discovery [21], and in 1990 there was an explosion on Comet Halley as it receded from the Sun [22].

On December 12, as the comet reached perihelion, Robert McNaught (see Chap. 4) said that there was no longer a danger either for 2126 or 2261 [23]. By March 24, 1993, two high-ranking officials of the SDI Office and the Air Force Space Command had refused to appear before the House Science, Space and Technology subcommittee, because the Pentagon "didn't want to see headlines that the Air Force was chasing space rocks. The subject looked like it had the potential for a high giggle factor when they are involved in so many larger issues." [24]

By late October 1992, the press was saying that the chances of an impact in 2126 were only one in 400—no problem, though some of the same journalists had considered one in two million to be dangerous 10 years before. Dr. Ken Russell, at the Anglo-Australian Telescope, was quoted as retorting, "A chance of one in 400 is not small when you are talking about the extinction of the human race," [23] which seemed a bit much for an object then estimated to be around 5 km across. If composed mostly of ice, it would have perhaps 1/40 the mass of the object that wiped out the dinosaurs. Yet now it seems Dr. Russell was right. More recent estimates of the diameter put it at 26 km [25], quite sufficient for what the film *Deep Impact* calls an 'Extinction Level Event'—see below.

Cometary Impacts

> A fearful star is the comet, and not easily appeased.
>
> —Pliny the Elder [33]

Both press reports of 1982 cited the Tunguska event of July 1908 as an example of what could happen. That explosion was in

FIG. 1.5 Trees felled by the Tunguska event

the multi-megaton range, flattened 80 million trees (Fig. 1.5) over an area the size of London, and threw enough dust into the upper atmosphere to produce spectacular sunsets for months. Deductions from records of the time suggest that the effect on the ozone layer on that latitude was comparable to a nuclear war. Yet no part of the object reached the ground. Trees remained standing at ground zero, though stripped of bark and branches, as they did at Hiroshima. There were no visible impact craters. The *Larousse Encyclopedia of Astronomy* says that a large number were formed, but that's a confusion with the Sikhote-Alin iron meteorite of 1947. Early reports referred to pits in the region, but these proved on excavation to be a common Arctic phenomenon, sink-holes in the permafrost.

In 2008 and 2009 the Space Settlers Society held a Tunguska centenary and 'Tunguska 101' seminars in Glasgow, organized by Andy Nimmo. One topic he raised was the recent suggestion that the elliptical Lake Cheko, 8 km from the blast center, might have been formed by a fragment of the incomer. Scientists from the University of Bologna suggested that its conical depth profile (exaggerated in Fig. 1.6a) could support this idea [26], but in reply scientists from Imperial College, London, pointed out that trees more than

FIG. 1.6 (a) Depth profile of Lake Cheko, exaggerated (© University of Bologna, 2001). (b) Undisturbed forest around Lake Cheko (© University of Bologna, 2001), http://www.-th.bo.infn.it/tunguska

a hundred years old were still standing around it (Fig. 1.6b), there was no ejecta field, and if it was formed by an impact, the incoming angle appeared to be too shallow [27]. If the lake was formed by a comet fragment or asteroid, presumably it's much older and its location is a coincidence. More recently it's been claimed that stones found in the Khushmo River might be fragments from the asteroid [28], if it was one, but they could be from any meteorite fall [29], and if there is an older impact feature in the area they might have been washed out from that.

The most recent thinking on it is that the object was a small comet that approached Earth unseen on the sunward side—a loosely structured object would fragment when the mass of atmosphere between it and the ground was equal to its own mass—and then supposedly all the ice evaporated and its kinetic energy was dumped into the atmosphere as heat, creating an explosion.

However, the Tunguska object was at least 60 m across, perhaps much larger, 3–6 km up and traveling at several km per second when it exploded. Could so much ice vaporize in a fraction of a second? The big thing about ice as a material is its recalcitrance. In the 1960s, the U. S. Coast Guard set out to destroy a small berg to see if they could keep hazards out of the shipping lanes. High explosives only blew chips off, and the incendiaries just glazed the surface without even altering the overall shape. The chastened Coast Guard had to admit that it wasn't really possible to destroy bergs in a hurry, even using nuclear weapons.

Harking back to those jets of gas from Comet Swift-Tuttle, Sir John Herschel reported only one [30], but Chacornac saw 13 jets over 17 days [31]. In 1978 Whipple announced a new analysis of comet rotations showing that in many cases only small areas of the nuclei were active. Computer modeling allowed the rotation rates to be determined—33 h in the case of Swift-Tuttle [32]—and Chacornac proved to be nearer the mark. The best fit with the observations, modeled by Zdenek Sekanina at the Jet Propulsion Laboratory, indicated seven active areas on the nucleus, none of them large [33]. The rotation periods for comets ranged from 4 h to 5 days, and implied that to hold together, the nuclei couldn't be loosely compacted but had to be frozen solid. For example the nucleus of Comet Giacobini-Zinner seems to be disc-shaped, with an equatorial radius eight times the polar one. If loosely structured, it would almost certainly come apart under solar heating, or atmosphere entry.

Until those astonishing tiles were invented for the space shuttle the only way to protect an incoming vehicle was by ablation—the surface layer of the heat shield vaporizing and carrying away the heat it has absorbed while the shockwave protects the material behind. Some materials achieve it naturally (Chinese re-entry capsules have shields of peanut fiber, and it turns out that Soviet ones were wooden all along). The tektite class of secondary meteorites have been ablated into aerodynamic shapes; larger stones are often covered with ice after they fall, even if red-hot at first, proving that the interiors are still at very low temperatures. The coast guard was right: ice would ablate in fireball conditions. Arthur Kantrowitz and others have envisaged ice-filled rockets, energized by laser or electron beams from the ground [34]. The same physics apply to an ice mass coming down through the atmosphere—even if it did break up when the sonic boom became trapped between it and the surface, if it remained intact down to 3 km it would fragment into pieces much too large to vaporize before they struck the ground below.

In 1976 an international program to photograph bright fireball meteors revealed that most are much bigger than had been thought, yet very fragile, disintegrating high above the ground [35]. The largest one photographed was estimated to mass 200 metric tons, yet be so fragile that it would have crumbled under Earth-surface gravity. In his book *Messages from the Stars* Ian Ridpath suggested very plausibly that the Tunguska object was similar in composition—possibly a comet nucleus from which all the ice had been driven off [36].

This could be the nature of many of the so-called Earth-grazing asteroids. Just 38 were known as of 1978, and more than 4,700 by 2012 [37], rising to 10,000 in 2013 [38]. All of them are liable to hit Earth, Moon or Venus within the next 100 million years [39]. If many of them are fragile objects, less threatening than their 1–10 km diameters would imply, then the situation might not be too bad. On the other hand, as ice sublimes off, the comets may form a protective crust of dust that can close over altogether, shielding a substantial mass of ice inside [40]. It seems that Encke's Comet may have been inactive until 1786, although it had been in its present short-period orbit since much earlier [41]. Dead comets have now been identified in the asteroid belt, and the near-Earth 'asteroid' Don Quixote has turned out to be one, so Earth-grazers masquerading as fluffy dust balls may actually be a great deal more dangerous than they seem. Either way, an airburst such as the Tunguska one in 1908 might have caused the conflagrations in New Zealand that wiped out the flightless Moa, contemporaneous with the first human settlement and the second wave on Easter Island c. AD 1200 [42].

The Nature of Comets

ISTI MIRANT STELLAM – these men are wondering at the star.

—Halley's Comet caption, from the Bayeux Tapestry

In 1680, 1843, 1880 and 1882 there were comets that passed extremely close to the surface of the Sun, and in 1888 Heinrich Kreutz showed that the last three had very similar orbits, suggesting that they were pieces of a larger comet that had broken up, possibly in 1106 (a great comet recorded in medieval chronicles) or one seen to break up by Ephorus in 371–372 BC. But when the Solar and Heliospheric Observatory (SOHO) replaced ISEE-3 at Earth-Sun L1 point in 1995, its cameras revealed that Kreutz comets are currently grazing the Sun on a near-daily basis (Fig. 1.7) [41].

FIG. 1.7 A 1995 SOHO coronagraph image of Sun-grazing comet (ESA)

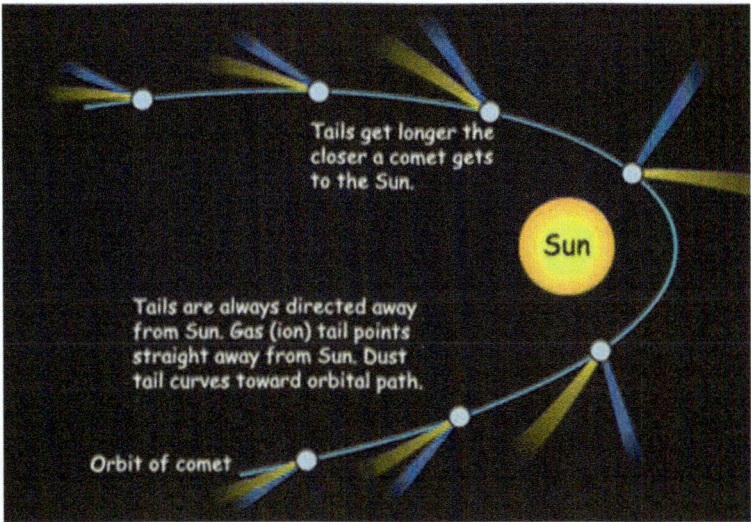

Tails get longer the closer a comet gets to the Sun.

Sun

Tails are always directed away from Sun. Gas (ion) tail points straight away from Sun. Dust tail curves toward orbital path.

Orbit of comet

FIG. 1.8 Comet pass around the Sun (NASA)

Almost all of them evaporate during perihelion passage, but Comet Lovejoy proved a major exception in 2011, providing spectacular views afterwards from Australia and from the International Space Station. It's entirely possible that more families of Kreutz comets will sweep past the Sun in future decades or centuries; at present SOHO or other solar probes might give us some warning if any of them heads towards Earth, but the spacecraft are not immortal. SOHO was dramatically saved after a major breakdown of attitude control in 1998, and it's been living on borrowed time ever since.

One common misconception is that tails follow comets through space. In fact, they always point away from the Sun, so comets draw away from the Sun moving tail-first, as in Fig. 1.8. But it's also not true that comets' tails are pushed away from the Sun by the solar wind. Although that outflow from the 'coronal holes' in the Sun's outer atmosphere has a major effect on Earth's magnetic field, the physical pressure it exerts on gas molecules, or on solar sails, is much less than the pressure of sunlight.

Most meteors are only specks of dust or small pebbles, and some of them at least come from collisions in the Asteroid Belt. But each time a comet passes close to the Sun, dust particles are driven off, sometimes forming a dust tail that separates from the gaseous one because dust grains move more slowly under sunlight

FIG. 1.9 Comet Hale-Bopp by Algol in Perseus (© Chris O'Kane, 1997)

pressure than gaseous ions. Comet Hale-Bopp, so prominent in 1997, had a strong dust tail (Fig. 1.9). As each dust particle has a slightly different velocity from the comet's, a band of dust forms along the comet's orbit. When Earth crosses the band we have a meteor shower, in which perspective makes the parallel tracks seem to come from a point in the sky termed the radiant, named after the constellation which holds it—hence the Perseid meteors above, in August, for example.

It's believed that the Solar System has a vast retinue of comets, occupying the so-called Oort Cloud at distances out to 2 light-years from the Sun. Dutch astronomer Jan Oort believed that these comets were debris from the breakup of the planet that formed the asteroids, but we now know that no such planet ever existed. Most astronomers believe that the cloud has been with us since the origin of the Solar System, though there are skeptics who insist that if so it would long since have been disrupted by passing stars (see below); however, various means have been suggested whereby the Sun can capture comets from interstellar space or from denser interstellar clouds. Isotope analyses from Halley's Comet may support those theorists, indicating a carbon-12/carbon-13 ratio very different from ratios found in the rest of the Solar System.

Wherever comets come from, every year a number of them with very long orbital periods swing past the Sun and recede again on long elliptical orbits. Some, however, pass close to major planets and have their speed reduced (Others are expelled from the Solar System altogether). These form 'families' of short-period comets. Halley's is one of the few members of Neptune's family. If it was one of Jupiter's many hostages, coming back every 3–5 years, it would have lost its ices and become fainter much more quickly.

The Oort Cloud of comets appears to be a sphere up to 2 light-years in radius (1.9 light-years out in the direction of Alpha Centauri), ranging from 13 billion km from the Sun out to at least 135 billion, and containing untold numbers of icy, dusty bodies whose origins and detailed compositions remain mysterious. Within the Oort Cloud the Kuiper Belt circles the Solar System beyond the orbit of Neptune and is made up of such objects as Pluto and Charon, Sedna, Eris and Dysnomia. This was the source of Kohoutek's Comet in 1973 and is probably also the source of the centaurs Chiron and Hidalgo, which are in erratic orbits among the outer planets, and of Phoebe, the captured outer moon of Saturn. We don't know if the Kuiper Belt has an outer edge or if it merges into the Oort Cloud, which extends to the gravitational limit of the Solar System (Fig. 1.10).

Astronomers use Earth's mean distance from the Sun as a convenient unit for discussing planetary and interplanetary orbits. The astronomical unit thus has a value of approximately 93 million miles. By definition, Earth's distance from the Sun is 1 au. Mercury's is 0.387 au, Venus's is 0.733, Mars's is 1.524, and Jupiter's is 5.2. The three main bands of the Asteroid Belt lie between 2.1 and 3.3 au, and the Kuiper Belt extends from 30 to at least 50 au (see Fig. 1.10 above). A substantial number of planetoids have now been discovered out there, leading to Pluto's reclassification as an asteroid rather than a planet.

At least we can be reasonably certain that the comet-like objects in the Kuiper Belt are left over from the formation of the Solar System. Indeed, F. L. Whipple suggested that the outer planets Uranus and Neptune had formed by accretion of comets, 800 million years after the rest of the Solar System had taken shape [11]. The event may even have been triggered by a passing brown dwarf star, before the cluster in which the Solar System formed was broken up [43].

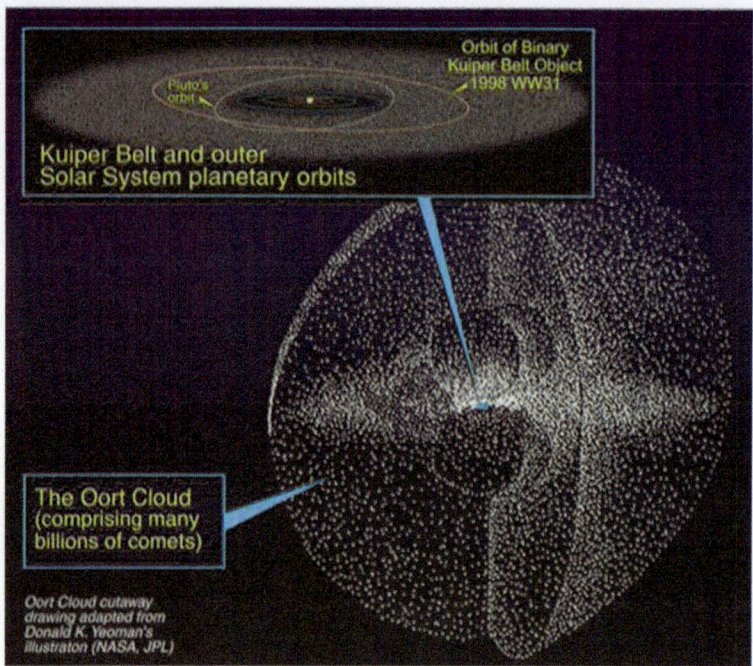

Fɪɢ. 1.10 The Kuiper Belt and Oort Cloud (NASA/JPL)

During the history of the Solar System, other stars may have passed within three quarters of a light-year of the Sun every 11 million years, on average [44]. The Soviet astronomer S. K. Vsekhs-vyatskiy calculated that as many as 10,000 stars may have passed within 0.6 parsecs of the Sun, during its history, and as they would have grazed the fringe of the Oort Cloud of comets at that distance, it would have completely disrupted it unless it was frequently or continually replenished [45]. The next star to enter the Oort Cloud is expected to be the red dwarf Gliese 710, 1.4 million years from now [46]. It's hard to see how our Kuiper Belt and Oort Cloud of comets could have remained stable during all that activity, unless we have exchanged comets with passing stars, or picked up new ones as we passed through the central plane of the galaxy [47].

The first close spacecraft encounter with a comet was Europe's Giotto in 1986, passing the nucleus at 500 km, backed up at greater distance by two Japanese and two Russian probes, and up-Sun by ICE, the International Cometary Explorer. Only Giotto

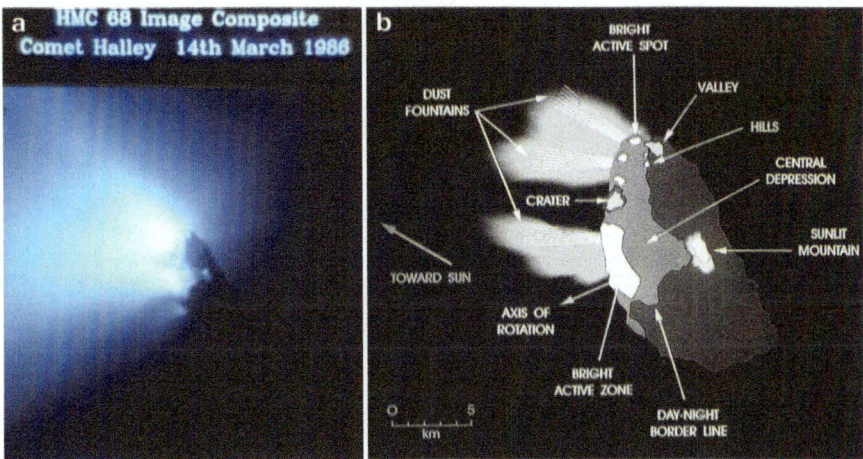

FIG. 1.11 (a) Halley's Comet nucleus, 1995 (ESA). (b) Map of Halley's Comet nucleus (ESA)

was armored, with a thick shield and an outer 'meteor bumper,' of the type suggested by Whipple in the 1950s for manned space stations. However, that was believed to have been completely destroyed in the hundreds of dust impacts sustained.

Although Sir Fred Hoyle and Prof. Chandra Wickramasinghe had predicted that the nucleus would be dark, covered in tar-like organic compounds, the general belief was that the nucleus would be icy, and Giotto's camera had been programmed to track the brightest object in the field of view. As a result it locked on to one of two gaseous jets. The nucleus turned out to be very black indeed and shaped rather like a peanut, with one circular feature that didn't appear to be an impact crater (Fig. 1.11a, b). Similar ones on Comet Wild 2's nucleus are definitely pits eroded by escaping gases, and the pits on Halley's Comet turned out to be about a kilometer across and several hundred m deep, possibly tunneling into the interior like Swiss cheese [48].

The dark nucleus was larger than expected, 15×8 km, and in a big surprise it was at 330° K, almost at room temperature at that distance from the Sun, so the crust must have had excellent insulating properties to prevent massive loss of the ice below [48]. Speaking in Glasgow, Dr. John Mason compared it to 'a cosmic choc-ice,' although less tasty methyl cyanide and hydrogen cyanide ices had also been detected. The percentage of hydrocarbon compounds

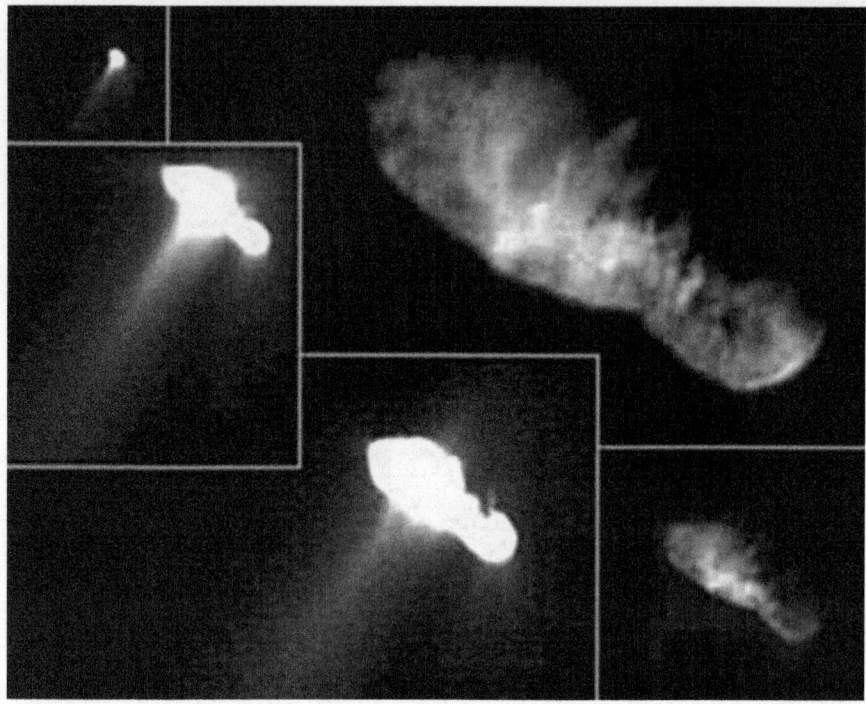

Fig. 1.12 Deep Space 1 ion-drive probe encounters Comet Borelly (NASA)

on the surface was ten times that of carbonaceous chondrite meteorites [49], and the depth of the crust suggested that the comet might have made as many as 3,000 previous passes around the Sun, entering the Solar System 200,000 years ago. With 200 million metric tons of material being lost on each pass, it could have 3,000 more to go [48]. No rocky core was detected, but with a total mass of 150 billion metric tons, its density of 0.2–0.7 g/c.c. was a mystery—too low for snow or ice [49], and probably indicating large internal voids.

In September 2001 the ion-drive mission Deep Space 1 passed Comet Borelly (Fig. 1.12), believed to be from the Kuiper Belt. Like Halley's the nucleus proved to be dark, this time shaped more like a bowling pin 8 km in length, with multiple small jets coming mostly from one side of it. The surface was extremely rugged, but smoother towards the middle of the elongated shape. Little water or hydrated compounds could be detected at the surface, perhaps suggesting that the crust is old and its volatiles have gone.

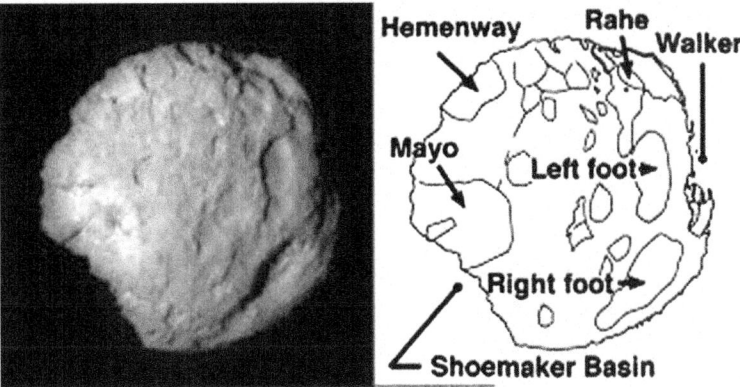

FIG. 1.13 Comet Wild 2 nucleus composite, 2004 (NASA)

In 2004 the Stardust probe visited Comet Wild 2, taking samples for return to Earth. The nucleus looked quite unlike Borelly's, a single rounded, elongated body, with multiple pits or craters, the larger ones flat-bottomed, and at least ten small but active jets (Fig. 1.13). The samples contained a wide range of organic compounds, including nitrogenous ones, but at first no hydrous silicates or carbonates containing water. Later iron and sulfur compounds were found that had formed in the presence of liquid water, while oxygen isotopes revealed that the mixture of rocky material had formed in different parts of the Solar System, some of it close to the Sun. How it got there, if it's from our Sun at all, remains a mystery.

In 2005 on the Fourth of July, appropriately enough, the comet Tempel 1 was struck at 37,000 mph by a copper projectile massing one third of a ton, released from the passing Deep Impact probe. Unlike any of the other comets imaged hitherto, this one is just a lumpy mass, with no obvious shape—strewn with craters and pits, and with very small jets, but with one strangely smooth area (Fig. 1.14), still present when the comet was revisited in 2011.

The explosion was equivalent to 4.8 metric tons of TNT, and the crater was expected to be seven stories deep and the size of a football field. After the flybys of comets Halley, Borelly and Wild 2, this should have been our first chance to study the interior

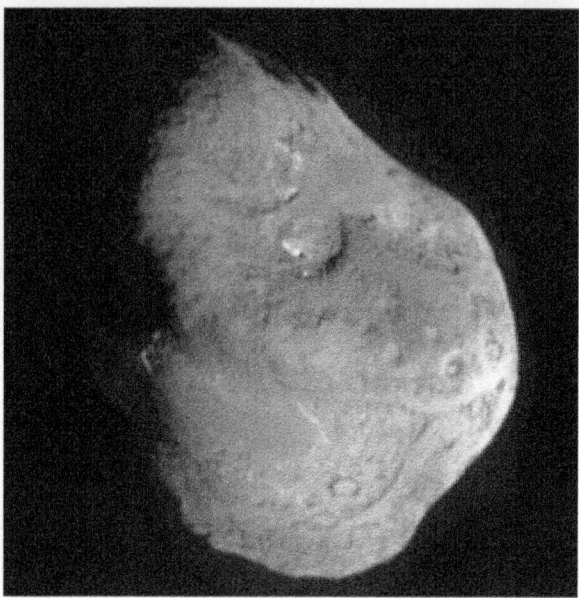

FIG. 1.14 Nucleus of Comet Tempel 1 (NASA)

of a comet. It should have grown brighter dramatically in binocu-
lars, perhaps even becoming visible to the naked eye and possibly fragmenting into two or more pieces. Instead, a huge plume
of water vapor and dust particles arose, with the consistency of
talcum powder, completely concealing the crater. It appears that
the plume consisted of wet clay, which was nothing like what was
expected. Because the dust cloud had prevented a view into the
interior of the comet, the Stardust mission was redirected to pass
Tempel 1 in February 2011. The impact crater was identified but
turned out to be insignificant. Density measurements suggested
huge voids inside the nucleus, with perhaps 75% of its volume
just empty space. It's not clear whether they've been formed by
outgassing, or whether the nucleus is just a loose cluster of icebergs below the dusty crust and clay.

Deep Impact was then retargeted to pass Comet Hartley 2 in
November 2010. The nucleus proved to be another big surprise,
a little like Comet Borelly at first glance, but with a definite
dumbbell shape and a smooth bar in between (Fig. 1.15). Similar features have been seen on the asteroids Eros and Itokawa
(Chap. 2), and on smaller moons of Saturn such as Calypso,

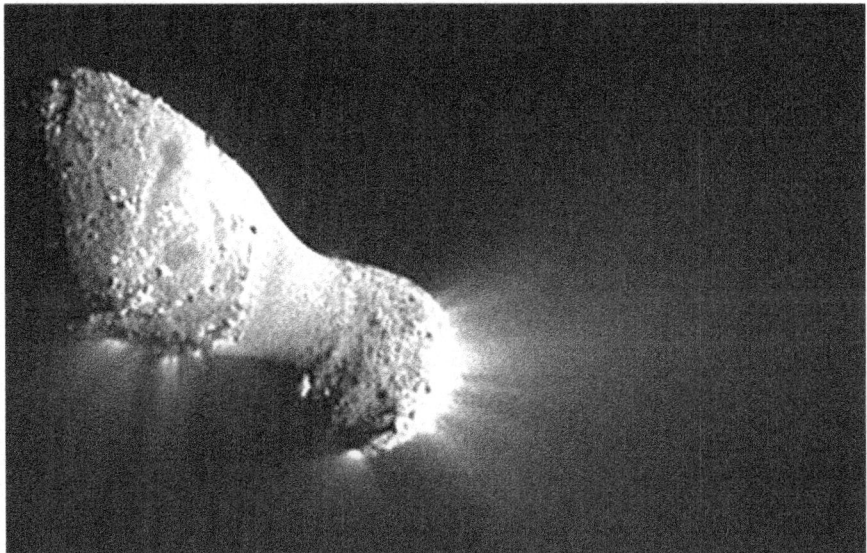

Fɪɢ. 1.15 Comet Hartley 2, EPOXI-Deep Impact 2010 (NASA)

Methone and Atlas. The former are thought to be due to dust grains migrating toward the asteroid's center of mass, the latter to ice stretching under tidal stresses. But it's not clear why Tempel 1 should have a similar smooth area when its shape, though lumpy, is more nearly spherical.

Carbon dioxide, dust and ice proved to be coming off from the same small locations on the Hartley 2 nucleus, whereas water vapor was coming from below the mysterious 'waist,' where the ice being heated by the Sun still lay below the surface. Even more remarkably, ice was coming off in the form of large numbers of fluffy snowballs, from golf ball to basketball size, propelled by carbon dioxide. Similar releases had been looked for, but not found, in the encounter with the much larger Tempel 1 [50].

But the truly remarkable result was that Comet Hartley 2's water emissions had the same isotope ratios of deuterium to ordinary hydrogen as in Earth's oceans. Comets previously studied had up to twice as much deuterium, but they all came from the Oort Cloud, whereas Hartley 2 was from the Kuiper Belt [51]. It revived the idea that Earth had gained its oceans from impactors, though perhaps from the Kuiper Belt rather than the Oort Cloud.

From the space missions to date, all we can say at the moment is that no two of the comets photographed seem to be the same. They can't even be grouped by shape, much less by composition or internal structure. If the first rule in protecting oneself is 'know your enemy', what we know so far is confusing indeed.

The Effects of Impacts

If it's as bad as you say, Mr. Öpik, perhaps you should be talking to the archbishop of Canterbury.

—Department of Trade & Industry spokesman to Lembit Öpik, MP

Edmund Halley applied Sir Isaac Newton's work on gravitation to observations of the comet of 1680, and realised the similarities in the orbital elements of comets seen in 1531, 1607 and 1682. If they were the same comet, he predicted that it would return in 1758–1759, "wherefore if it should return according to our prediction about the year 1758, impartial posterity will not refuse to acknowledge that this was first discovered by an Englishman." And indeed it gained him the recognition that he hoped for, with the comet named after him although he wasn't its discoverer.

Halley's Comet is in retrograde orbit around the Sun, against the motion of the Earth and the other planets. To achieve a rendezvous with it, a conventional rocket vehicle would have had to launch in 1978 and make a very tight swing around Jupiter [52]. Halley's determination of 'his' comet's orbit reveals not just the potential difficulties of intercepting a comet but also the nature of the threat. The damage caused by an impact is proportional to the kinetic energy released as heat, and that energy is given by the equation $K. E. = \frac{1}{2} m v^2$, where m is the mass of the projectile and v is its velocity. For a given diameter, the mass of an icy body like a comet nucleus is approximately one-tenth that of a similar one composed of nickel-iron, like the asteroids that formed the Barringer crater in the United States or the Vredevoort Ring in Africa (see Chap. 2), with intermediate values for different varieties of rock.

The velocity term involves not just the speed but the direction of motion. Earth's orbital velocity around the Sun is approximately

18.5 miles per second, and the majority of Earth-grazing asteroids are in direct orbit around the Sun, fairly close to the plane of the ecliptic. Those with orbits similar to Earth's will therefore come in with low relative velocities, whether they overtake Earth or are overtaken by it; but even they will gain seven miles per second in the final plunge inward from the distance of the Moon, equal to Earth's escape velocity.

The greater the orbital eccentricity and/or the orbital inclination of the impactor, the greater the incoming velocity of the object relative to Earth will be, up to a maximum of around 25 miles per second. Short-period comets, captured within the inner Solar System by encounters with Jupiter, will have a similar range of velocities but will pack only one-tenth of the punch. However, the Solar System's escape velocity, at this distance from the Sun, is approximately 20 miles per second. Comets from the Oort Cloud, the Kuiper Belt or from the comet 'families' of the outer planets can come from any direction, even from the opposite direction, such as Halley's, and could hit at anything up to 65 miles per second. Since the impact energy is proportional to the *square* of the relative velocity, and an incoming comet could easily have three times the velocity of a nickel-iron asteroid, it could release similar impact energies and that's why we're still unsure whether the dinosaurs were finished by an asteroid or a comet.

Actual comet impacts, in the last few thousand years, have still to be identified with certainty in the historical record, but some of these may have been considerably more destructive than the Tunguska-type ones. It seems to have been Laplace, in *The System of the World* (1799), who first recognized the likely effects of impacts upon Earth, but subsequently a view developed which ruled out catastrophic events, including impact events, in the history of the Earth. The author, for one, grew up with the belief that Earth's atmosphere stopped almost everything thrown at it. Until the 1960s most people greatly underestimated the hazard from incoming asteroids and comets, and greatly overestimated the size that would be needed to cause serious harm. An early revelation came in Isaac Asimov's article "The Rocks of Damocles" (March 1966) [53]. Asimov demonstrated that a 'city-buster' asteroid could be as small in mass as a few dozen tons, about the size of a large desk if composed of nickel-iron. A two-part article in *Analog* by

Ralph A. Hall had previously demonstrated that a similar object a mile across would sterilize a continent. Even the X-rays from the plasma around the falling object would be lethal. Impact craters are generally around 10–20 times the diameter of the impactor; but even though the crater would punch through Earth's crust to the magma, much of the energy released would be radiated into space as the hole filled up [54].

However, worse things happen at sea. Tsunamis up to three miles high would radiate away from the impact, and though they'd become lower within a 100 miles, they'd rear up again in coastal waters at least to the 1,000-ft height seen in the film *Deep Impact*. The pressure of the sea water outside would push the half-molten ring wall into the crater, and eventually it would quench the rising magma, but not before huge amounts of water vapor, solid matter and energy had been released into the atmosphere, creating a storm that would cover at least an entire hemisphere while dust blanketed the whole planet. J. E. Enever's article "Giant Meteor Impact" (*Analog*, March 1966) had a cover by Chesley Bonestell showing an airliner caught in the winds drawing it into the firestorm [55]. When Dr. Asimov did a new edition of "The Rocks of Damocles," the Bonestell cover was a counterpart showing a cruise liner on fire, borne on the crest of a tsunami 30 miles from the rim, with the air blast following and molten ejecta plunging around it [56].

It took a long time to appreciate the danger of big impacts because Earth's surface is so dynamic. The effects of weathering, continental drift and earthquakes is to conceal the crater's impact origin. Gosse's Bluff in Australia (Fig. 1.16), a large impact ring formed 142 million years ago near Alice Springs, has been so smoothed out over the last 142–143 million years that it was first spotted from orbit [57], in the 1960s Gemini program, and many more examples have been found since. The Manicougan Ring in Canada is one of the largest (Fig. 1.17), caused by a 3-mile impactor 215.5 million years ago and seemingly one of a chain, probably from a cometary breakup, but still larger features such as Hudson's Bay and the Gulf of Mexico have come under suspicion.

Drs. Victor Clube and Bill Napier have suggested that around 3000 BC a 'super-comet' broke up in the inner Solar System, filling the sky with smaller comets and meteor showers, which

FIG. 1.16 Gosse's Bluff crater, central Australia (NASA)

stimulated the dramatic large-scale astronomically aligned structures all over the world in the next 1,000 years—perhaps in attempts to get advance warning of impact events and their environmental consequences. In their view the present population of short-term comets and meteor showers is larger even today than can be explained by capture and break-up of smaller comets from the Oort Cloud on the fringes of the Solar System [47]. Dr. Duncan Steel has further suggested that so much dust was released in the break-up that the two cones of the zodiacal light might extend up the sky to join at the gegenschein, so making the ecliptic visible as a glowing arc across the night sky [57]. Present-day Earth-grazing asteroids and meteor showers may be remnants of the

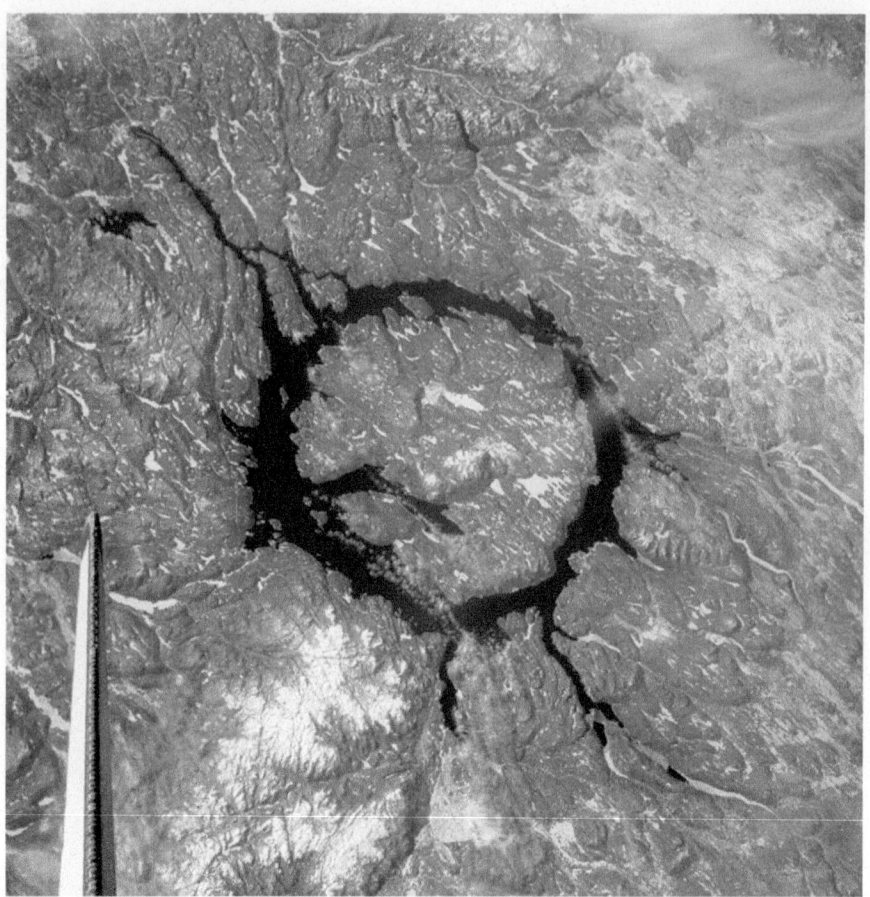

Fig. 1.17 Manicougan Ring in Canada from STS-9 (NASA)

object, including the Tunguska object and the Beta Taurid meteors, in June during daylight.

If it was large enough to be differentiated by internal heating, fragments from its core could have struck Earth as stony or even metallic meteorites. The Henbury craters in Australia were formed around 2700 BC [58]. The original flood, in Sumerian mythology, could have been generated by one of those impacts, in the Persian Gulf. In 2354–45 BC there was an abrupt cooling in global climate, and the oldest and most detailed accounts of the biblical flood, which appear to describe an impact and tsunami, date from c. 2250 BC [59]—see Chap. 2. Tree ring and related data from Ireland show massive rainfall, 20 years of near-zero plant growth, the collapse of Neolithic farming and a population crash c. 2350 BC. Remarkably the *Annals of the Four Masters* relate that Ireland was a waste-land for 30 years up to 2344 BC , and Archbishop Ussher's biblical chronology put the date of biblical deluge at 2349 BC [61], though it's probably a coincidence because other Bible studies then put it much further back.

In Psalm 18, after David calls to God for help:

> Then Earth shook and trembled; the foundations also of the hills moved and were shaken, because he was wroth. There went up a smoke out of his nostrils, and fire out of his mouth devoured: coals were kindled by it. He bowed the heavens also, and came down: and darkness was under his feet. And he rode upon a cherub, and did fly: yea, he did fly upon the wings of the wind. He made darkness his secret place; his pavilion round about him were dark waters and thick clouds of the skies. At the brightness that was before him his thick clouds passed, hail stones and coals of fire. The Lord also thundered in the heavens, and the Highest gave his voice: hail stones and coals of fire. Yea, he sent out his arrows, and scattered them; and he shot out lightnings, and discomfited them. Then the channels of waters were seen, and the foundations of the world were discovered at thy rebuke, O Lord, at the blast of the breath from thy nostrils.

We've seen episodes like those in the films *Deep Impact* and *Armageddon* much more recently, and Professor Mike Baillie suggests in his book *Exodus to Arthur* that similar imagery in Psalm 74 and the book of Isaiah relates to impact events [59]. The Psalmist might have called on the pre-biblical Sumerian, Hittite and Egyptian versions of the flood, as above, which have all the characteristics of an impact—probably in the Tigris marshes around 2300 BC (Chap. 2). But the same imagery is found in 2 Samuel 22 and is thought to be contemporary with I Chronicles 21:16: "When David looked up and saw the angel of the Lord standing between earth and heaven, and in his hand a drawn sword stretched out over Jerusalem, he and the elders, clothed in sackcloth, fell prostrate to the ground"—very probably a comet recorded in Chinese annals for the 970s or 960s BC.

Starting in around AD 534, and continuing at least until 550, there was a major cooling in global climate, followed by famine, plagues, mass migration and war in Mesopotamia, China and the British Isles. The Chinese imperial capital of Loyang was abandoned, with political collapse in northern China and economic chaos in the south [62]. Droughts and crop failures triggered Mongol migrations westward [63], the collapse of the Mesoamerican civilization at Teotihuacan was contemporary with the rise of the Muslim religion in the Middle East, and the rise of the Huari empire in Peru, opportunistically built on climate change [64]. In 540 a new bubonic plague appeared in Antioch and Constantinople [65], spreading through Pelusium and Alexandria in 541 to central Africa [66]. Possibly worse than the Black Death [67], but now thought to be an earlier version of it, in Constantinople in 542 the so-called Justinian Plague was killing 10,000 per day [65, 68], one-third to half of the population. It reached Britain in 544 [59, 69], by which time it is estimated to have killed 25% of the population south of the Alps. By 547 it had swept through the British kingdom, killing King Maclgwyn and leaving the Anglo-Saxons dominant [70].

The climatic precursors of those events were equally dramatic. After the 'fire across England' in 534, "the Sun became dim, 'its darkness lasted for 18 months [62, 71]." Starting in 536 there was diminution of sunlight for 14 months [72]. There was 'dry fog' or a 'dust veil' in the Mediterranean [67], while Byzantine records [73] describe a cold summer, a blue sun, and obscuration of the night sky in which even Canopus was hard to see [59]. In Britain severe cold set in during 536, with a brief recovery 537–540, but renewed in 541 with lasting effects to 550 [74, 75]. There was a marked narrowing of tree rings worldwide [66], with growth halted in some places [59], while in China there was snow in August 537 at Qingzhou, and four provinces were hit by famine, with 80% deaths in some places [68].

Many of those events could be explained by a volcanic super-eruption, possibly of Krakatoa. But the correlation is uncertain, and the evidence of microscopic diamonds in tree rings, volcanic deposits and ice cores suggests a major impact event [62]. There are multiple reports of one or more comets in 537–540 [59, 76], with major meteor showers [77], but the most telling is the account of a comet in the west [78], "as if the sky was aflame [76]," identified with Lugh Lamh Fada, the Celtic sky-god manifested as Lug of the Long Arm—due to his skill as an artificer, but in this context, brilliant and rising in the west, surely a comet if ever there was one. It seems highly likely that the problems of the second half of the sixth century, and their knock-on effects even today, can be laid at that comet's door: "Then arose Breas, the son of Balor, and he said: 'It is a wonder to me that the Sun should rise in the west today, and in the east every other day.' 'It were better that it was so,' said the Druids. 'Would that it were no more than the Sun!' 'What else is it?' said he. 'The radiance of Lug of the Long Arm,' said they." *The Fate of the Children of Tuirenn* [79].

In AD 1178 Gervase of Canterbury reported a major event on the Moon, though he didn't see it himself. Gervase, who carefully recorded events such as aurorae and mock-suns, wrote:

> In this year, on the Lord's day before the nativity of John the Baptist, after sunset, on the first day of the new Moon [June 18 or 19], there appeared a wonderful sign, five or more men witnessing it. For the new Moon was bright, extending the horns of its new form to the east, and behold suddenly the upper horn was divided in two. From the middle of this division there rose up a burning torch, hurling flame, burning coals and sparks for a long way. Meanwhile the body of the Moon which was below it was twisted as if anxiously, and as it is borne by the words of those who retold it to me and saw it with their own eyes, the Moon waved like a wounded snake. After this it returned to its normal state [80].

Carl Sagan interpreted this event as an impact in his TV series and book *Cosmos* [81]. The 'doubling' would be caused by a fountain-like spray of dust in vacuum and low gravity, like the plumes of the volcanoes on Jupiter's moon Io (Fig. 1.18). It's been calculated that debris could have been thrown over 1,200 km, so the impacting body was traveling at 20 km per second or more; about 1% of material ejected would have escaped from the Moon altogether [82]. The plumes prove that the sighting wasn't a meteor in Earth's atmosphere, in line of sight with the Moon, although that idea has been resurrected lately. But Gervase goes on to provide further details that Sagan didn't include, and which wouldn't have made sense until March 1993:

Fig. 1.18 Lunar impacts, 1178 (© Sydney Jordan, 2000) [83]

It repeated this change twelve times and more, namely the various torments of fire, as if it endured again what had already happened, and returned to its former state. And after these and such changes, it was made as if darkened from one horn all the way to the other. The very men who saw these things with their own eyes retold them to myself who writes them, willing to give their oath or to swear to it, that to the above they have added nothing false [80].

In 2001 the accuracy of this account was challenged by Paul Withers, who suggested that the Moon wasn't visible on June 18 [84]. Analysis by Graeme Waddington showed that it was, and Peter Nockolds showed that it was visible on the nineteenth even in the Holy Land, where its very thin crescent would have marked the beginning of the Moslem month [85]. Because it occurred in June, this event could have been caused by a member of the Beta Taurid meteor stream, the intense shower of daylight meteors first detected by radar in the Second World War. During it the seismometers left on the Moon by the Apollo astronauts detected several

Comet P/Shoemaker-Levy 9 (1993e)
Hubble Space Telescope
Wide Field Planetary Camera 2

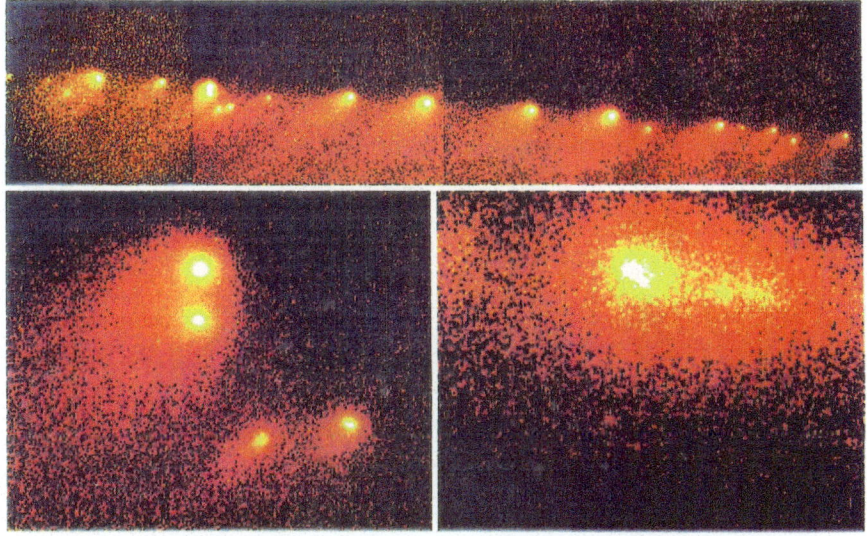

Region near Brightest Nucleus   Region near Brightest Nucleus
January 1994   July 1993
After Servicing Mission   Before Servicing Mission

Fig. 1.19 Hubble images of Comet Shoemaker-Levy 9, 1993 and 1994 (ESA)

large impacts, and the Tunguska impact in Siberia, in June 1908, may have been another example. Precise calculations, allowing for the Julian calendar then used, put the Moon's setting at Canterbury only 45 min after sunset. So the 12 impacts were in rapid succession—and we know now what sort of event that was.

Comet Shoemaker-Levy (SL-9) passed close by Jupiter in July 1992, and split into 22 fragments that were discovered in March the following year (Fig. 1.19). They hit Jupiter in July 1994, on the far side of the planet from Earth, but the flares were seen by the Galileo space probe and over the rim of Jupiter by the Hubble Space Telescope. As Jupiter's rotation brought the impact points into view, optical and infrared telescopes saw features larger than Earth forming in Jupiter's bitterly cold cloud layers (Fig. 1.20). Even at Airdrie Observatory, where the author was Assistant Curator at the time, with a 6-in. refracting telescope we could clearly

Fɪɢ. **1.20** Hubble image of SL-9 impact scar, several times larger than Earth (ESA)

see the impact scars a week later (Fig. 1.21). The whole sequence was much more spectacular than expected—with correspondingly scary implications about past and future impacts on Earth.

The Shoemaker-Levy 9 impacts renewed interest in overlapping crater chains on Ganymede and Callisto, two of Jupiter's large moons that preserve much of the record of bombardment during their history (Fig. 1.22) (On Io impact craters are erased by active volcanoes; Europa is covered by ice and overlying liquid water, and only small, recent craters are visible). Callisto has 12 or 13 of these chains and Ganymede three.

On the Moon, Mars and Mercury, chains are formed by secondary impacts of debris from major collisions, but on Ganymede and Callisto there are 15 chains for which no parent impacts can be found [86]. The biggest is 620 km long, with 25 craters roughly 25 km across, formed by bodies less than 4 km across.

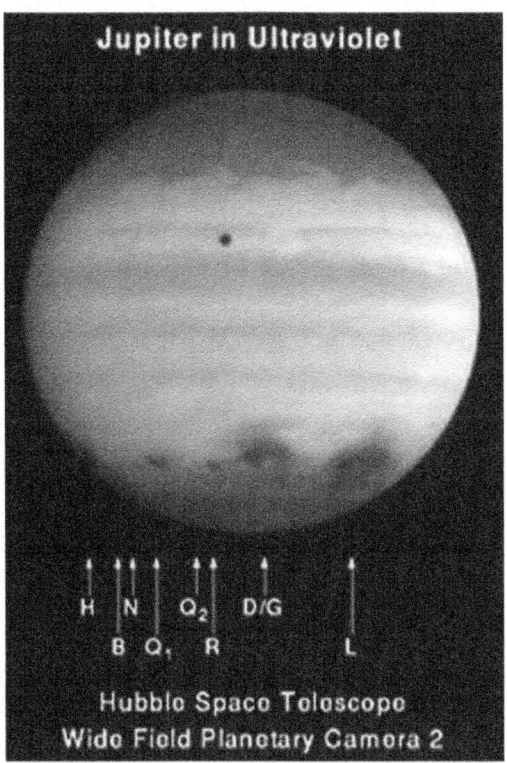

FIG. 1.21 Hubble ultraviolet image of impact scars, similar to the view in Earth-based telescopes (NASA)

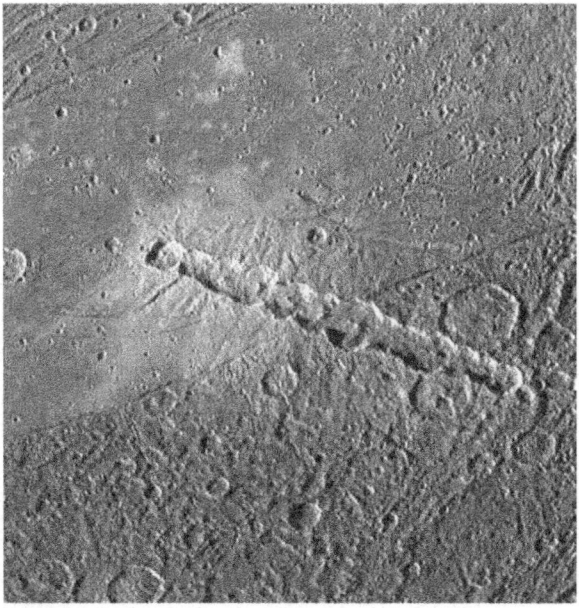

FIG. 1.22 'Parentless' Enki crater chain on Ganymede, from the Galileo spacecraft (NASA image)

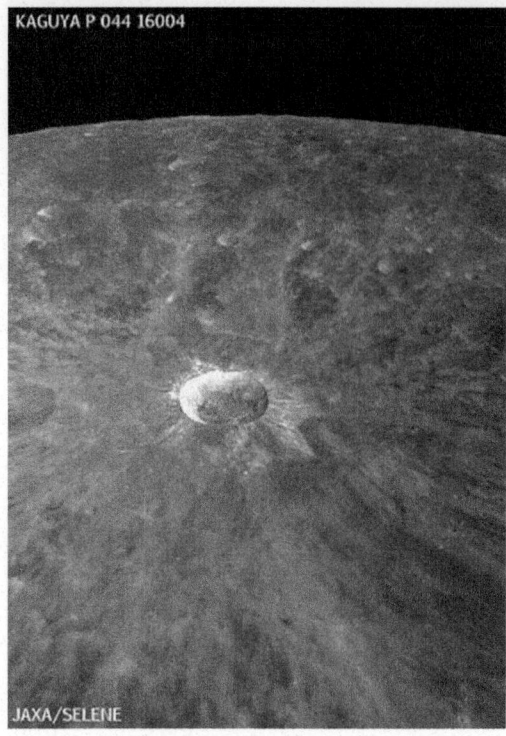

KAGUYA P 044 16004

JAXA/SELENE

FIG. 1.23 Lunar crater Giordano Bruno (Kayuga-Selene image, © JAXA, 2009)

Our Moon has secondary chains, e.g., from the craters Davy and Humboldt, but they begin next to their 'parents.' There are at least two 'parentless' crater chains on the lunar far side, north of the crater Tsiolkovsky but not pointing to it [87]. One is right on the lunar equator, the other at 5° Lunar North, and perhaps they were SL-9-type events. But they look old, not showing bright against the lunar landscape.

The 1178 impacts described by Gervase of Canterbury apparently formed the 20-km crater Giordano Bruno (Fig. 1.23), discovered by the USSR's Luna III probe that first photographed the lunar far side in 1959 [88]. Its existence had been deduced from the rays of impact debris spreading out from it (Fig. 1.24). They reach the hemisphere facing Earth [89]. The Bruno rays are as extensive as the much larger crater Tycho's, thought to be 60 million years old, and since ray systems fade with time, Giordano Bruno is much

FIG. 1.24 Apollo 11 lunar approach, 1969, crater Giordano Bruno at upper right (NASA)

more recent. If we take literally Gervase's statement that the flare divided the upper horn of the Moon in two, that puts the impacts around 45° Lunar North, which fits.

One major discovery of the Apollo missions was that the Moon's crust was apparently lacking in volatiles, such as carbon, hydrogen and nitrogen compounds. For this and other reasons it's thought that the proto-Earth collided with another protoplanet whose ripped-off, superheated crust formed the Moon. To give the Moon a lasting artificial earthlike atmosphere would require a comet-like body 80 km in diameter [90], or 400 comets the size of the Shoemaker-Levy parent. Smaller impacts might create a temporary atmosphere, darkening the Moon by the end of the event as described at Canterbury. In 1960, if not sooner, it was suggested that released volatiles could collect in 'cold traps' on the permanently shadowed floors of craters near the lunar poles. The total collecting area would be very small, less than half of 1% of the lunar surface—though even that could hold enough water to cover the Moon to a depth of one meter. On the other hand, the small collecting area could mean that the traps would fill rapidly and any ice in them would now be very old [91].

The Clementine probe in 1994 found signs that there was indeed ice at the south pole, in small craters within the giant impact basin Aitken, which is 2,500 km across and at least 12 km deep [92]. First results from Lunar Prospector in 1998 supported that, but then suggested that at both poles there were much larger deposits of ice, very thinly mixed with dust [93]. Lunar dust is a very effective insulator, which could explain why the ice grains survived even in sunlit areas; it, too, would have been laid down in 1178-type events, and the darkening which the Canterbury monks observed would indicate a temporary atmosphere thick enough to support the dust and carry it around the Moon. Later results from Lunar Prospector cast doubt on the frost interpretation, however, indicating that the ice was indeed concentrated in deep subsurface deposits, and there was still more of it. Arguments about whether there was ice, in what form and where it was from continued until 2009, when the L-Cross probe impacted near the south pole and raised a plume of water vapor whose isotopic composition showed it was definitely from comets; and at the same time India's Chandrayaan-1 lunar orbiter discovered at least 600 million metric tons of water ice in shadowed craters at the north pole [94]. The more details of Gervase's account come to make sense, the less likely it is that the story is invented.

In 1991 radar scanning by the Very Large Array in New Mexico found evidence that there is ice at Mercury's north pole. It shouldn't have caused that much surprise because the Scottish astronomer V. A. Firsoff had calculated that ice caps on Mercury were possible [95]. Patrick Moore wrote, "I admit to being profoundly skeptical. It does not seem likely that there has ever been water on Mercury, and without water there can be no ice [96]." Presumably it, too, is water ice deposited out of temporary atmospheres formed by comet impacts, as its existence has now been decisively confirmed by the Messenger probe orbiting Mercury.

Another reason for Paul Withers to doubt the impact was that he'd found no records of debris falling to Earth. The *Chronicon Anglicanum* of Ralph of Coggeshall, considered a reliable source, doesn't mention either the lunar event or the eclipse of 1178, but he describes the finding of St. Alban's relics that year. To it another hand has added the words "*et lapides pluebant*—and stones rained down [97]."

About 1% of the ejecta, hurled up from the lunar surface, would exceed the Moon's escape velocity [82]. The 'sparks' that were seen flying off into space must have been big, and there are two asteroids in orbits that could have originated in lunar impacts. The one designated 1991 VG is about 10 m across; 1999 CG9 is even larger, between 220 and 380 m in diameter [98].

However, some would have less than the combined escape velocity for Earth and the Moon [82]. Because they were from the Moon's trailing hemisphere, some of those could fall on Earth as secondary impacts. Since there were eclipses that year the Moon was near the ecliptic, and in June, its declination (terrestrial latitude, projected on to the sky) would be 23–24° North. Debris from Giordano Bruno crater, at 37°.7 Lunar North, could fall on Earth, but it depends on the angles of impact and of the secondaries' ejection from the lunar surface. Around 1200, there were big meteorite falls in Nebraska [99] and perhaps in New Zealand [42]. They might have been secondary impacts, or pieces of comet that missed the Moon altogether: Nebraska is 10° south of southern England, and New Zealand close to the antipodes. At the time there were major disruptions of population in Polynesia and South America; Emilio Spedicato, of the University of Bergamo, suggested that they too were due to impacts [100], and that would tie in with legends of the origins of the Incas, whose founders were said to have fought with rocks that tore holes in the landscape [101].

On a larger scale, the impact of a comet nucleus ten miles across, the size of Halley's, or an asteroid eight miles across, would be as bad as the Chicxulub event 65 million years ago, which not only wiped out the dinosaurs but made 90% of all species extinct. Ultraviolet light, X-rays and gamma rays from the plasma surrounding its atmosphere entry would have killed virtually anything in the line of sight. The shockwave of its impact threw everything on Earth that wasn't rooted to the ground 10 ft up in the air (Fig. 1.25). Our tiny mammalian ancestors fell unharmed and scuttled away, but the dinosaurs broke every bone in their bodies.

Twenty minutes after the impact, the rain of molten ejecta had killed every living thing in North America; the tsunamis might

FIG. 1.25 The Chicxulub impact as related by the *trees*, which *were* rooted to the ground (© Sydney Jordan, "I Talk to the Trees," story by Duncan Lunan, Lance McLane strip, *The Daily Record*, Glasgow, 1984)

have reached a mile in height, but it's thought they were limited to 100 m because the impact was in comparatively shallow water—bad enough, when most of Florida and the Mississippi basin was still shallow sea. The pyroclastic flow riding on the air blast was still raging through Central America, but within the hour everything on the land surface of Earth would be ablaze as the ejecta enveloped the planet. The crater was punched right through the crust into the magma, and as the magma rose, it met the sea falling in from above and vaporized it. The sea won, as it always does, but not before so much vapor and energy had been pumped into the atmosphere that the storm covered a hemisphere. The ozone layer was gone in a heartbeat, but that was a lesser factor because darkness was total worldwide for up to 2 years, during which acid rain came down everywhere. Dust suspended in the upper atmosphere, and poisonous nitrogen and sulfur compounds formed below by the object's passage through the lower atmosphere, all contributed to the extinctions. Over 90% of all living species were eradicated, but the little mammals in the burrows lived because they could hibernate, they were nocturnal, and there was lots of carrion above ground to sustain them.

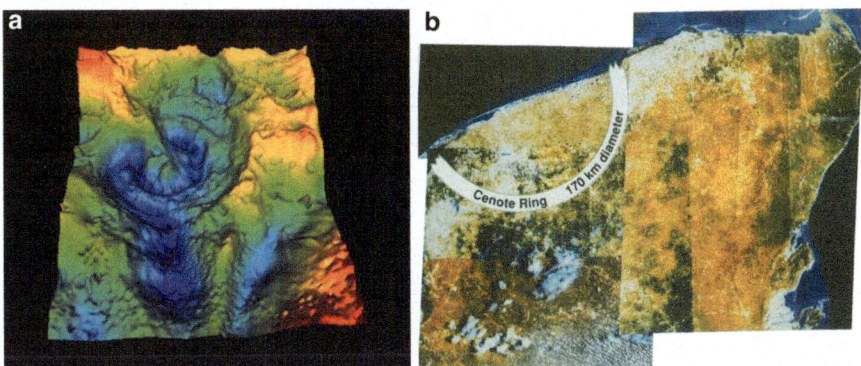

FIG. 1.26 (a) The Chicxulub impact feature (NASA). (b) Cenote ring in the Yucatán, mapped by space shuttle radar (NASA)

The impact crater was 120–160 miles in diameter, depending on just where one places the outer rim now that it's overlaid by a mile of sediment (Fig. 1.26a), but the full Chicxulub impact feature is easily 200 miles across. It takes in the circular pits in Yucatán called cenotes, giant wells where rising fresh water has cut away the limestone from below: the Maya used them for sacrifices, both of treasure and of people. They lie along the stress rings that were produced by the impact, concentric with the impact point offshore (Fig. 1.26b), and the offshore cenotes are a danger to shipping, especially submarines, because the rising fresh water is less dense than the seawater around it. In an episode of the TV show *Seaquest DSV*, at the time when Robert Ballard was still scientific advisor to the program, the submarine inadvertently entered one of the fresh water columns, and because it was trimmed for salt water, it plunged through the seafloor into the aquifer below.

The Chicxulub impact may have been one of a series, and there have been many other destructive events before and since. Two within the last million years, both on land, coincided with reversal of Earth's magnetic field, and Carl Sagan suggested that was no coincidence [102]. Since the crater of a large impact would punch right through Earth's crust, maybe the shockwaves radiating into the interior would also disrupt the currents that generate the magnetic field, and it's a 50-50 chance which way 'up' it will regenerate. Sagan suggested that in that case, sea impacts would explain why the Great Dyings of marine species in Earth's history seem to coincide with magnetic field reversals, when it's not obvious why sea creatures would be affected by other effects of reversals such as temporary thinning of the ozone layer. If it is impacts that cause the reversals, then the average interval between events big enough to do serious damage is 170,000 years, and we may count ourselves lucky as a species that there hasn't been one for 700,000 years.

However, if Comet Swift-Tuttle's nucleus is indeed 26 km across, up to 30 times the mass of Halley's, then the consequences of an impact could be anywhere from 8 to 30 times as severe as the Chicxulub one. Halley's Comet would strike Earth with a release of 10^{23} joules, 20 million megatons [103]. In his novel *The Hammer of God*, in which a smaller impact is turned into a close graze, Arthur C. Clarke argued that the 120-year orbital period determined for Swift-Tuttle was originally correct, and it changed to 130 years due to the action of the jets; the threat of a 2126 impact had been real, but it had also changed for the same reason. After the fictitious threat 'Kali' had been dealt with by a whisker, he wrote, "And Comet Swift-Tuttle was already accelerating towards perihelion. There was plenty of time for it to change its mind again" [104].

It has to be said that the more massive the nucleus, the less effect that its jets are likely to have on its trajectory. But at our present state of readiness, such an object could come out of the Sun and be upon us with no more warning than there was of the Tunguska object's approach.

If an event like a Swift-Tuttle impact were to happen today, then the effects would indeed be analogous to the quotation from *The Star* at the beginning of this chapter. If the first mention of the star described the approach of a comet, the second mention of it 'hotter now and larger and brighter than the sun' would correspond to the incandescent plume over the impact site. The blast wave would be like the star's motion over the rotating Earth, trailing 'thunder-storms like the hem of a robe.' The spreading dust and the trails of the falling ejecta would provide the 'pitiless brilliance' of the 'incandescent sky,' an impact in water would provide the tsunami, and the helpless terror and the 'flight no whither' are all too easy to imagine.

And then death. Wells was the master of saying a great deal in very few words.

2. Asteroids

I surmise (again) that possibly numbers of such small bodies that have not matter enough in them to hurt one another by attraction, or to disturb the planets, may possibly be running through the great vacancies, left perhaps for them, between the other planets, especially Mars and Jupiter.

– William Herschel, *Observations on Two Newly Discovered Bodies (Vesta and Ceres)* [1]

This may serve as a specimen of the dreams in which astronomers, like other speculators, occasionally and harmlessly indulge.

– Sir John Herschel, on the possibility of more than four asteroids [2]

When we look at a scale map of the Solar System, it's immediately obvious that there's a disproportionately large gap between the orbits of Mars and Jupiter. After he had deduced the relative sizes of the planetary orbits, Johannes Kepler modestly wrote, "Between Mars and Jupiter I have put a planet." The gap is dramatically illustrated by the image of Earth, Moon and Jupiter taken by the Mars Global Surveyor, and the corresponding orbital geometry (Fig. 2.1a, b).

In the late eighteenth century an international group of astronomers started a search, calling themselves the Celestial Police, and on January 1, 1801, Piazzi found Ceres, the first of the asteroids to be discovered. It was immediately obvious that it wasn't large enough to be classed as a true planet, so the search continued. Even though Ceres is the largest of the asteroids, it's less than 1,000 km in diameter, and the surface gravity can't be much more than 0.03 g.

Wilhelm Olbers discovered Pallas, the second known asteroid, on March 28, 1802, and Harding of Lilienthal discovered Juno on September 2, 1804. Vesta, the fourth to be discovered in the Main

D. Lunan, *Incoming Asteroid!: What Could We Do About It?*,
Astronomers' Universe, DOI 10.1007/978-1-4614-8749-4_2,
© Springer Science+Business Media, LLC 2014

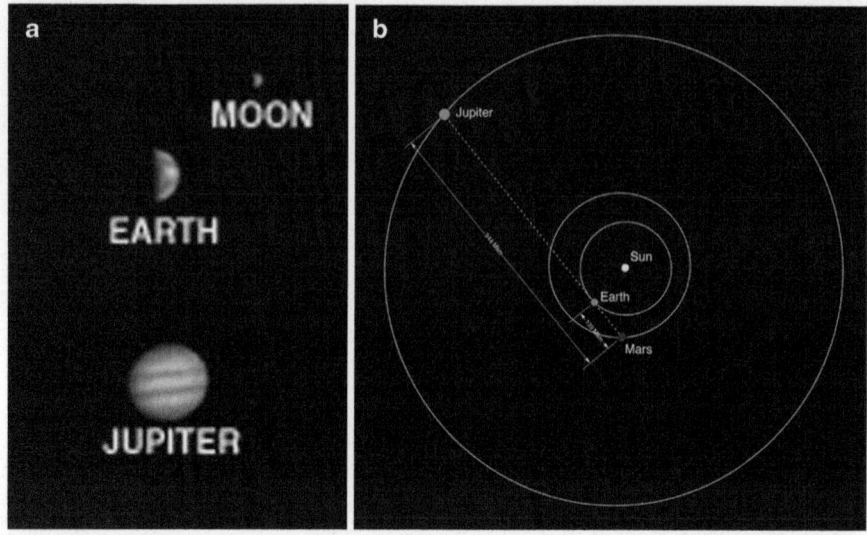

Fig. 2.1 (a) Jupiter in conjunction with Earth and the Moon, imaged by Mars Global Surveyor, May 2003 (NASA). (b) Orbits and positions of Earth, Mars and Jupiter, May 2003 (NASA)

Belt, was found by Olbers on March 29, 1807, and since then many thousands have been found, more than the Herschels dreamed of in their wildest moments. Vesta is visually the brightest, at the limit of naked-eye visibility when it's overtaken by Earth. About the size of Arizona, in the southern hemisphere it has a huge crater 465 km across and 12 km deep, as wide as Vesta itself; the equivalent on Earth would be the size of the Pacific basin. More than 50 smaller asteroids with similar compositions, 'Vestoids,' were formed in the collision, and many fragments blasted off it make up a family of asteroids [3]. Some of them reach Earth as meteorites.

At first it was thought that there had originally been a planet between Mars and Jupiter, where the Asteroid Belt is now (Fig. 2.2). Instead, we now know that Jupiter's gravitational pull caused the protoplanets in that area of the Solar System to collide with too much violence for their fragments to coalesce. Some of those protoplanets were large enough to have been heated internally by radioactive decay, causing them to be gravitationally differentiated, with crusts, mantles and cores—pieces of which now reach Earth as stony meteorites, stony-irons and metallic ones of nickel-iron.

Protoplanets in the outer region also had significant concentrations of water and possibly organic compounds. So in the multiple

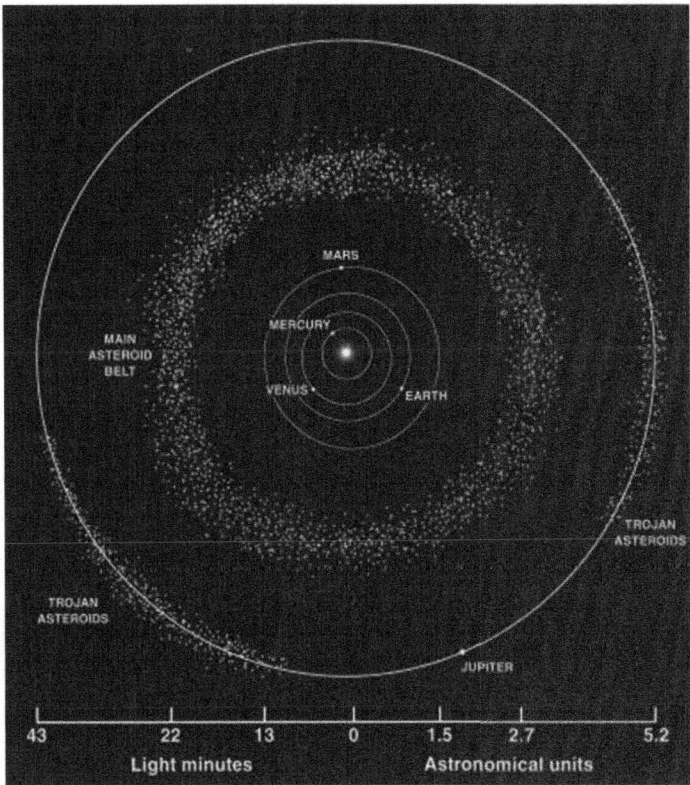

FIG. 2.2 The Asteroid Main Belt and Trojan groups (NASA)

collisions asteroids of many different types were formed, and although there is a general difference in composition between the inner and outer belts, all kinds are represented, as far as we know, in the population of asteroids whose orbits have been perturbed sufficiently to pass near Earth and occasionally collide with it.

Generally, carbonaceous asteroids come from the outer edge of the belt, stony ones from the middle and metallic ones from the inner edge. Some of them have undergone virtually no evolution from the primal days of the Solar System, and they contain the decay products of radioactive transuranic elements that were formed in the supernovae whose shockwaves caused the original Solar System nebula to collapse. The oldest material is often found in chondrules, which are fragments of the earliest condensates, embedded in more recent rock. The fragments collected from the asteroid that exploded over Chelyabinsk in Russia on February 15,

FIG. 2.3 Dawn among the asteroids (NASA)

2013, showed it to have been a rocky chondritic body, with a 10% content of nickel-iron, shot through with veins of once-molten material from a collision and fragmentation event, much earlier in its history.

Writers, artists and film producers (even today) like to portray dense asteroid fields (Fig. 2.3). The pulls of the planets have separated the Main Belt into three main bands and a lesser one, separated by 'Kirkwood gaps' (Fig. 2.4), and although occasional collisions between asteroids produce streams of asteroids in near-identical orbits (Fig. 2.5), there's so much space between them that normally one could spend a lifetime on an asteroid without ever having another come within naked-eye range. If the Asteroid Belt were really as dense as portrayed, maybe it would lend weight (literally) to the idea that the belt is the debris of a shattered planet. In fact the total mass of the belt is less than 10% of Earth's, possibly much less.

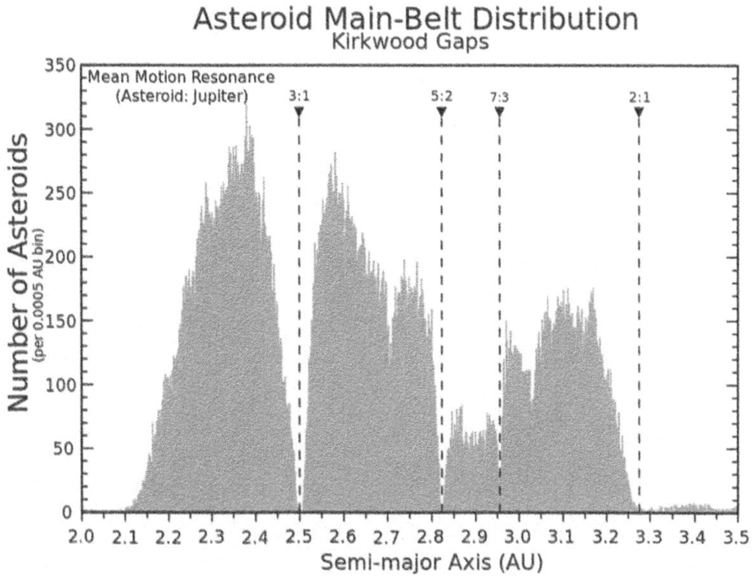

Asteroid Main-Belt Distribution
Kirkwood Gaps

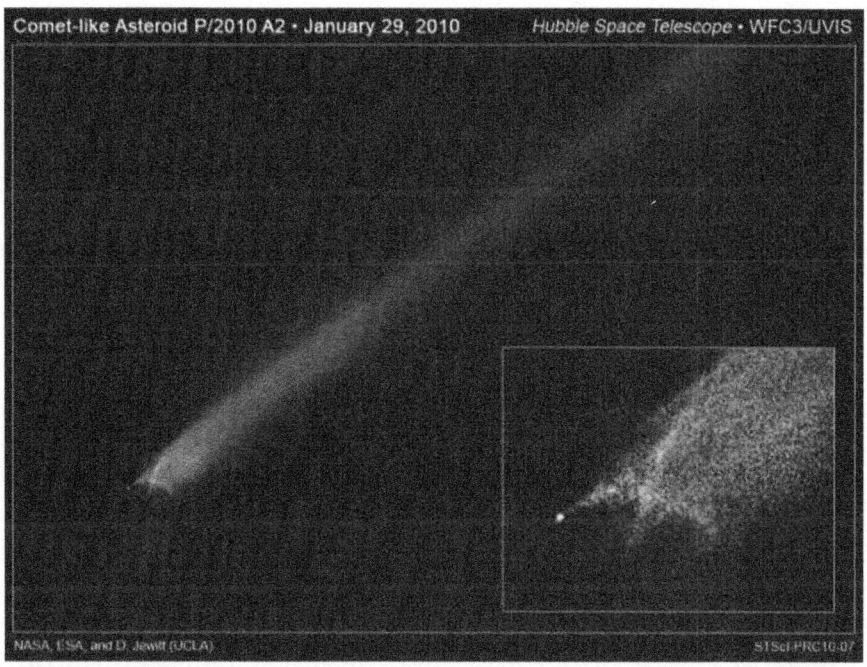

FIG. 2.4 The Kirkwood gaps (NASA)

FIG. 2.5 Hubble Space Telescope image of asteroid 2010 A2 LINEAR collision (NASA)

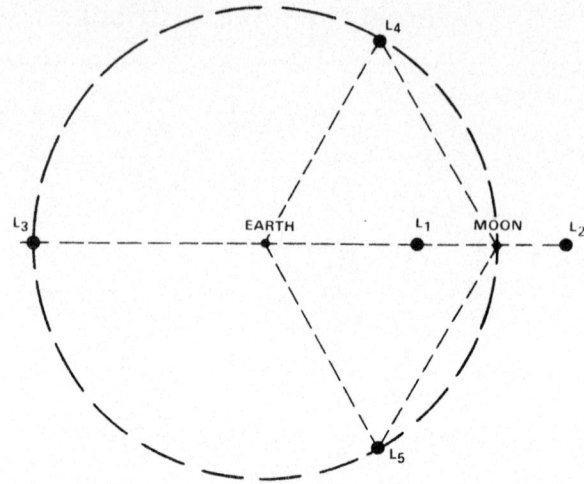

FIG. 2.6 Lagrange points in Earth-Moon system (NASA)

Thule is the Greek and Roman word for the north, as in Ultima Thule = furthest north, and it was allocated to an asteroid at 4.3 astronomical units, which was thought to mark the outer limit of the Asteroid Belt, until the Trojan asteroids were found sharing the orbit of Jupiter. The Trojan asteroids are in the same orbit around the Sun as Jupiter itself, equidistant from the planet and the Sun. They circulate around the fourth and fifth position calculated by Lagrange as special solutions to the problem of gravitational attraction between three bodies, which has no general solution even today. Lagrange points are generated whenever one massive body orbits another, for example Earth and the Moon (Fig. 2.6), Earth and Sun (whose L1 and L2 points were mentioned in Chap. 1), or Jupiter and the Sun. An asteroid has recently been discovered occupying the Sun-Earth L4 point, Mars has at least seven such companions, Neptune at least four [4], and Uranus has at least one [5]. Several of the moons of Saturn have companions at L4 and L5, which are also known as 'equilateral points' because the satellites maintain a fixed triangular relationship with the planet and the larger moon.

Orbits around equilateral points are more stable than those around L1 and L2, which are conditions of unstable equilibrium in which artificial satellites require regular station-keeping. Originally, the asteroids around one of the Jupiter equilaterals

FIG. 2.7 Venus from Toro during sunward pass (© Andy Paterson, 2000)

were to be called Trojans and the ones in the other were to be Greeks, all named after characters in the *Iliad*, but not all astronomers have full classical educations and the two groups quickly became mixed up.

In recent years a number of asteroids have been found circling the Sun in orbits that are resonant with Earth's, producing recurring close encounters. Cruithne (1986 TO) is an example [6], one designated 2002 VE68 has a similar bound relationship to Venus and Toro has a recurring pattern of encounters with Earth and Venus (Fig. 2.7). Some of them follow highly complex, unstable 'horseshoe' orbits, from which they will soon drift away, and Uranus has at least three companion asteroids like that [5]. Although these small bodies are of great interest for future space missions (see Chap. 8), none of them pose immediate threats to Earth, although that could change in the long term due to the pulls of the other planets.

(Note that asteroids are designated by the year of discovery, followed by a letter saying in which of 24 half-months it was found, and another letter indicating the day, followed by a number if more than one was found on the same day. As the discovery rate goes up, such designations are becoming increasingly cumbersome and are eventually replaced by single numbers in a full catalog. Names are usually proposed by discoverers, and ratified by the International Astronomical Union after due consideration. At least one participant in this project, the late Prof. Archie Roy, has already been honored—5806 Archieroy has been official since May 1995.)

The majority of meteorites come from the Asteroid Belt, where protoplanets formed and shattered in the early history of the Solar System. Since they had a range of chemical compositions, and had differentiated internally to different extents, the fragments contain a bewildering range of materials that took a long time to interpret. The important class of carbonaceous chondrites, containing organic compounds and the oldest, unaltered solid material, come from the outer Asteroid Belt, or from cometary nuclei, and as noted earlier, some 'Earth-grazing' asteroids, which come this far in towards the Sun, may be the nuclei of dead comets.

The Dawn probe (Fig. 2.3), launched in September 2007, was boldly targeted to orbit both Vesta and Ceres. Little was known of either, even the best images from the Hubble Space Telescope being little more than silhouettes, though a lot could be guessed about Vesta from comparison with meteorites, and there were great hopes that the crater at the south pole would let us see the interior of it.

Nothing in the Solar System has turned out to be what we expected. The author's New Worlds for Old (1979) was subtitled The New Look of the Solar System [7], and every space mission changes that 'look' again. Dawn entered orbit around Vesta in July 2011 and the asteroid has turned out to be far more complex and interesting than anyone had imagined. Hubble images of Vesta showed a marked point at the south pole, which had been thought to be a ridge between craters. But instead it's a mountain, 15 miles high—more like Olympus Mons on Mars than Mt. Everest, to which it was compared—planked on the pole as if by a gentle impact. Curving around it was a long, scalloped line of overlapping rings from two very large, superimposed impacts, and on the other side, there were roughly parallel grooves in the shape of a chevron. At first glance they resembled features on Miranda, the innermost of the five large moons of Uranus, which has been blown apart by a

FIG. 2.8 Dawn spacecraft image of Vesta, showing equatorial grooves and 'snowman' craters (NASA)

huge impact, with the fragments reassembled in the wrong order, like pieces of a jigsaw forced into places where they don't belong. But further evidence now shows that Vesta, like Lutetia below, is a single piece—intact since the origin of the Solar System, large enough to have been differentiated internally by radioactive heating, but with its surface much altered by collisions. Standing on the equator, three craters of increasing size in a row resemble a negative image of a snowman. Also on the equator huge parallel grooves were found, like those on Phobos (see below), but the largest of them, 10 km across, is as wide as Phobos itself (Fig. 2.8).

Craters with dark markings proved to be the impact points of carbonaceous asteroids, confirmed by releases of hydrogen from dark deposits on the surface, with pits like those formed by escapes of water from below the surface of Mars. There was even apparent evidence of erosion by water action, presumably in the presence of temporary atmospheres. And among many unexplained features, there was a pyramidal peak. Similar ones in terrestrial deserts are shaped by wind-blown sand, as presumably are the larger ones a kilometer wide found on Mars—but this was a jaw-dropping 5 km.

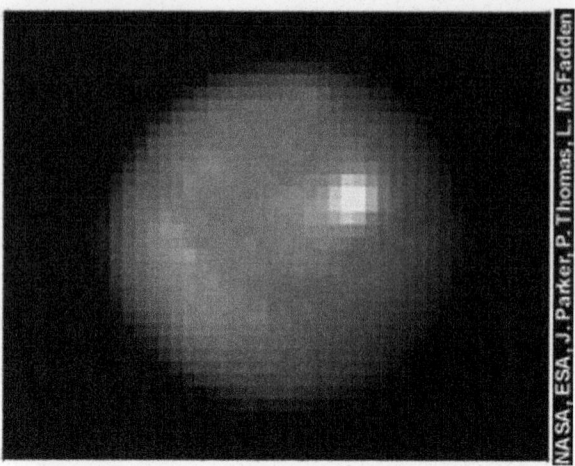

FIG. 2.9 Hubble Space Telescope image of Ceres (NASA/ESA)

It all promises great things when Dawn gets to Ceres. After over a year of taking pictures from orbit at various heights over Vesta, Dawn is due to arrive at Ceres in 2015. The largest of the asteroids has a very intriguing feature (Fig. 2.9), a bright spot reflecting sunlight as if it's water—presumably ice. But there have been suggestions of liquid water, which isn't possible at that distance from the Sun, still less in vacuum—unless it's under glass! After what we've seen on Vesta, it's tempting to add that nothing would come as a surprise—but we can be sure that there will be.

Encounters with Asteroids

> ...the ways by which men arrive at knowledge of the celestial things are hardly less wonderful than the nature of the things themselves.
>
> – Johannes Kepler

From 1971 to 2010, it was believed that the first asteroids to be photographed by spacecraft were Phobos and Deimos, the two tiny moons of Mars discovered by Asaph Hall in 1877. However, the close flyby of Phobos by Europe's Mars Express probe in 2010 detected hydrated rocks of types that had already been located on the Martian surface. The startling conclusion was that Phobos and

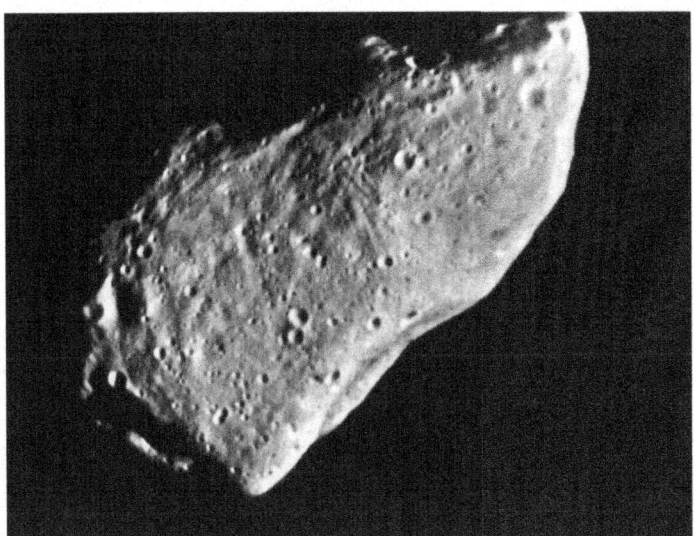

FIG. 2.10 Galileo spacecraft image of the asteroid Gaspra, 1991 (NASA)

presumably Deimos were formed from material blasted off the surface of Mars in big impacts—like the formation of Earth's Moon, but on a smaller scale. It ruled out the idea that the moons of Mars are captured asteroids, which until then had been widely believed.

So the first true asteroid pictures, after all, were the ones which the Galileo spacecraft obtained of the asteroid Gaspra on its way to Jupiter in October 1991. Gaspra is a stony asteroid about 12 km in diameter near the inner edge of the Main Belt. It proved to have a faceted appearance (Fig. 2.10), having been broken off from a larger body about 200–300 million years ago, and the only real surprise was that enhanced contrast showed up many hundreds of small craters, partly masked, it seemed, by thick regolith 'soil' composed of broken rock, as mysterious as the surface layers of Phobos and Deimos. It's not yet understood how such small bodies can accumulate regolith, when the debris from impacts should be blasted off into space and they don't have sufficient gravity to pull it back.

In Galileo's photographs of another asteroid, Ida, in August 1993, there was an unexpected satellite named Dactyl. Until the Ida flyby, professional astronomers insisted that asteroids were too small to have satellites. Amateur observers and meteor experts were less surprised because there had been a number of occasions

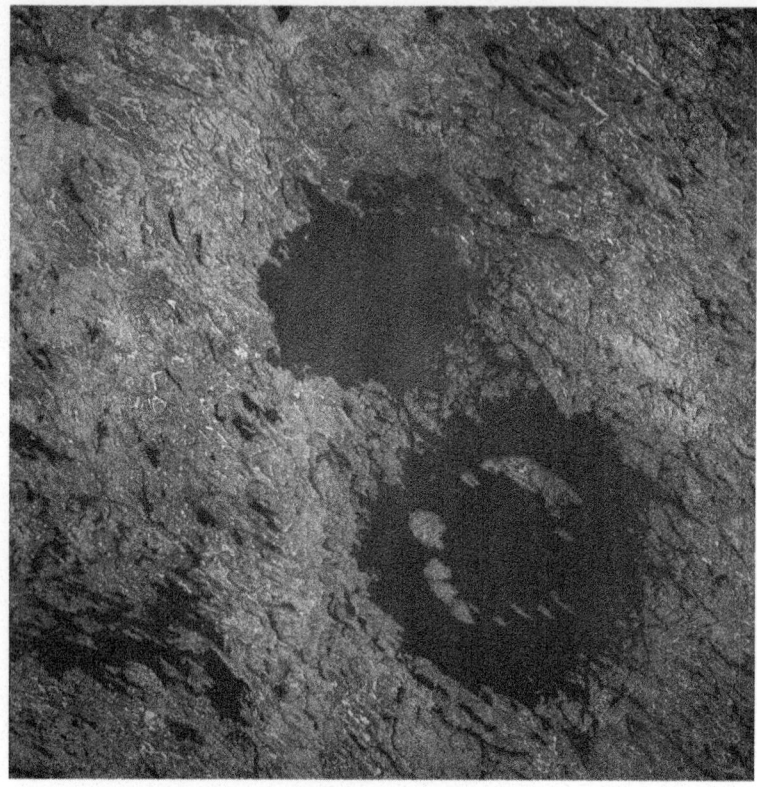

FIG. 2.11 Clearwater Lakes, Canada (NASA)

when asteroids passed in front of stars, and amateurs reported double occultations. In Canada there's a matched pair of impact craters called the Clearwater Lakes (Fig. 2.11), and a number of the asteroids were known to have dumbbell shapes, though nobody could explain how such small bodies might bump together gently enough to stick. But perhaps it does happen, because Ida and Dactyl are quite different in composition (Fig. 2.12), so Dactyl isn't a fragment detached from Ida by a collision, although capture of one asteroid by another is even less likely. Nevertheless, large numbers of binary asteroids have now been found, and several with more than one satellite.

It was thought that dumbbells have been formed by mergers, and that seemed to be confirmed by a deep valley found on Ida, plus a second one on Eros found by the NEAR-Shoemaker probe (Fig. 2.13). Nevertheless, despite having a huge saddle-shaped bite

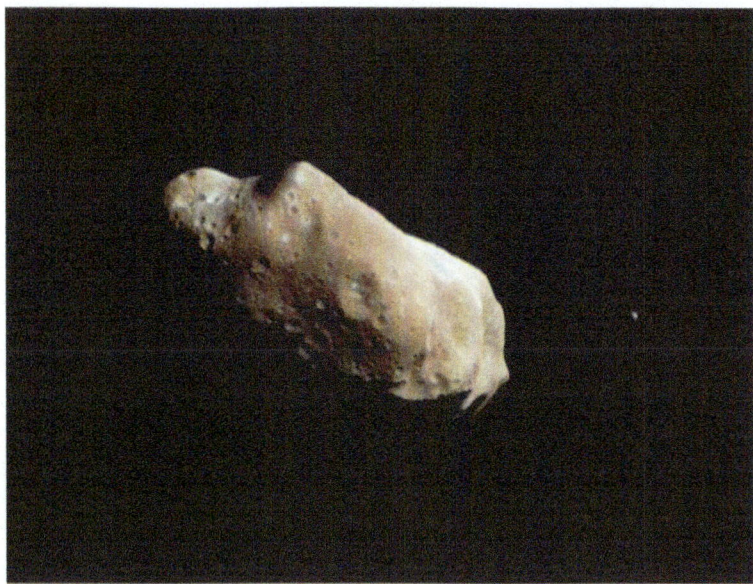

Fɪɢ. **2.12** Ida and Dactyl, 1993, in enhanced color to show differences in composition (NASA)

Fɪɢ. **2.13** NEAR-Shoemaker image of stony asteroid Eros in 2000 (NASA)

FIG. **2.14** Asteroid Steins from Rosetta, 2008 (ESA)

out of it, Eros is still a single object. In 2008 the Rosetta probe
flew by asteroid Steins, and found it to be so faceted by multiple
collisions that it looked like a gemstone, a diamond in the sky
(Fig. 2.14). The biggest crater was nearly half the asteroid's diameter at 2.1 km, suggesting that it had to have a really solid structure.
But the Main Belt asteroid Annefrank, visited in 1992 by the Stardust probe, shows genuine variations in composition and appears
to be a contact binary, formed by merger of two or more objects.

In June 1997 the Near Earth Asteroid Rendezvous mission,
renamed NEAR-Shoemaker in honor of the late co-discover of
Comet Shoemaker-Levy (Chap. 1), flew past Mathilde, a carbonaceous Main Belt asteroid 50 km in diameter, in a comparatively
eccentric orbit, ranging out to the belt's outer edge. The biggest
difference from Gaspra and Ida was that Mathilde showed very
large craters, and their sharp edges indicated that a lot of material had been removed by 'spalling,' in which the crust peels away
from the impact site. Its low density indicated large voids within
the asteroid, yet there was little variation in the appearance of its
dark surface, suggesting a very even composition (Fig. 2.15). There
was an even bigger surprise in 2010 when the Rosetta probe passed
the asteroid Lutetia. With a diameter of about 100 km Lutetia was

Fig. 2.15 NEAR-Shoemaker image of carbonaceous asteroid Mathilde (NASA)

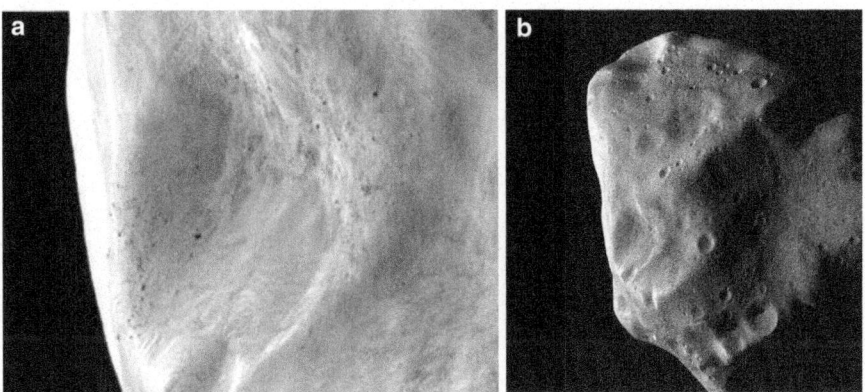

Fig. 2.16 (a) Lutetia regolith and boulders in close-up. (ESA). (b) Asteroid Lutetia imaged by Rosetta, 2010, with parallel grooves at bottom center (ESA)

much larger than the other asteroids surveyed so far, and again a 45-km crater suggested that it was solid, although surface features suggested extensive internal fracturing. Lutetia had been thought to be metallic and may have a metal core (see Chap. 8), though the outer layers appear to be metal-rich carbonaceous chondrite. Its regolith cover was 600 m thick, again strewn with boulders (Fig. 2.16a), and intriguingly it showed parallel grooves (Fig. 2.16b). Similar ones are found on Phobos, but straight; perhaps both are due to the flow processes now being found on many other small bodies. It appears that even Lutetia is not a collision fragment but a body that has remained intact since the formation of the Solar System, initially possessing a molten core.

Near Earth Asteroids

What you don't know about won't hurt you.

– common misconception

In the last few years there has been a big increase in the number of asteroids discovered and classified (see Chap. 4), and it is becoming apparent that most of the ones approaching Earth are the products of collisions in the main belt, rather than simply being perturbed in our direction by Jupiter. The first near-Earth asteroid to be visited was Eros, the 433rd asteroid to be discovered, by Witt, at Berlin in 1898. It immediately became important because it can approach to within 23 million km of Earth (Fig. 2.17) and could be used to gain a more accurate value of the astronomical unit, Earth's mean distance from the Sun, which gives the scale of the whole Solar System. In 1900 Von Oppolzer noticed big changes in magnitude as the asteroid rotated, indicating an elongated shape, about 37 km long and 16 km wide. With a rotation of only 5 h 16 min, that would give it a gravity of about one-hundredth of Earth's at the poles, but only one thousandth at the equator because of centrifugal force. In their book *Islands in Space, the Challenge of the Planetoids* (1966), Cole and Cox argued that the United States should abandon its target of a Moon landing before 1970 and go for a mission to an Earth-grazing asteroid instead [10]. Some of them could be reached with less fuel expenditure than the Moon itself, but their mission plan would take weeks longer than a Moon landing and put a severe demand on life support. They proposed landing Apollo tail-first on the target, and the difficulty of that was discussed, using Eros as an example, in the author's *Man and the Planets* [9].

Despite its dumbbell shape it has a uniform composition, and another mystery is that its surface was dotted with small boulders, although its surface gravity is too low to pull impact debris back. One possible explanation was that they've been left exposed

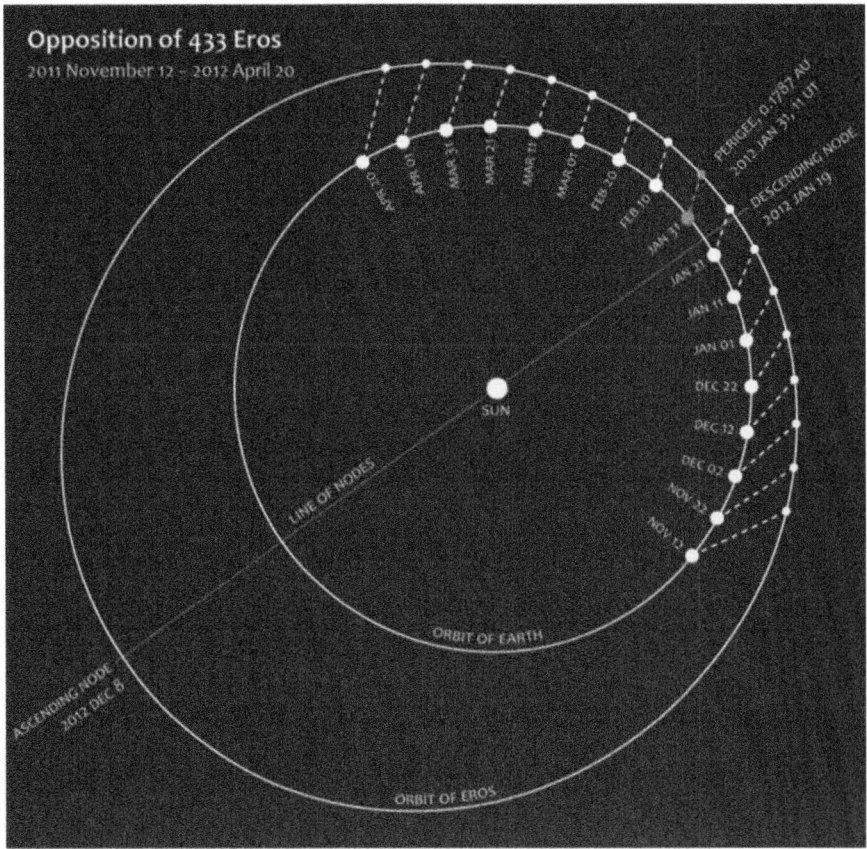

Fɪɢ. **2.17** Orbit of Eros, passing Earth in 2011–2012 (NASA)

as softer material was worn away by thermal and micrometeorite erosion, but grooves on the surface indicated that regolith material was flowing downhill towards the center of mass, even in the very low gravities of 0.001 g or less (Fig. 2.18). At the end of its mission NEAR-Shoemaker touched down on Eros. It wasn't designed to do it, but the landing proved much easier than expected, and contact with Earth continued for two more weeks, though no pictures could be transmitted from the surface.

Eros is technically an Amor-class asteroid, meaning that its orbit brings it relatively close to Earth's but doesn't actually cross it (Fig. 2.19). Apollo-class asteroids do cross Earth's orbit, Aten-class ones orbit entirely within the orbit of Mars, and Atira-class move entirely within Earth's orbit. All these are what we used

FIG. 2.18 Boulders and grooves on Eros (NASA)

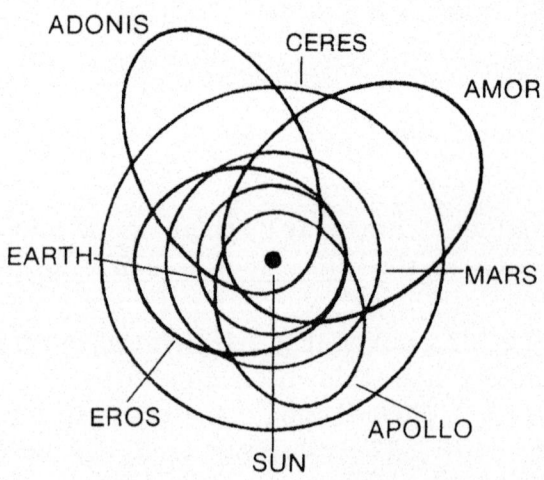

Nature of the orbits of some characteristic asteroids.

FIG. 2.19 Orbits of Earth, Mars and some asteroids (NASA)

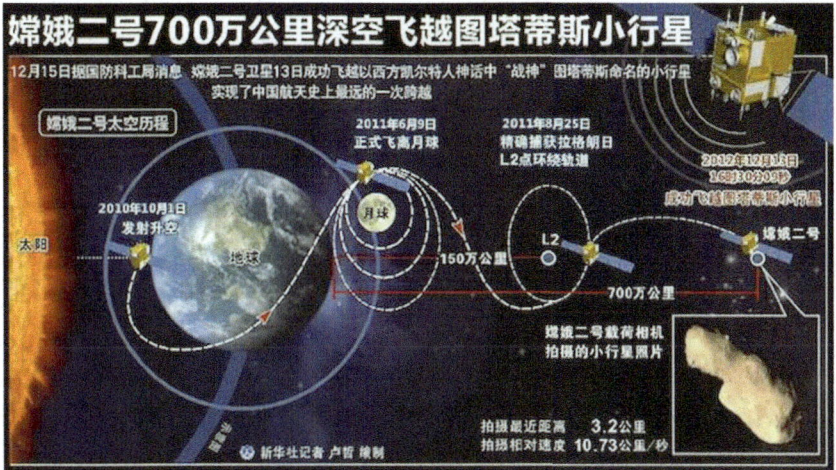

嫦娥二号700万公里深空飞越图塔蒂斯小行星

12月15日据国防科工局消息 嫦娥二号卫星13日成功飞越以西方凯尔特人神话中 "战神" 图塔蒂斯命名的小行星 实现了中国航天史上最远的一次跨越

嫦娥二号太空历程

2011年6月9日
正式飞离月球

2011年8月25日
精确捕获拉格朗日
L2点环绕轨道

2012年12月13日
16时30分09秒
成功飞越图塔蒂斯小行星

2010年10月1日
发射升空

太阳

地球

月球

L2

塔城二号

150万公里

700万公里

嫦娥二号载荷相机
拍摄的小行星照片

拍摄最近距离   3.2公里
拍摄相对速度 10.73公里/秒

新华社记者 卢哲 绘制

Fig. 2.20 Chang'e 2 mission including Toutatis flyby (© news.xinhuanet. com, 2012)

to call Earth-grazers, now more neutrally known as near Earth objects (NEOs). Most pose no immediate threat to Earth, although it can be shown mathematically that all of them will hit Earth, the Moon or Venus over the next 100 million years [10]. As most of the NEOs are stony in composition, it can be said that Eros is our first close-up glimpse of the threat.

At the time of writing the most recent encounter with an asteroid was in December 2012, when the Chinese space probe Chang'e 2, diverted from orbit around the Moon via the L2 point (Fig. 2.20), passed an asteroid called Toutatis at 3.2 km. Toutatis is a stony Apollo asteroid, 4.5 km long by 1.9 km across, in a resonant orbit with Earth and Jupiter and so creates recurring encounters with Earth every 4 years. It had previously been imaged by radar from Arecibo and other sites and found to be a dumbbell (Fig. 2.21a), almost certainly two merged objects and perhaps a 'rubble pile.' The Chang'e 2 images showed the asteroid as more angular than had been supposed, but with the same regolith, small craters, boulders and smooth areas seen by other missions, though there seemed to be some major variations in the composition of the surface (Fig. 2.21b).

In November 2005 the Japanese Hayabusa probe used low-thrust propulsion for a rendezvous with Apollo asteroid Itokawa.

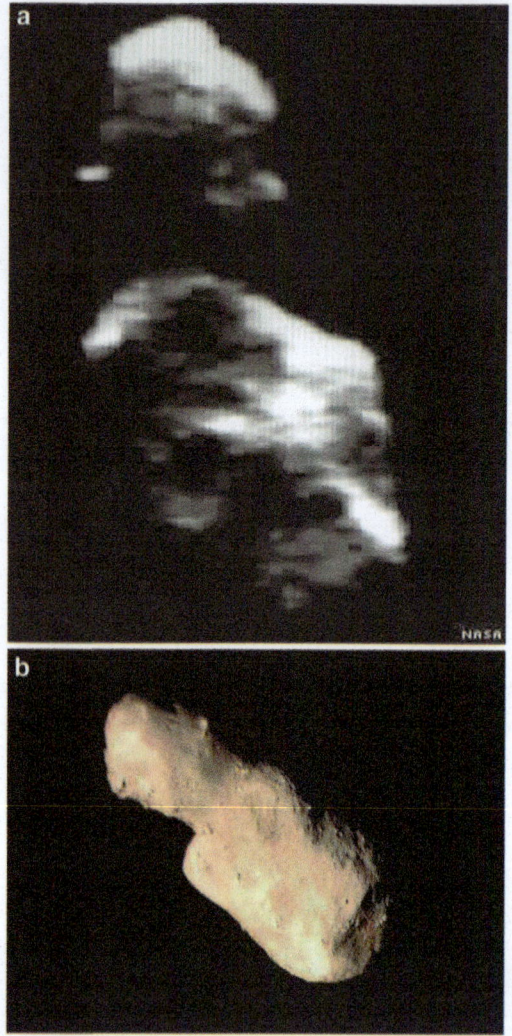

Fig. 2.21 (a) Toutatis radar image (NASA). (b) Toutatis from Chang'e 2 (Montage © Daniel Macháček, The Planetary Society, images © Chinese Academy of Sciences, 2012)

Two landing attempts were only partly successful, but some dust samples were gathered and brought back to Earth in 2010. They resembled rocky chondritic meteorites, incorporating material from very early in the Solar System, but Itokawa appeared to have split away from a larger body about 8 million years ago. Again there was a two-lobed structure, but this time, the surface was strewn with boulders and the 'neck' was remarkably smooth,

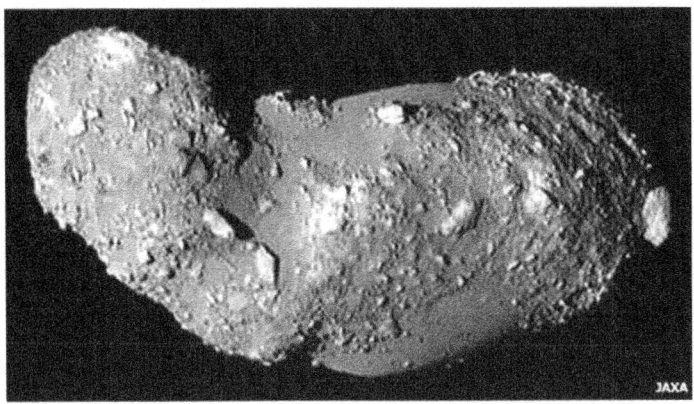

FIG. 2.22 Asteroid Itokawa from Hayabusa (© JAXA, 2005)

making it look more like the nuclei of Borelly and Hartley 2, presumably due to similar processes at work (Fig. 2.22). The low density suggested that internally Itokawa was little more than a rubble pile, much more fragile than Steins, for example. A tiny space hopper probe called Miranda was released at the wrong time and failed to make contact with the asteroid; however the JAXA space agency proposes a Hayabusa 2 mission for launch in 2014 and sample return in December 2020, this time with an impactor fired into the surface by shaped charge.

In late 2016 NASA plans to launch a mission called OSIRIS-Rex (Origins, Spectral Interpretation, Resource Identification, Security, Regolith Explorer). The target is a carbonaceous Apollo asteroid roughly 500 m in diameter, designated (101955) 1999 RQ36; a competition to name it was held in 2012 and the name Bennu, from Egyptian mythology, was the winner. This name was suggested by Mike Puzio, a 9-year-old from North Carolina [11]. The mission profile involves rendezvous and sample gathering, separation from the asteroid in 2021 and sample return to Earth in late 2023. The security aspect is that Bennu is currently considered to be the most dangerous known asteroid, with a 1:1,800 chance of impacting Earth in the late twenty-second century, particularly in the year 2188. So this will be the first mission to a potentially hazardous asteroid (PHA), and well worth doing for that reason alone.

Carbonaceous bodies generally fragment in the atmosphere before hitting the ground, and already from optical and radar studies there's reason to think that Bennu may have an aggregate structure, possibly with large voids within it (though the apparent solidity of Mathilde gives grounds for caution here). But while Bennu may not be a *very* hazardous asteroid, the 'resource identification' may be very important for human expansion beyond Earth (see Chap. 8).

Incoming Asteroids

> There comes a time when every scientist, even God, has to write off an experiment.
>
> – P.D. James [12]

At the other end of the size scale from the giant asteroids, which pose no threat to us, neither does the dust that causes the zodiacal light (Fig. 2.23) and the gegenschein (Fig. 2.24).

FIG. 2.23 Zodiacal light over Cerro Paranal, Chile (ESO)

Fig. 2.24 Gegenschein over the European Southern Observatory (ESO)

Coming partly from the Asteroid Belt, partly from comets passing through the inner Solar System, it spirals towards the Sun due to the Poynting-Robertson effect, in which light from the Sun exercises a slight but significant braking influence. If that was all that came our way from the Asteroid Belt, there would be no cause for concern; but unfortunately larger objects can be perturbed by mutual encounters and collisions, or by the pull of Jupiter, into orbits that throw them, too, in our direction.

On February 15, 2013, the 50-m asteroid 2012-DA14 passed Earth from north to south, moving within the ring of geosynchronous communications satellites 36,000 km above the equator (Fig. 2.25). The angle at which it passed us was a resultant of Earth's orbital velocity and its own, because its orbital plane was actually quite close to the ecliptic. Previous close passes had included 2003 SQ222, about the size of a small house, which came within 88,000 km of Earth on September 27 that year—the closest approach of a natural object then recorded, and at just over twice the distance of the geostationary satellites (There was a time when

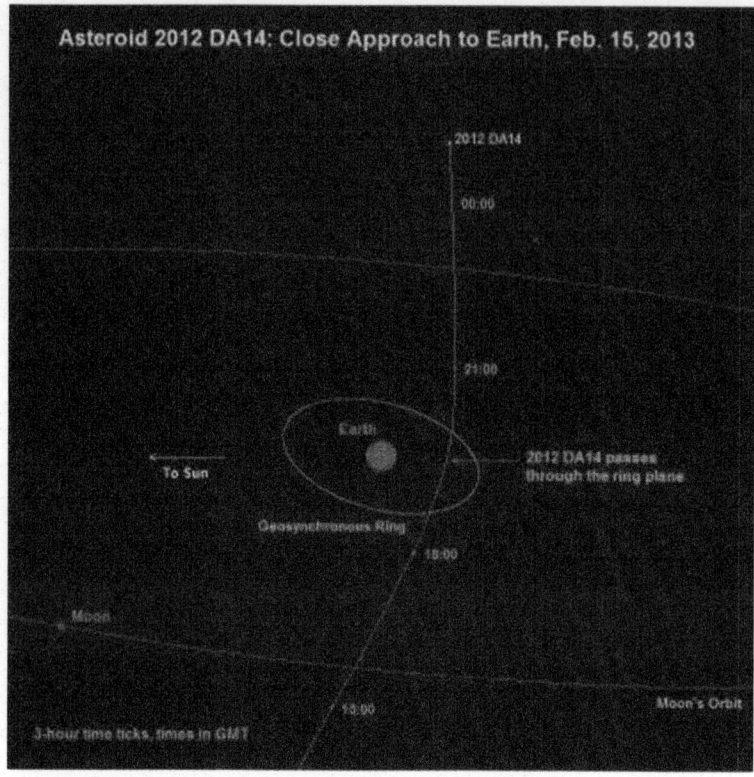

Fig. 2.25 Trajectory of asteroid 2012 DA 14 (NASA)

passing distances were related to the orbit of the Moon, but the Comsat ring has become the new yardstick).

Like so many that narrowly miss us, the asteroid came from inside Earth's orbit and wasn't spotted until after it went past. It was first seen on September 28 by the Lowell Observatory Near-Earth Object Search program in Arizona. Brian Marsden (see Chap. 1) calculated that its orbit is eccentric, with a period of just 1.85 years. Less than 10 m in diameter, it might have made a spectacular fireball had it entered the atmosphere—or it might have destroyed a city, depending on what its composition was, from a ball of dust to a stone to a solid lump of nickel-iron.

This object passed Earth at about 2300 GMT, only 10 h after a bright fireball and meteorite fell in the Orissa region of India, but the two events were not connected. The previous record for closest approach of an asteroid—108,000 km, measured from the center of Earth—was set in 1994 by another 10-m object named 1994 XM1, and the third closest was object 2002 MN, about 80 m in diameter, at 120,000 km [13]. At that size an impact would have caused a huge amount of damage.

Again by chance, on the same day that 2012-DA14 flew past in 2013, a smaller Apollo asteroid about 15 m in diameter, with a mass of about 7,000 metric tons, entered the atmosphere at a 20° angle, traveling east to west at about 19 km per second, and exploded 14–20 km above the city of Chelyabinsk in Russia with a force that was thought at the time to be at least 300 kilotons of TNT. Later estimates put the blast at 440 kilotons, roughly 30 times the yield of the atomic bomb over Hiroshima [14]. Nearly 1,500 people were injured, most of them by flying glass broken by the sonic boom.

At the 2003 seminar in this book project, Al Gore was quoted as having said that a Tunguska-type event over a city in the Midwest might have to happen before the asteroid threat was taken seriously. At the 2013 Planetary Defense Conference, one speaker suggested that the Chelyabinsk one was an ideal wake-up call—brilliant, damaging, but not a tragedy, over an area not too heavily populated, but where everybody had dashboard cameras (seemingly for social reasons, to do with false accident claims and accusations of traffic offenses).

Clearly this was one of Isaac Asimov's 'city-busters' (Chap. 1), and the consequences would have been far worse if it had hit the ground. Even so, its comparatively low velocity was again a resultant of the asteroid's speed and Earth's, having been overtaken by the planet [15]. Asimov calculated that at typical asteroid velocities, a city could be destroyed by a nickel-iron object massing only dozens of tons, the size of a large desk [16]. One of those fell at Sikhote-Alin in Siberia in 1947, again breaking up in the air before impact but forming hundreds of small craters at ground level. Observations from the Mars Reconnaissance Orbiter have now confirmed that Mars, nearer the Asteroid Belt and with a much thinner atmosphere, takes around 200 such hits every year (Fig. 2.26a, b) [17].

In *New Worlds for Old* in 1979, the author wrote "From that point of view, probably the most terrifying photos of the decade appeared in *Sky & Telescope* for July 1974. They show a meteorite,

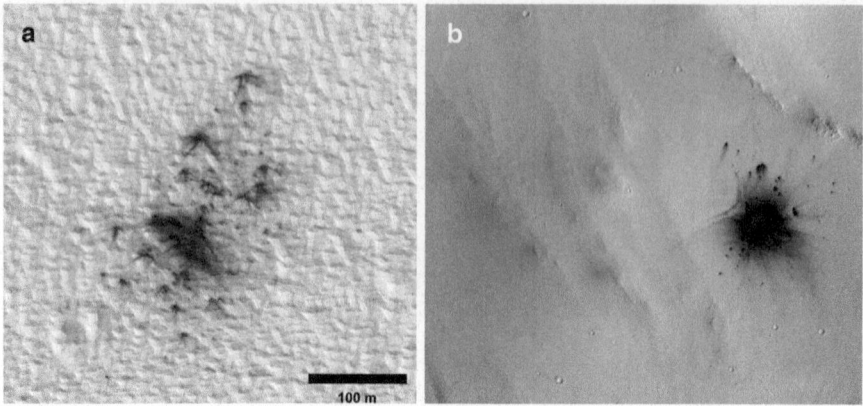

FIG. 2.26 (a) A fresh crater field, less than 5 years old, found by the Mars Reconnaissance Observer. (NASA). (b) A fresh impact with secondary craters (NASA)

whose mass may have been as high as several thousand metric tons, passing through Earth's atmosphere without striking the ground. It passed within 58 km of the surface at 15 km/s, causing sonic booms; its flight was from Utah, over Montana, and out into space again over Alberta. The Flying Saucer movement claims it for a spaceship; would that it had been anything so innocuous" [7]. We don't know its composition for certain, but it held together while subject to strong atmospheric forces. This suggested that it was nickel-iron, even at the time; comparison with the Chelyabinsk object makes that a lot more likely. *New Worlds for Old* continued, "In a very low-key assessment of what could have happened, Luigi G. Jacchia wrote: 'It seemed strange not to have any report of an impact.... A body capable of producing a fireball having the observed brilliance impacted with the energy of an atomic bomb, and seismic disturbances should have been recorded.' [18] Military and political ones, too, I fear.... In pursuing 'deterrence' based on nuclear weapons, one of the craziest aspects of a crazy philosophy is that we have reduced the impact mass necessary to bring about our annihilation by about nine orders of magnitude." The fear of global nuclear conflict has receded with the ending of the Cold War, but Prof. Colin McInnes has published a chilling fictional version of what might ensue from a similar strike on the border between India and Pakistan [19].

As a possible example of a city-buster, Asimov had considered the biblical flood, which he suggested might have been due to a city-buster scale impact in the Persian Gulf. In the 1920s, it caused a sensation when C. Leonard Wooley discovered a flood layer at Ur in Mesopotamia.

> During the seasons 1927–8 and 1928–9 our work on the prehistoric graveyard had resulted in the excavation of a huge pit some 200 feet across and between 30 and 40 feet deep.... The shafts went deeper, and suddenly the character of the soil changed. Instead of the stratified pottery and rubbish we were in perfectly clean clay, uniform throughout, the texture of which showed that it had been laid there by water. The workmen declared that we had come to the bottom of everything, and at first, looking at the sides of the shaft, I was disposed to agree with them, but then I saw that we were too high up. It was difficult to believe that the island on which the first settlement was built stood up so much above what must have been the level of the marsh, and after working out the measurements I sent the men back to work to deepen the hole. The clean clay continued without change... until it had attained a thickness of a little over 8 feet. Then, as suddenly as it had begun, it stopped, and we were once more in layers of rubbish full of stone implements, flint cores from which the implements had been flaked off, and pottery... Taking into consideration all the facts, there could be no doubt that the flood of which we had thus found the only possible evidence was the Flood of Sumerian history and legend, the Flood on which is based the story of Noah.... This deluge was not universal, but a local disaster confined to the lower valley of the Tigris and Euphrates, affecting an area perhaps 400 miles and 100 miles across; but for the inhabitants of the valley that was the whole world! [20].

The oldest surviving account of the biblical flood is in the Sumerian *Epic of Gilgamesh* and dates from c. 2250 BC [21]. It's the most detailed account and clearly attributes the event to an impact. Versions of the legend are found in Egypt, the Hittite kingdom, India and China [22], and the biblical one—which was picked up by Jewish exiles in the Babylonian captivity—is the only one that leaves out the impact [21].

Not knowing that, in *The Rocks of Damocles* Asimov nevertheless suggested that there had been one. It began with "a cloud no bigger than a man's hand" (a distant mushroom?), *then* a tsunami ("the same day were all the fountains of the great deep broken up," and the Ark was carried inland to Ararat), and only after that the sky grew dark and "the windows of heaven were opened" with torrential rain (Genesis 7, 11). But the Sumerian account has a heat-flash, an incandescent rising cloud with ejecta ("the Annunaki lifted up their torches, setting the land ablaze with their glare"), a groundshock ("the god of the underworld tore out the posts of the world-dam"), an air-blast, and only then the tsunami and the deluge.

Hittite legend says the flood was caused by the Moon falling to Earth (descending fireball) [21], and the Egyptian Coptic account says it began with fire from the constellation Leo, while divine personages stalked the land striking down the populace with iron maces [23]. The ancient Egyptians knew iron only from meteorites, and the Leonid meteors still provide spectacular displays every 33 years (Chap. 1). It's been suggested that parts of the story of Samson are a confused account of a Leonid fire-storm, although other writers associate him with Orion and its January meteors [24].

The Henbury craters in Australia were formed by a nickel-iron asteroid that broke up at low altitude, around 2700 BC. The flood could have been generated by a similar impact, perhaps in the Persian Gulf as Asimov suggested. In 2354–45 BC there was an abrupt cooling in global climate, and there is now evidence that the impact may instead have been in the Iraqi marshes and a century before the Gilgamesh text [25]. As previously noted, it's remarkable that one nineteenth-century estimate put the date of biblical deluge at 2349 BC, though it's probably a coincidence because other Bible studies of the time put it much further back [26].

Around 2000 BC a so-called bouncing asteroid (probably twinned, as many asteroids are) created a double crater at Campo Cielo in Argentina; this was the largest impact of modern times, with an energy release of about 300 megatons [27]. All this ties in with Victor Clube and Bill Napier's belief in an ongoing series of events from the break-up of a 'super-comet' in the inner Solar System around 3000 BC (Chap. 1).

Alan Bond and Mark Hempsell have presented evidence that a low density 1-km rock asteroid with a mass of 800 million metric

Fɪɢ. **2.27** Meteor crater, Arizona (Shane Torgerson, Wikipedia Commons)

tons passed over Sumeria at 14 km/s on June 29, 3123 BC, clipping the Gamskogel ridge and impacting the Köfels area in the Austrian Tyrol, triggering a massive landslide that erased the main crater, while smaller fragments caused other impact features in the area:

> As the object travelled up the Adriatic Sea…and across the Alps the supersonic shock would have caused considerable destruction on the ground beneath the trajectory. The impact…would release energy equivalent to 1.4×10^{10} metric tons TNT. [The plume] would rise… to some 900 km before falling over the Levant and Sinai causing considerable destruction over a wide area.…There would have been many direct casualties, near 100% mortality over areas of thousands of square kilometres in both the Alps and the Near East. There would also have been a severe global climate change that caused further death and social disruption [28].

Although newspaper reports very frequently cite the Tunguska event when discussing the effects of impacts, a much better example is Meteor Crater in Arizona, also known as Barringer Crater and before that as Canyon Diablo Crater (Fig. 2.27), with a diameter of 1,186 km (0.737 miles). The event took place about 50,000 years ago, and the nickel-iron impactor was about 50 m across, with an energy release on the order of 10 megatons.

At first the crater was thought to be a volcanic feature, but in 1903 Daniel Barringer suggested it was caused by an impact and was proved to be correct by Eugene Shoemaker in 1960, identifying the types of shock features in surrounding rock that have since been used to identify impact features worldwide. Barringer's attempts to excavate iron ore from the crater were less successful, because most of it had vaporized on impact, and he had drastically overestimated the mass, now thought to have been about 50,000 tons.

A great deal of underestimating the destructive effect of impacts stems from a painting by Chesley Bonestell, Plate LXI of *The Conquest of Space* by Willy Ley, in which he superimposed a similar crater on Manhattan Island. He showed parts of the city on fire and some of the bridges broken, but roads, piers and other features were still recognizable [29]. Compare that with the description by Larry Dean Marshall, a paleontologist on the Kansas University investigating team, of the 'football-field-size' object mentioned in Chap. 1 that hit Earth between 500 and 1,000 years ago at Merna, 20 miles west of Broken Bow, Nebraska. The impact created a depression a mile across, originally 500 ft deep and even now 70 ft below the local surface. "If you were in the vicinity and looked at it, you'd be blinded. And there would be a tremendous roll of thunder followed by a shock wave" [30]. The heat would have been intense enough to ignite anything for about a 20-mile radius, and after the blast wave nothing manmade would be recognizable in the Bonestell painting.

Australia's terrain has preserved some very fine impact features, and more are being discovered even today. One of the better known, Wolfe Creek in Western Australia, wasn't discovered until 1947, when an oil survey party flew over it, although it's 875 m across (Fig. 2.28). A 1980s tourist book described it as "the second biggest meteorite crater on Earth" [31], which was long out of date even then, unless you add the words 'immediately recognizable.' Wolfe Creek was formed about 300,000 years ago by another iron meteorite, with a mass of about 50,000 tons.

As noted in Chap. 1, a problem with larger, older craters is the difficulty in assigning them to asteroid or cometary impacts.

Fig. 2.28 Wolfe Creek Crater, Kimberleys, Western Australia (© W. Pederson, Australian News and Information Bureau Photograph)

But there's no ambiguity about the Nordlinger Ries, discussed in Chap. 3, nor about the Sudbury impact crater in Canada, 1.85 billion years old, which is the source of much of the world's nickel. The Vredevoort Dome in South Africa, which until recently had the distinction of being the oldest known impact feature at 2.023 billion years (Fig. 2.29), has now been surpassed by a 100-km (62 mile) wide crater near the Maniitsoq region of western Greenland, believed to be 3 billion years old and also thought to be due to a nickel-iron body, 30 km in diameter, twice the size of the Vredevoort object (Fig. 2.30) [32].

J. E. Enever's 1966 article, cited in Chap. 1, assumed that the Vredevoort impactor was a nickel-iron body about a mile in diameter [33]. If his calculations of the energy released were correct, as well as the 15 km diameter that is now estimated, this would suggest it was a rocky body rather than metallic. Enever went on to suggest that a big marine impact could have wiped out the dinosaurs, one of the biggest 'megadeaths' in the history of evolution.

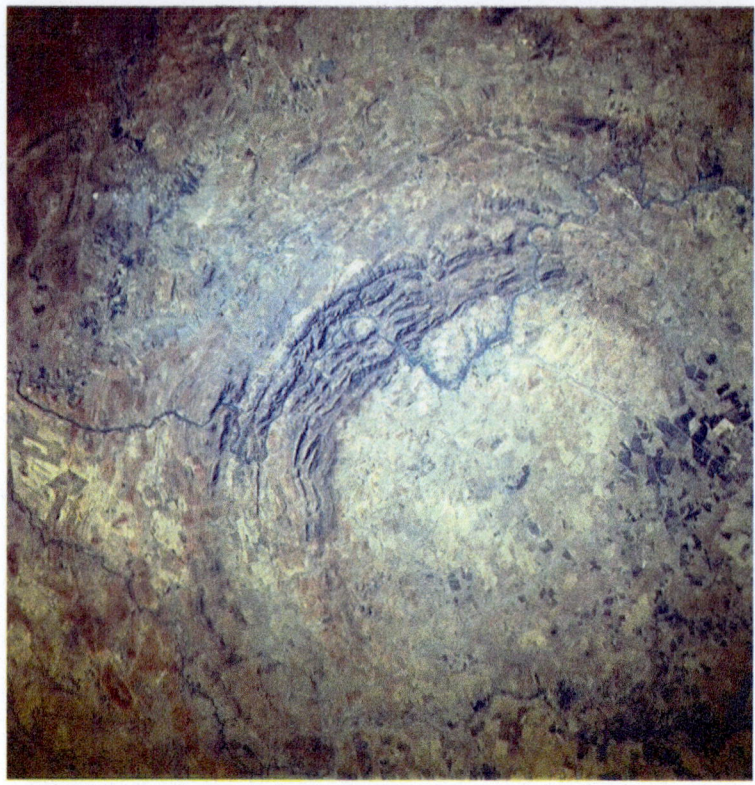

FIG. 2.29 Vredevoort Dome multiple ring impact feature, South Africa. STS-51i, August 1985 (NASA)

As noted earlier, a big enough impact on sea or land will punch a crater right through Earth's crust into the magma, but in a land impact most of the energy is re-radiated back into space, and only one continent is devastated. But a sea impact generates tsunamis of colossal size; then, as the sea tries to quench the rising magma, a terrible storm ensues whose effects, together with the dust thrown into the upper atmosphere, may be enough to block off all sunlight from Earth's surface—perhaps for years. In his novel *The Hermes Fall*, John Baxter generally underestimates the effects of the marine impact portrayed, but he does paint a chilling picture of a storm unlike anything in the experience of the human race [34].

Then in 1980 came the discovery of a worldwide stratum enriched with iridium, too much to explain by volcanic activity, coinciding with the dinosaurs' disappearance and also containing

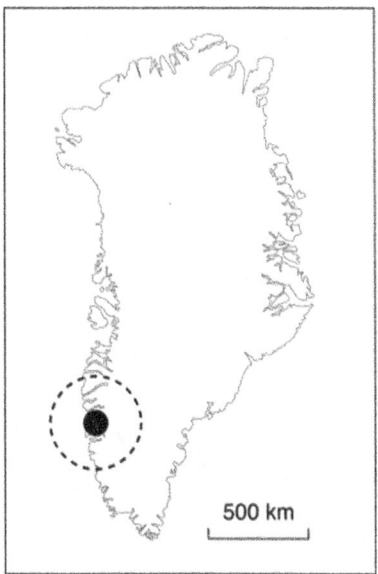

Fig. 2.30 Location of Maniitsoq impact feature, Greenland (© Geological Survey of Denmark and Greenland (GUES), 2012)

evidence of conflagration over much of the globe. Earth's air is balanced between having too little oxygen for animal metabolism and having so much that everything becomes super-inflammable. Air bubbles in amber, formed shortly before the disappearance, show a high oxygen content. It may have enabled the pterosaurs to fly despite their huge wingspans, giving them a high metabolic rate, but the consequences were drastic. The iridium layer in the geological strata world-wide at the Cretaceous-Tertiary boundary contains a micro-thin layer of carbon, showing that the impact was followed by fires over enormous areas. A similar event 135 million years ago gave flowering plants their first advantage over ferns [36].

   At first it was argued that there was no crater on Earth of the appropriate size and date, but the Chicxulub feature off the coast of Yucatán emerged as a strong candidate [37]. Below the marine deposits covering it, shattered and melted rock has been found matching the composition of the sediments at the Cretaceous/Tertiary (K/T) boundary. The *Lunar and Planetary Information Bulletin* reported: "Prior to the analyses of the basement rock at the crater, other workers suggested that several simultaneous impacts might be needed to explain the material at the boundary layer....

Chicxulub seems to be able to account for it all — singlehandedly."
An initial uncertainty of several hundred thousand years in the dating has been narrowed down to less than 32,000 years, making it virtually certain that the events were causally linked [38].

We still don't know whether the incomer was a comet or an asteroid, but we have enough evidence of big asteroid impacts above to say that it could have been either. It seems that the iridium levels in deep-sea sediments may be higher than on land, and if that's due to a process of concentration, the lower levels found elsewhere may indicate that it was a comet after all [39].

However, the good news is that it's estimated that over 90% of the current potentially hazardous asteroids (PHAs) over 1 km in diameter have now been detected, and none of those pose near-term threats [40]. But there are literally thousands in that size range down to 100 m (see Chap. 4), and a strike by any of those could have global consequences.

Playing the Odds

> Most of Target Earth is Target Ocean... there is at least a 1% chance that all of the cities around the Pacific rim will be obliterated by an asteroid-induced tsunami within the next century.
>
> – Dr. Duncan Steel [41]

"One of the most feckless arguments on this subject is that 'the chances are millions to one.' As just over half the people who have played Russian Roulette can testify, the small chance that a given chamber will go off is a mathematical fiction: similarly for the asteroid and Earth there's one day that counts and the rest are immaterial. Nor can anything, at present, alter the odds on the day itself — asteroids hardly ever have misfires…" [9].

As John G. Kramer pointed out in *Analog*, at the time of Comet Swift-Tuttle's return in 1992, how impact risks are assessed is a matter of definition. Three years earlier, David Morrison and Clark Chapman had calculated that the chances of dying in an asteroid impact are slight, on a day to day basis, but so many deaths would result that the risk over 50 years is greater than from airplane accidents, tornadoes or electrocution [42]. Duncan Steel recalculated the odds for his book *Rogue Asteroids and Doomsday Comets*, concluding that death by asteroid was twice as likely as they had estimated. For U. S. residents, that makes it less likely than death by automobile accident (at the top of the list), homicide, fire, accidental shooting or electrocution (in that order), but more likely than an airplane accident, a flood or a tornado—against all of which people take out insurance [41].

A still more abstract calculation put the global insurable loss per year, averaged over 50 years, at $20 billion, updated to $28 billion by 2001. For the United States alone the figure would be $750 million per year, £150 million for the United Kingdom—about the cost of a small housing estate, as rocket engineer Roy Dommett pointed out [43]. In relation to that, the annual cost of a proposed Spaceguard system of six dedicated 2-m telescopes, at $300 million over 25 years, to identify all possible hazards (see Chap. 4), could be compared to insuring an automobile for a dollar [41]. Here's another way to look at it. In any given year, a regionally destructive impact is only 150 times less likely than a major earthquake in Japan, only 500 times less likely than a flood in Bangladesh, and death by impact on a global scale is 10 times more likely than a regional one [44].

In July 1997 a conference at Cambridge concluded that because of the tsunami hazard, the risk to the UK from impacts was greater than that from Russian nuclear reactors or from plane crashes such as Lockerbie. Whether the risk evaluation was actuarial or statistical, it well exceeded the allowable limits of current Health and Safety legislation: "If anyone owned Near Earth Objects, they'd be in jail." If there's a 1-km impact every 100,000 years, which kills 25% of the world population, the UK's statistical share of that is £12.5 million, at £850,000 per life, so the cost per year in lives alone is £123 million, before considering the infrastructure, property, heritage and commercial losses. Comparison with the actual costs of the 9/11 terrorist attacks suggested those could increase the loss by a factor of 7–8.

Nigel Holloway, a risk analyst at the UK Atomic Energy Authority, assessed the risks as being at the limit of tolerance, adding, "It is some time since astronomers have been called upon to serve in a directly useful fashion at the expense of their more theoretical aspirations. The discovery of the NEO risk changes that" [44]. Bringing the tsunami probability into the equation made the total risk package far beyond acceptable, putting even Tunguska-type events on a par with Chernobyl-type reactors. Even a 100-m impact in the Atlantic could wreck the UK economy.

If the threat was proven and the timescale known, as our scenario supposes, the most likely reaction in the financial world was summed up in the "Alex" cartoon of the *Daily Telegraph*'s financial section. After reading in an issue of *Science* that an asteroid impact could wipe out a continent and cause a mass extinction, Alex and his colleague are reassured by the thought that the universities are producing science graduates who "might be able to figure out a way to save us from being wiped out... brilliant scientific brains... engineers, physicists and those in the field of pure maths. Then we can lure them off into the derivatives market for tons of money as usual..."

> "Quite. We *still* pay more than NASA, and we'll want those eggheads to work out quick ways of shifting our risk exposure into a safe hemisphere if Earth gets hit [45]".

Surprisingly perhaps, some of the people who insure against risk took a different view. One might think that the predicted consequences of an actual event might outweigh the actuarial ones, calculated over 80 years. If 25–50% of the human race died in a 1-km impact event, how many claims would be filed, and how many paid out? To quote Tom Lehrer, "No one will have endurance to collect on his insurance." But to this author's surprise, seemingly those arguments did carry weight and cause the impact hazard to be taken more seriously in government and financial circles. At the 2003 conference, under the heading 'New U. S. Initiatives,' Jay Tate reported that the goals for all related programs were to be stated in terms of statistical risk and cost/benefit analysis [49].

This section of this book began with the question, 'Is there a danger?' The answer is yes. But as Lesley Wright pointed out, to act on a danger one has to recognize it. Babies will try anything. Adults believe others' accounts, even of events they've never experienced or can't experience. Men believe women's accounts of childbirth. But acceptance of the threat from NEOs remains abstract. To make it concrete we'd have to ask the dinosaurs, because only they have experienced it (Actually we should ask the crocodiles, because they survived it). Anything that makes an event easy to recall enhances public belief in its frequency; the more media coverage it receives, the more its frequency is overestimated, and vice versa. The risk of fatal cancer is overestimated, relative to diabetes, although statistically they're about equal; likewise death by fire and death by drowning.

Movies tell us that governments can deal with impactors, given a few months' warning (Both *Deep Impact* and *Armageddon* assume secret government space vehicles, just waiting to be called upon). What we don't know can kill us, but so can what we refuse to know. The 'ostrich effect' is especially pronounced with cancer. But fear, too, can be disempowering, leading to disjunction in response to the danger. The spacing in time between major impact events leads to a misunderstanding of the probability—the odds may be millions to one, but somebody wins the lottery every week [47].

Governments tend to think in 4–5 year cycles when assessing cost effectiveness. Investment in detecting and countering NEOs, when there is no apparent threat, is unlikely to win votes. The counter-argument will always be that the money would be better spent on health issues, or any other immediate concern—as witness John Braithwaite's question in the Preface. Changing government policy is a big hurdle to cross. The lesson of the last 20 years is that it won't be easy to arrange protection for Earth without a quantifiable menace to give it the necessary urgency. All we can do is keep watching, in hopes of finding something that is going to hit us, but not too soon to do anything about it—which brings us to the topic of Chap. 3.

Part II
Incoming!

3. A Designer Hazard

Bethan: But the moon smiles on them always,
The moon smiles back at them.
Sara: All of the world
Calls that a smile. I call it a frightened stare
At something crawling up behind us.

—Glyn Maxwell, *Wolfpit* [1]

In the 2000s, one of the major sources for information about impact hazards was the Cambridge-Conference Network, edited by Dr. Benny Peiser, who collected published work on all aspects of the topic from all over the world. On February 7, 2000, he printed an essay by Brigadier General S. Pete Worden, USAF, Deputy Director for Command and Control Headquarters in the Pentagon [2]. Emphasizing that the essay stated his personal opinion, Gen. Worden wrote, "Should we worry now about mitigating the NEO hazard? I would say no, until a *bona fide* threat emerges." And he concluded, "When, and not until, we find a likely threat, is the time to work hard on mitigation."

In February 2002 Bill Ramsay issued the challenge to ASTRA: "What would we do, in that case?" And specifically, to focus the discussion, "What would we do if we knew there was going to be an impact in ten years' time?" 10 years was picked because it's not so far off that governments can leave action to the administration that succeeds them, nor so close in time that nothing can be done. It took 7 years to put man on the Moon after President John F. Kennedy set his deadline in 1962 that it should be achieved "before this decade is out." We need to put national governments, international organizations, design teams and the aerospace industry under similar pressure.

"After all," Gen. Worden had added, "we can't reliably divert an NEO until we know more about its structure." But even before that, the author pointed out at the first meeting of Ramsay's

D. Lunan, *Incoming Asteroid!: What Could We Do About It?*,
Astronomers' Universe, DOI 10.1007/978-1-4614-8749-4_3,
© Springer Science+Business Media, LLC 2014

project, we need to know what kind of object it is, what orbit it is in and what opportunities, if any, we will have to reach it, deflect it or destroy it. Even though our aim was to look at the political realities of the situation, they would depend to a great extent on the technical options that were open. So we couldn't discuss all possible cases or combinations of circumstances. If we came up with a plausible scenario to deal with one specific hazard, then we and others could look afterwards at how it would change if the circumstances were different.

Expanding on the reasoning above, to make this project realistic we need a 'negative Goldilocks' scenario—not that our incoming hazard will be 'just right' for deflection, but that nothing about it will be too unfavorable, nor make the task too easy. The object we are to deflect must be:

- Not too easy to reach, nor yet so difficult that it couldn't be done in the time; not so small, nor so fragile, that we can easily turn it aside or break it up, but not so big that no steps we take will make much difference;
- Of a plausible composition, for an object of reasonable size in such an orbit, and not of material that makes the task impossibly difficult in the 10-year timescale allowed, nor of a material that does us any special favors.

Inevitably, this led to the accusation that we were designing the asteroid to meet our scenario, rather than the other way around. But there was no choice.

Comet or asteroid? This was the first question to be answered. As we've seen, comets coming from the Kuiper Belt or the Oort Cloud will be traveling at very high impact velocities and are likely to be detected only a year before collision, at most. At the moment of writing, we've had a near-perfect example of what we couldn't cope with in the latest Comet McNaught, C 2013 A1, which initially was given a one in 2000 chance of hitting Mars on October 15, 2014 (Fig. 3.1). The comet, 1–3 km in diameter, was only discovered on January 3, 2013, at Siding Spring Observatory in Australia, though earlier images were found dating back to October 4, 2012. Mars Express, Mars Odyssey and Mars Reconnaissance Orbiter should all still be operational in orbit and will return spectacular pictures even if it's 'merely' a very close encounter. MAVEN, the next Mars probe, should arrived shortly before the

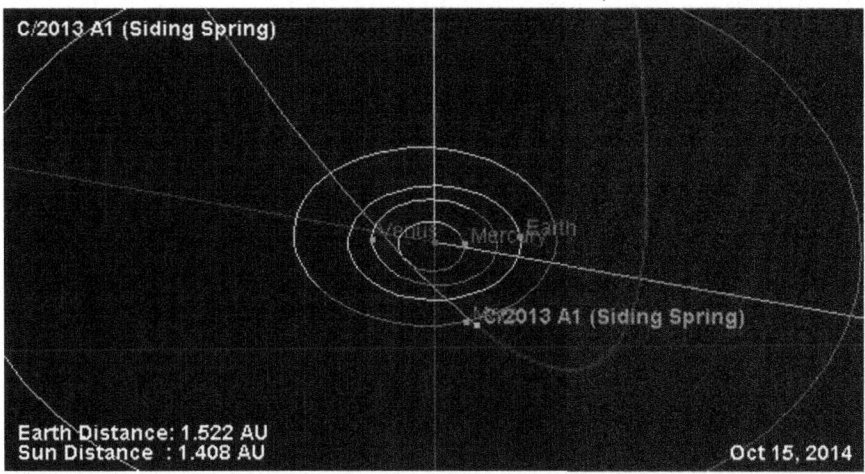

Fig. **3.1** Comet McNaught, 2013, A1 (Siding Spring) at Mars, October 2014 (NASA/JPL)

event, but unfortunately its instruments may not be operational in time. Further observations put the closest approach within 68,000 miles, but reduced the chance of impact to 1:120,000.

"If it does hit Mars, it would deliver as much energy as 35 million megatons of TNT," said Don Yeomans of NASA's Near-Earth Object Program at JPL. That would be about 10% of the energy released by the Chicxulub impactor, enough to cause major climate change on Mars. The atmosphere would thicken and fill with dust, so temperatures could go up or down; the solar-powered Opportunity rover might be lost, but it's thought that Curiosity, which is nuclear-powered, would survive [3]. That does rather depend on how close the impact point is to the rover's location in Gale Crater. It's not generally realized that in the absence of oceans, Mars has more land surface area than Earth has, but it still has roughly half the diameter, half the gravity and one-eighth the volume of Earth, so the destruction caused by the impact would be much more widespread than it would be on Earth.

With so little warning of an object of that size, at the present stage of our technology, it's not likely that there would be time to prepare any interception attempt other than a nuclear weapon launched during the final approach to Earth. That leaves very little to discuss except whether it's likely to succeed or not, and, since it's not very likely to succeed, the question becomes what we could do to try to ensure that some self-supporting element of the human race could survive with the cultural resources to rebuild our civilization. Dr. Arthur Hodkin's contribution to this project in Chap. 7 addresses that issue.

Periodic comets (usually identified with P/before the name(s) of the discover or discoverers) are a different matter because with one of them, there's likely to be at least a few years and maybe a few decades in which to take action. Each of the periodic comets, which regularly return during human history, belongs to a family of comets that have been captured by the gravitational field of one of the giant planets, Jupiter, Saturn, Uranus or Neptune. (Halley's Comet is one of the few members of Neptune's family.) Not surprisingly, the largest family is Jupiter's, also known as short-period comets. Their orbital periods are typically around 3.5 years, and because they come back to the Sun so frequently, all the current ones have lost much of their ice content already and are too faint to be visible to the naked eye unless they pass very close to Earth.

As a result they weren't recognized as a class of comet until quite recently. The classic example is Encke's Comet, which was first observed in 1786 by Pierre Méchain [4]. Its 1795 return was observed by Caroline Herschel, and in 1805 it was seen by Axis Bouvard, but it was still not recognized as periodic. In 1818 Pons observed it and recognized its periodicity in 1819; since then it's been monitored in great detail.

Clube and Napier thought their super-comet (Chap. 1) broke up around 3000 BC, giving rise to the huge Taurid meteor shower and to many short-period comets, and they thought that displays in the sky and impacts on Earth could account for the sudden rise of positional astronomy, all over the ancient world, between 3000 and 2700 BC [5]. It's now known that around 2700 BC the Strohl meteor stream split from Comet Encke, and around AD 1000 the North Taurid meteors split off likewise. But it's also known now that the asteroid Oljato split from Encke's parent object around 8000 BC, and as it is in the main Taurid stream, which indicates the parent body had been around and shedding material for even longer [6].

The main Taurid meteor stream, the Beta Taurids, is crossed by Earth in its orbit every June. It bombards the daylight side of

Earth, so its existence wasn't discovered until wartime radars began scanning for V2 rockets launched towards London from the continent. The Beta Taurids are the most intense meteor shower of the year, and it's been said that if only it came by night, scientists would have no problem getting the public and politicians to take the impact threat seriously. For there seem to be other large objects among the Taurids: candidate members include two large objects that hit the Moon while the seismometers left by the Apollo astronauts were still operating, the Tunguska object of June 1908, and the 12 or more near-simultaneous impacts on the far side of the Moon in June 1178 (Chap. 1).

After the multiple impacts on Jupiter by Comet Shoemaker-Levy 9 in 1994, that event is easier to understand, but there are still puzzles associated with it. Comet Shoemaker-Levy had broken into more than two dozen fragments during a previous encounter with Jupiter, while being trapped in a very long orbit around Jupiter, which brought it back for the collisions. The exposed surfaces of the fragments became active even at that distance from the Sun, and the stream of debris dispersed until the impacts were on average 24 h apart. Outgassing from Encke's Comet has noticeable effects on its orbit even now. The 1178 AD objects might have been separated by a previous close pass of Earth. In 1173 "two moons were seen in the sky" in England, while from Scotland a star was seen in the west, visible even in daylight, and by night other stars could be seen around it [7]. At this distance from the Sun, one would expect fragments to separate rapidly. Yet the impacts reported by monks at Canterbury occurred in less than 45 min, and it appears that they all struck the Moon in the same place, forming the crater Giordano Bruno instead of a chain like the ones found on Jupiter's moons Europa and Callisto.

So if we were threatened by a comet, and we had a decade to deal with it, the most likely candidate would be a short-period one, probably of the Comet Encke family, and the one that threatened us might have become a family in turn. When we looked at what that might entail, even with a single impactor, we realized that for our purpose, there is far too much variation in the structure and composition of comets. In the film *Deep Impact*, the astronauts land on the night side of the comet's nucleus, but are caught by violent outgassing as the side they're on rotates into sunlight. The snapshots we have of comets from spacecraft suggest that gas emissions are more localized, coming from only small active zones on the nucleus, but the nuclei have also turned out to be much darker than expected and covered with thick dust, perhaps even with tar, while the interiors remain mysterious but highly diverse.

As noted in Chap. 1, the orbital eccentricities, inclination and even the direction of a comet's motions create big potential problems for comet rendezvous missions. The rotations of the nuclei and changes of orientation such as Comet Encke's add still more to the complexity of the operation. In *The Comet, the Cairn and the Capsule*, the author took major poetic license with the imaginary comet's rotation, which could not have remained virtually Sun-synchronous through perihelion passage as the plot required [8]. To eliminate the unknowns, we decided, what we wanted to cope with should be more solid and predictable—in other words, an asteroid.

The two main clusters of asteroids lie in the main Asteroid Belt, between the orbits of Mars and Jupiter, and the Kuiper Belt, beyond the orbit of Neptune. The indications are that the reddish Kuiper Belt objects may resemble icy comets more than rocky or metal asteroids, and that if they are deflected into the inner Solar System they may become comets, perhaps highly dusty ones like Comet Kohoutek. The same may be true of asteroids in the outer main belt, which also tend to be reddish in color. But most of the main belt objects have formed by the collisions of around eight 'protoplanets' in the early history of the Solar System, apparently prevented from merging into a single planet because the growing pull of Jupiter made their collisions too violent for that. Those protoplanets were of differing sizes and compositions, and had evolved to different extents due to internal heating by radioactive decay, some of them differentiating to form metallic cores, others remaining more mixed internally even if they formed differentiated crusts. The wide variety of stony, stony-metallic and metallic meteorites caused puzzlement for a long time, but it's now clear that they are pieces of different objects.

Like the comets of the Oort Cloud and the ambiguous objects of the Kuiper Belt, main belt asteroids can undergo collisions or mutual perturbations that send them into the inner Solar System.

Many of them adopt elliptical orbits still reaching out to the main belt or even beyond, categorized according to whether or not they actually cross Earth's orbit and go nearer to the Sun. (The asteroid Icarus goes so close that at perihelion, the nearest point to the Sun, it's thought to glow red-hot!) But there's another puzzling category of objects in orbits very similar to Earth's own. Some of the smaller ones may be fragments blasted off the Moon by bigger impacts. The Canterbury monks saw 'sparks' flying off in 1178, but the ones in orbits like ours show the full range of types, up to and including the very dense nickel-iron objects that would be very destructive if they hit us.

The good news is that those objects come by comparatively seldom, typically once every 20 years or more; if they have to be reached by spacecraft, it takes less energy than going to the Moon—*when they're going past* (see below); and dynamical analysis indicates that if one is going to hit us, it will make multiple passes of Earth first to give us warning. In the critical pass it must go through a 'keyhole' only 100 km across, in order to hit Earth the next time, and at that point a deflection of just a few miles could be enough to prevent the impact, at least for many more years if not altogether [9].

We know that the Encke's Comet family also includes asteroidal objects such as Oljato. After considerable discussion, we decided that our imagined threat would be in an orbit of that type. But we didn't want it to be an easy object to dispose of, like the Tunguska one that appears to have been part of the Encke family. Dr. Anatoly Zaitzev believes that such a loosely structured object is the most likely thing to hit us in the immediate future, and that it could be disintegrated by a nuclear warhead with a yield as low as 1 megaton (and if not disintegrated, it would at least be so damaged that it would break up on its way into Earth's atmosphere) [10].

If such a threat was identified, and we could be sure that was all there was to it, stopping it in that way would be well worth doing. If the Tunguska impactor had come 3 h later, it could have exploded over St. Petersburg; 6 h later, over London. Its major blast damage covered an area that would fit within the M25 motorway around London (Fig. 3.2), or New York City [11]. Hundreds of thousands would have died, millions would have been injured— a much bigger event than the Hiroshima bomb, with about 200

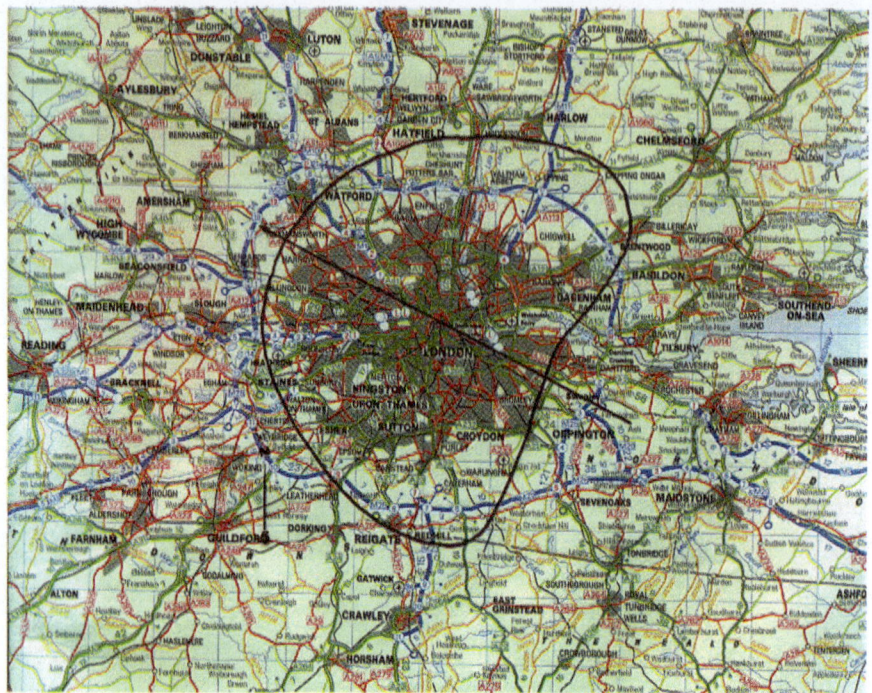

FIG. 3.2 Tunguska major destruction superimposed on central London (© Spaceguard UK, 2002)

times the explosive yield. It's thought that it depleted atmospheric ozone over the northern hemisphere by 30–50% [12] and may have quenched it altogether for a year on the impact latitude. But Steel also believes that Comet Encke was inactive until the eighteenth century, hiding a large amount of ice within it [11], and if we send a nuclear warhead to such a 'dead comet' masquerading as a dust ball, we may bite off a lot more than we can chew.

Dr. Zaitzev's specific proposal, part of his plan for a planetary defense system, is that two spacecraft should be positioned at Earth-Sun Lagrange 4 and 5 points in the orbit of Earth, as sentries for the planet. The Zenit booster, manufactured in the Ukraine and currently used by the Sea Launch corporation, is capable of launching 1.5 megaton warheads into asteroid intercept trajectories. From Earth-Sun L4 and L5 points, a system constantly scanning for threats to Earth has two relatively small areas of sky to scan, less than 60° across and much smaller than the global 360°

Fig. 3.3 Chondritic meteorites (NASA)

that an Earth-based watch would require. That would allow 3–5 days' warning of an incoming impactor, and a dedicated Zenit could be scrambled to intercept one in just 1.5 h, with a second launch from the same pad 5 h later.

To give ourselves a serious, but not impossible, task we decided we must have a more solid object. On the other hand, it seemed unfair to give ourselves the worst possible threat, which would be a nickel-iron asteroid massing billions of tons. Such an extremely dense and refractory object might prove highly resistant to any of the deflection techniques we had in mind.

So we decided, in the end, that our threat was made of stone. The choice appears to be justified by the percentages of compositions found in Earth-grazing asteroids, reflected in the meteorites that fall on Earth. About 4% are iron meteorites, 1% are stony-iron mixtures (from incompletely differentiated parent bodies) and 95% are stones, 81% of which are chondrites [13], two-third of those Type LL chondrites (Fig. 3.3) [14]. Of course, our meteorite sample of what's up there is biased by Earth's atmosphere, which destroys most of the lighter material on the way in, but it's what does reach the surface that concerns us most here.

So we decided that the object should have a diameter of 1 km, to make it detectable but not too difficult to detect. Pursuing the same negative Goldilocks reasoning we decided not to go for a dust ball or a carbonaceous chondrite (too easy) or a nickel-iron object (too hard), but a chondritic rock with some nickel-iron content. If it's homogenous, it would originally have been part of a body larger than 100 km in diameter.

Using negative Goldilocks, we had arrived at the orbit of our hazard and its composition by different routes, and Gordon Ross suggested that we'd produced an incompatible result—a relatively dense stony object in an Encke's Comet-type orbit. It would be reasonable to expect Tunguska-type objects in that orbit, and although Oljato appears to be a solid asteroid, it's more probably a Tunguska-type dust ball. But the situation is more complex than that.

The year 2700 BC is the date around which the Strohl meteor stream split from Comet Encke and also the approximate date of the impact event that formed the Henbury craters near Alice Springs in Australia. At that time 14 craters were formed after the break-up near the ground of an incoming object made of nickel-iron. The two events may be unrelated, but if so, it's a remarkable coincidence. Since we know now that the Encke family parent began to break up at least 5,000 years earlier than Clube and Napier first supposed—and since they themselves thought that 'super-comets' are captured from interstellar dust and gas clouds as the Sun periodically plunges through the plane of the disc of the Milky Way—then, perhaps the parent body was even larger and more differentiated than they imagined.

Or perhaps, our object has been perturbed out of the main belt, initially by encounters with other asteroids and then by the influence of Jupiter. There are many satellites of Jupiter that appear to be captured asteroids, plus the two families of 'Trojans.' So it's entirely possible that a stray stony object from the Asteroid Belt could be perturbed into the type of orbit the Encke family follow. The asteroid Eros, which was visited, orbited and finally landed on by NASA's NEAR-Schumacher space probe, is a stony object in an orbit that approaches Earth's on the outside, and which could evolve to cross it.

To divide approximately into 10 years, the author suggested that we give our hazard an orbital period of 3.33 years around the Sun, like Encke's Comet (Fig. 3.4). That means that, if we detect it and establish a strong likelihood that it will hit Earth in 10 years' time, there will be two opportunities to rendezvous with it at perihelion, its closest approach to the Sun. The first will be 3.33 years after discovery, the second 6.66 years after discovery, and there will be a last chance to take action, however desperate, on its

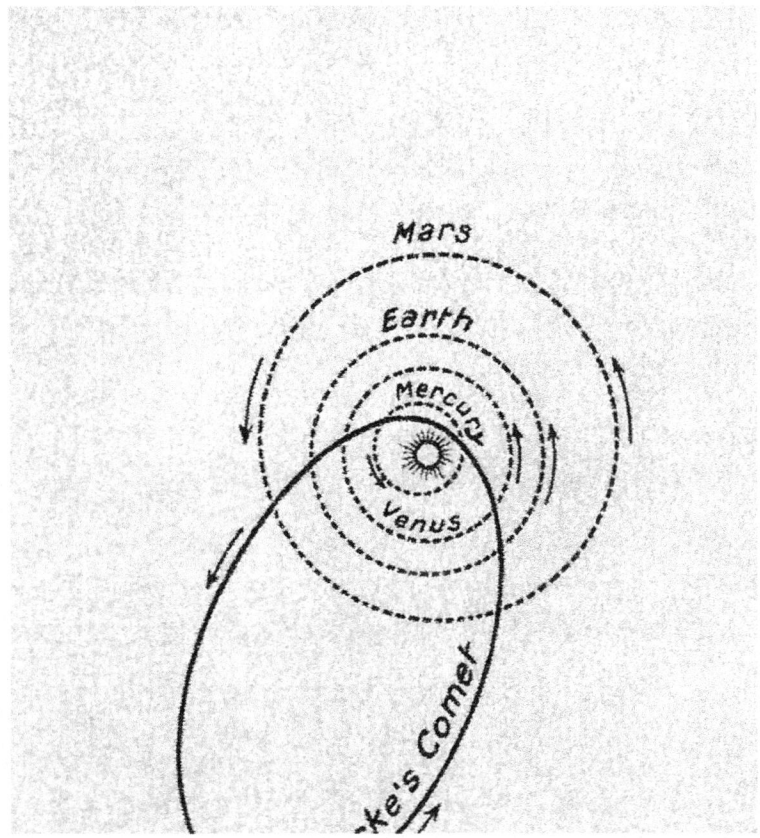

FIG. 3.4 Orbit of Encke's Comet [15]

third pass around the Sun, when it will try to go through the space occupied by Planet Earth at the time. Bill Ramsay characterized this as 'the three-hit scenario'. A major attraction for our purposes is that it focuses attention on what can be done in those three 'windows,' rather than a prolonged argument about the possibility of launches at other times.

Dr. David Asher of Armagh supplied some very useful information on the known compositions of asteroids and confirmed that a stony asteroid could be found in an orbit such as Encke's Comet's, also pointing out that the orbital elements could change comparatively fast. He cited 2000 UV13, in whose discovery he took part [16]. With a perihelion distance of 0.91 au and aphelion at 5.12 au, diameter 5–12 km, last close to Earth in AD 600, it's not currently hazardous, but expected to become so in 3,500 due to perturbation by Jupiter. 1915 Quetzalcoatl has a perihelion at 1.09 au and aphelion at 3.99, so it doesn't currently cross Earth's orbit, but is expected to start doing so in 2020. 1580 Betulia's perihelion varies between 1.2 and 0.4 au, with aphelion at 3.88 au, and eight intersections with Earth over 5,000 years.

Meanwhile, Gordon Ross had raised another objection to that orbit. He wanted to differentiate the question of deflecting the asteroid from the issue of the propulsion system required to get there, and because Encke's Comet approaches closer to the Sun than the planet Mercury, an exotic system such as ion drive or solar sailing would be needed to get there (see Chap. 5), even with an unmanned mission. In the 1970s the Mariner ten probe reached Mercury with a conventional rocket launch, but that was because the planet Venus was in the right place to provide gravity assist to the probe. We can't guarantee that Venus will be placed to help us even once during the 10 years we have, let alone twice.

Encke's Comet, however, approaches the Sun as closely as 0.342 au, swinging out to 3.8. Since Gordon's remit was to consider the use of solar power to deflect the asteroid, he also worried that the range of incoming solar energy would be too high near perihelion, when it would be more than nine times higher than at Earth's orbital distance, and too low at aphelion, furthest from the Sun, when it would be over nine times less than here. His calculations indicated the *average* solar input would be adequate for the system he had in mind (Chap. 5), but he was concerned that the engineering loads on the system covered too great a range. The same point would apply even more strongly to manned missions, placing extreme demands on life-support systems.

Shortly before our conference at the Spaceguard Center in October 2003, there was an international stir about a passing NEO designated 2003 QQ47. 2003 QQ47 will return to the vicinity of Earth in 11 years, and at first it seemed there might be a risk of collision, though that was quickly ruled out as the analysis of its motion continued. In size and composition it was very like our imaginary impactor, and Andy Nimmo of ASTRA suggested that we might substitute it for Goldilocks. But on consideration we decided against it, because it would change our scenario too much at such short notice; in particular it would eliminate the three-hit scenario. Objects like QQ47, in orbits very close to Earth's, are only easy to reach when they're passing us. Even if QQ47 were menacing us it would not be anywhere near Earth again in the near future, and exotic propulsion would be needed to reach it at any time during the 11-year period.

If we keep the 3.33 year launch windows, however, we can still change the characteristics of the orbit, and, interestingly enough, that doesn't change the total amount of sunlight which the asteroid received over its 'year.' After trying a variety of alternative orbits, we found ourselves with this more reasonable range of options.

Perihelion (AU)	Aphelion (AU)	Period (years)
0.7	3.7	3.33
0.9	3.5	3.33
1.0	3.4	3.33
1.7	2.7	3.33

To avoid any arguments about what can or can't be done with existing rocket technology, for the moment we've settled on the 0.9–3.5 au orbit. The characteristic velocity of such a mission, also called the total delta-v requirement, i.e. the sum of all the velocity changes which the spacecraft has to make during it, are roughly equivalent to a flyby mission to Mars, well within the scope of existing rocket launchers such as the European Space Agency's Ariane V and Japan's H-2. We'll look at that question in more detail in Chap. 4. Moving the asteroid's perihelion nearer to Earth's orbit would make the elliptical shape of the orbit fatter (decrease its eccentricity), moving the points where it crosses Earth's orbit further apart, but wouldn't affect the general transfer orbit times discussed in Chap. 4.

On January 8, 2011, an asteroid in a somewhat similar orbit was found and classified 2011 AG5 (Fig. 3.5). It posed no threat in 2023 or 2028, but was first reckoned to have a 1 in 625 chance of hitting Earth on Feb. 5, 2040 [17]. With a diameter of 140 m and that type of orbit, it just qualified for classification as a PHA (potentially hazardous asteroid), although its composition is currently

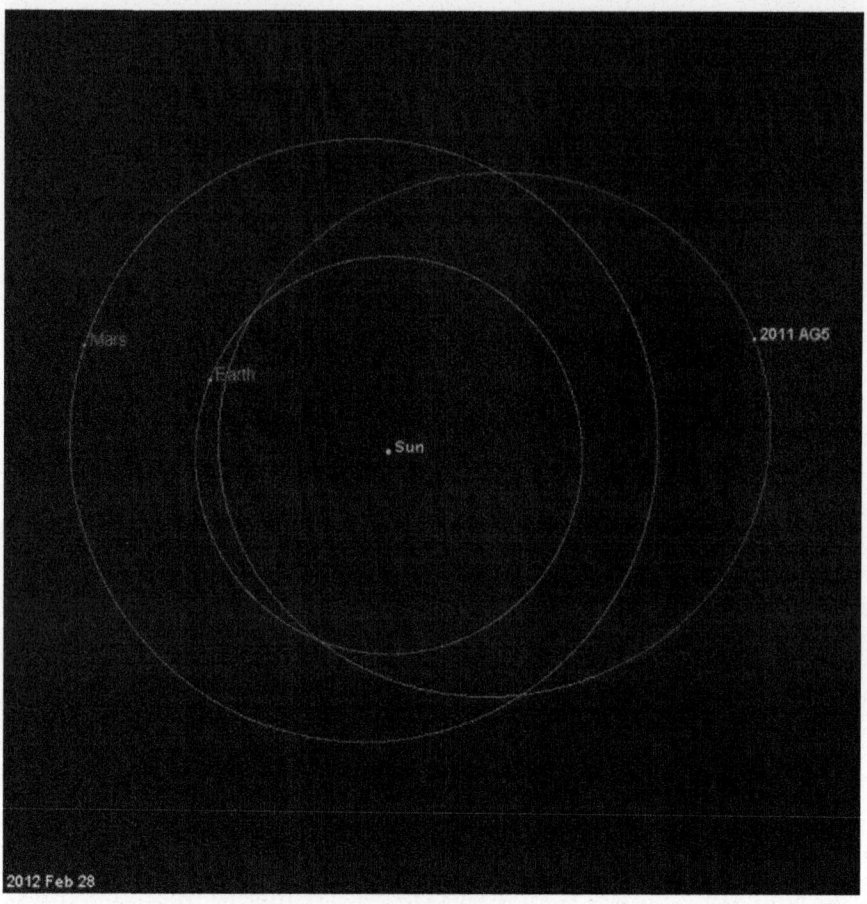

FIG. 3.5 Orbit of asteroid 2011 AG5 (NASA)

unknown and the impact could have been in the 100-megaton range. Calculations were to be refined at a distant pass in September 2013, but by the end of 2012 the threat had already been disproved [18]. Nevertheless, threats like our Goldilocks object are still out there.

One such, 2012 LZ1, was discovered by Robert McNaught on June 10, 2012, passing at 5.3 million km. At first it was thought to be 'only' 500 m across, but radar scans from Arecibo proved it to be twice that size (Fig. 3.6) [19]. It poses no threat at its next pass in 2053, nor at any time in the next 750 years, but its size puts it in the PHA category.

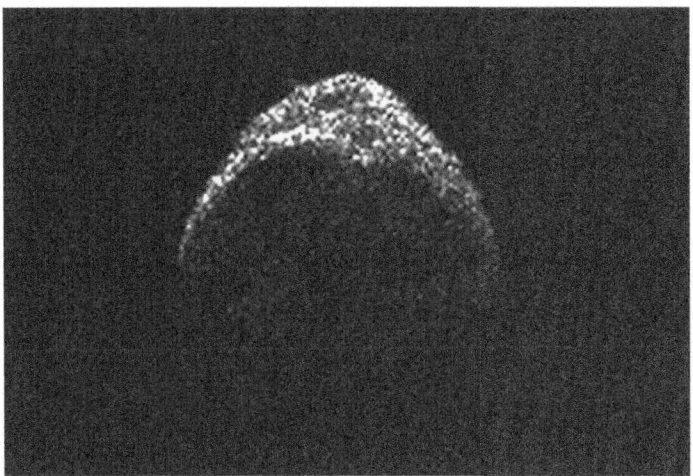

FIG. 3.6 Arecibo radar image of asteroid 2012 LZ1, 1 km in diameter

Readers who have been following the impact issue over the last 10 years or so may be wondering why we didn't focus on the asteroid Apophis (2004 MN4), which has been used as the basis for many other studies. When it was discovered in 2004, there was calculated to be a 2.7% likelihood that Apophis would hit Earth in 2029, and if it didn't, there could be a still closer pass in 2036 [20]. With the limitations of the 2004 data, the chances of an impact in 2036 couldn't be determined until 2029 (Fig. 3.7). Former Apollo astronaut Rusty Schweickart was among those who argued strongly for a space probe mission to put a transponder on Apophis, to establish by 2017 whether there was a serious danger [21].

There were a number of reasons for not focusing our discussions on Apophis. One obvious one was that our project had been running for 2 years before Apophis was discovered; another was that it seemed to be a relatively easy target (stress *relatively*), in an orbit significantly closer to Earth's own, with a relatively long timescale for action, compared to the Goldilocks object; and still another was that there was a lot of uncertainty about its size, with estimates of its diameter ranging between 100 and 400 m, but all of them much smaller than the hazard we had postulated. (As of January 2013, the current estimate was 325 ± 15 m, determined by ESA's Herschel space telescope [22].) Its composition was equally

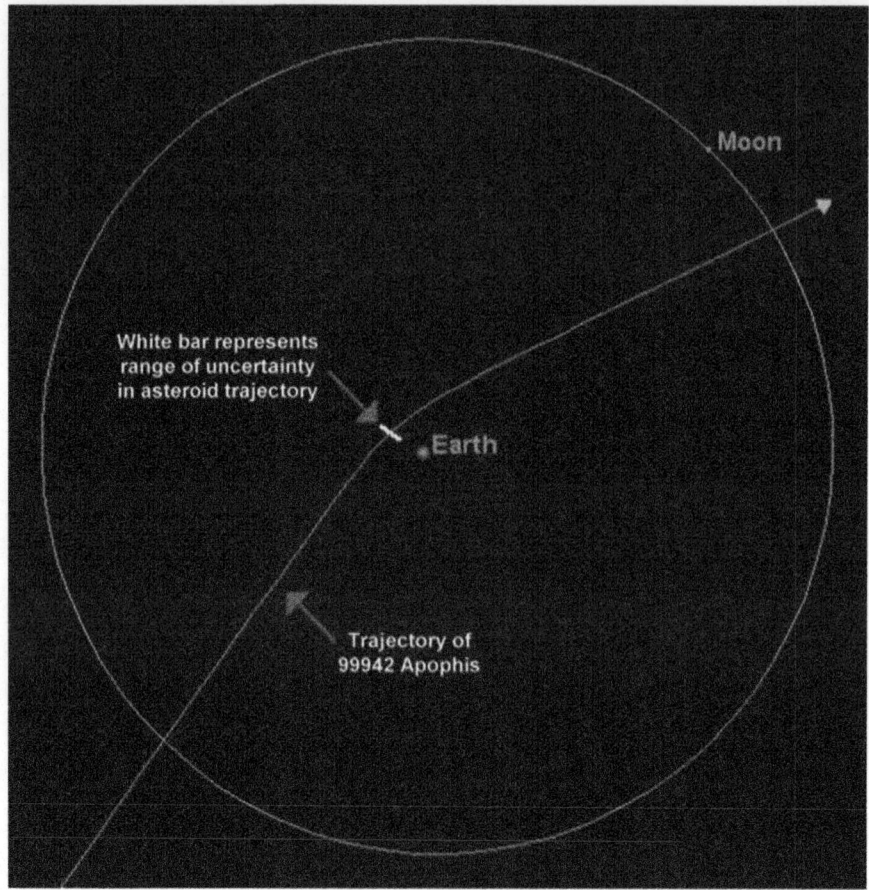

The figure labels (within image): Moon; White bar represents range of uncertainty in asteroid trajectory; Earth; Trajectory of 99942 Apophis

FIG. 3.7 Uncertainty in the 2029 trajectory of Apophis, as of 2009 (NASA)

uncertain, whereas we could talk with certainty about Goldilocks' size and composition because we had made them up.

Finally, it seemed likely that further observations would pin down the Apophis orbit before 2029, or even 2017, just because it was attracting so much attention. Sure enough, earlier detections were tracked down, and the odds of a 2029 impact came down to 1:45,000, then to 1:250,000 [20], and then to zero [22], while the chances of an impact in 2068 remain at less than 1:300,000. If we had shifted our attention to Apophis we would have wasted our time, because it's no longer a credible hazard on anything like the 10-year timescale Ramsay originally postulated.

If Goldilocks Hits

Death is like an arrow already in flight, and your life only lasts
until it hits you.

—Georg Hermes [23], German theologian (1775–1831)

If a 1-km rock asteroid like Goldilocks strikes Earth, it may not
be an extinction level event, as *Deep Impact* calls such impacts, but
it could be a disaster on a level never experienced by civilization.
The only eyewitness accounts are from far enough away for the wit-
nesses not to be directly affected, like the Sumerian record of the
Köfels event and the Hittite legend of the Moon falling to Earth—
neither of which left mute testimony to the devastation in the form
of craters, due to the types of terrain at the impact points.

Sea impacts in this size range are surprisingly difficult to
pinpoint. As Jay Tate of Spaceguard UK pointed out at the 2012
Glasgow seminar, a 1-km strike in deep enough water may not
penetrate the crust, and if so may not form a sea-floor crater at all.
No candidate feature has been found to correlate with Australian
Aborigine stories of a 'great white wave' around AD 1500—thought
to refer to the coming of European settlers, until boulders weighing up to 100 t were found wedged 33 m above sea level and more
than 100 m inland on the coast of New South Wales, along with
gravel dunes 130 m above the water [24]. They may tie in with
related stories of fire in the sky, and possibly with the great fires in
New Zealand cited by Duncan Steel (see Chap. 1). If so it would be
an impact event and not a Tunguska-style airburst, though Steel
believes that even Tunguska-type airbursts can generate killer tsunamis. "About the worst place for an impact of any size is in the
open ocean. The sea is deadly efficient at transporting energy from
one place to another. And I do mean deadly" [11].

It seemed that there was a candidate sea impact in the Eltanin
crater, formed 2.15 million years ago by an asteroid impact in the
Pacific off the tip of South America. This impact generated tsunamis for which evidence is found all around the Pacific rim, starting
over 50 m high and still 5–10 m high when they reached Alaska
24 h later. The event lies at the transition from the Pliocene to
the Pleistocene era, and may well have triggered a planetary ice
age. The impactor is usually said to have been in the 0.8–4.0 km

Fig. 3.8 The Eltanin impact site, 2012 (NASA)

size range, and the authors of the paper consulted gave it as 1 km [25]. But when the crater was first discovered, the upper figure was preferred [11], and since the sea-floor feature is up to 80 km across and surrounding concentric features range out to 200 km (Fig. 3.8), possibly larger than the Chicxulub crater, a 4 km impactor seems more convincing unless the impact velocity was very high for an asteroid.

On land, a 5.5-km impact crater was recently discovered during a geological survey of mineral and water resources near the town of Decorah in Iowa (Fig. 3.9) [26]. It's one of a chain of craters stretching across the United States from Oklahoma to Lake Superior, apparently caused over 470,000 years ago by the break-up of an incoming L-type chondritic asteroid. The Decorah fragment was about 200 m across, and the impact energy was around 1,000 megatons, so the original incoming body was well up in the Goldilocks range. Although its existence had been suggested by an amateur geologist named John Young 10 years earlier, the crater is so deeply buried that it could only be verified by aerial surveys, confirmed by the existence of shocked quartz below the 15 m of surface shale. It's not the only buried crater in Iowa. There's also Manson Crater in Humboldt county, 15 km wide and 74 million years old, more like a Goldilocks object that came to ground in one piece. "[It] would've leveled trees for a radius of 180 km," noted the news report, citing one of the least of the probable consequences.

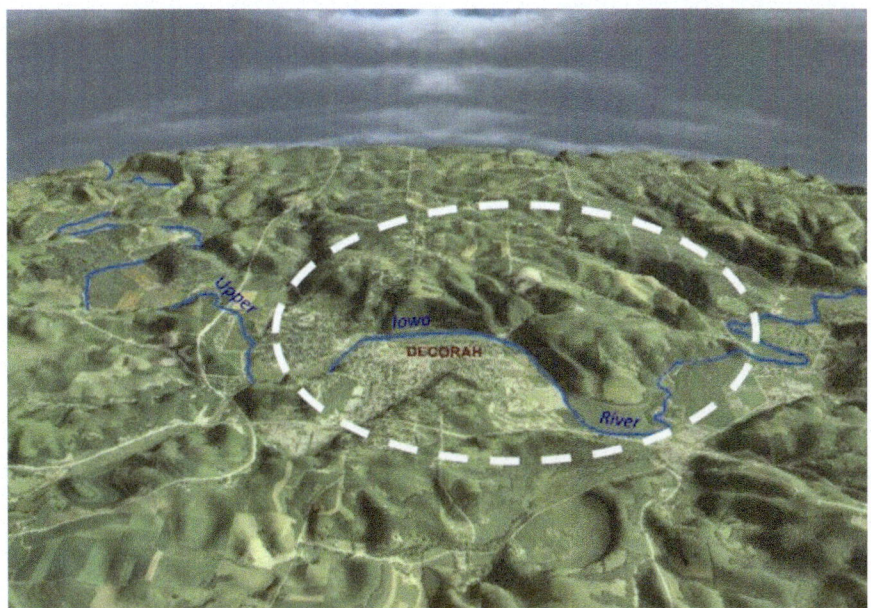

Fɪɢ. **3.9** Decorah impact feature, Northeast Iowa (© U. S. Geological Survey, 2013)

For a more recent event on the Goldilocks scale, whose consequences are more visible today, we can look to the Nördlinger Ries feature in Germany (Fig. 3.10). The original crater rim was 24 km across, formed 14.3–14.5 million years ago in the Miocene period, and today the city of Nördlingen nestles within it like the village of Avebury within the great Neolithic ring in Wiltshire, England, but on a far greater scale. It was identified as an impact feature by Eugene Shoemaker, and among the shocked features he discovered were the millions of microscopic diamonds, formed when the impactor—about 1.5 km in diameter—struck a graphite deposit. These diamonds are found today in many of the city's stone buildings. The asteroid came in at 35–50°, at around 20 km/s (45,000 miles per hour), and the energy release was about 1.8 million times that of the Hiroshima bomb. This was another paired impact, like the Clearwater Lakes, and the second asteroid was about 150 m across. A field of secondary meteorites fell in what is now the Czech Republic; these Moldavite tektites are green glass, formed in the impact, and are in demand as tokens of love.

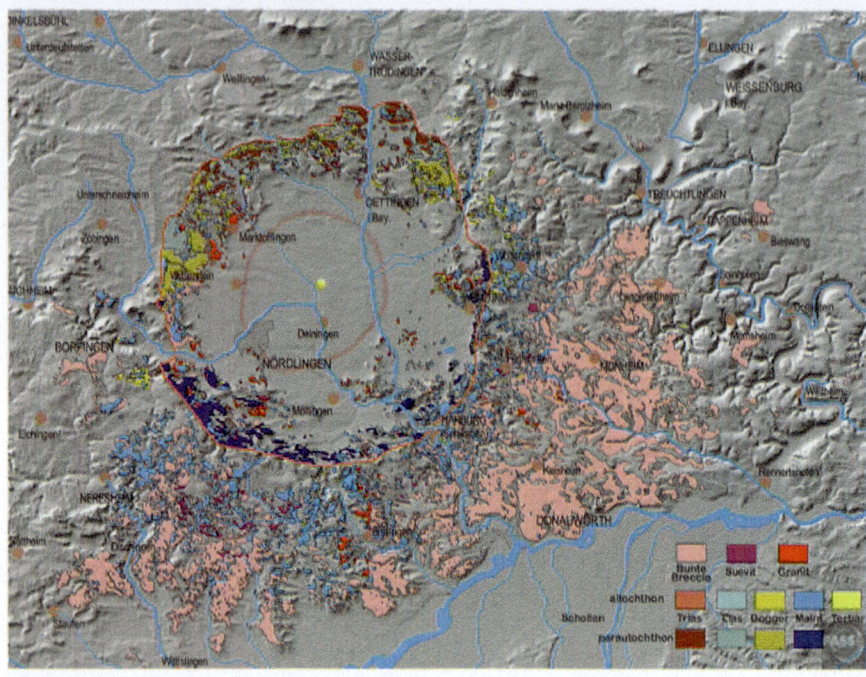

FIG. 3.10 Nordlinger Ries map (NASA)

Statistics give little impression of what this event must have been like, but a picture is worth a thousand words. For Sydney Jordan's comic strip "Lance McLane," in 1982, the author devised a story featuring a Goldilocks-scale impact in the Pacific Ocean (Figs. 3.11a, b, c). The impact occurred while the characters were testing a craft called VESSEL, designed for human exploration on Venus; as it happens, the water pressure on the seafloor off Easter Island, a kilometer down, is a good approximation to the 91 atmospheric pressure at the surface of Venus itself. VESSEL, the VEnus Surface Sustained Exploration Lander, had been designed to go into areas of volcanic eruption and high lightning activity [27]; maybe, just maybe, it could survive being sucked into a newly formed crater, which went through the seafloor into the magma beneath. Getting it out again would be another matter; but then, this was fiction.

As the characters began efforts to rescue one another, the ground shock, blast waves and tsunamis would be taking their toll, radiating out from the Easter Island epicenter. The waves from a

Fig. 3.11 (a, b and c). A Goldilocks-scale impact off Easter Island in the late twenty-first century (© Sydney Jordan, 'The Phoenix at Easter,' story by Duncan Lunan, *Lance McLane* comic strip, *The Daily Record*, 1982–1983)

Fig. 3.11 (continued)

1-km impact would be 4–5 m high in the open sea, 300 m high on the continental shelf. In the next story, Sydney Jordan tried to give some idea of the devastation, but wisely restricted his portrayal to the effects on and around just one of the Pacific islands. To show the effects across the whole Pacific basin would have taxed even his artistic powers.

Further afield, the effects would be global—perhaps not the worldwide rain of molten rock and conflagration that followed the Chicxulub impact, but depending on the angle of impact, quite probably intense local bombardments like the formation of the Moldavite tektite field. On land and sea, with 6 months to a year of darkness, the food chain would collapse from the bottom up; afterwards, the whole Earth would be colder as snowfall and bare ground reflected more sunlight back into space. Climatic changes might be temporary, but not if they triggered new ice ages, as the Eltanin impact may have done. In his 1998 Airdrie lecture, Jay Tate estimated that one-third of the human race could die: "The individual risk in a given year may be comparable to that of a plane crash or a tornado, but the difference is that no one comes to the rescue."

Speaking at the 2001 Charterhouse Conference, Dr. Benny Peiser pointed out that statistically, over the 5 million years of human evolution, there might have been as many as 50 1-km impacts—say 25 on a conservative estimate, 15 of them over 1.7 km diameter and with global environmental effects. In the same period there had been 14 different hominid species (now 15, with the discovery of extinct 'hobbits' in Indonesia), and anthropologists don't know why they are gone. When the human populations were so small, some of them could have been taken out by still smaller, 500-m 'continent-busters' [28].

The apparent similarity of the numbers may be mere coincidence, and even if true, the conjecture may be unprovable. But its implications deserve more than a moment's thought, and should focus our minds on what might be done to prevent a recurrence.

4. Detection and Reaction

Awake! For morning in the bowl of night
Has flung the stone that puts the stars to flight.

—Edward Fitgerald, *The Rubaiyat of Omar Khayyam*

Asteroid 1989-FC, later named Asclepius, caused some stir when it passed Earth at 400,000 miles in March 1989. Coming from sunward, the asteroid was unseen until it had passed by [1]. In March 1990 the magazine *Space-faring Gazette* reported that its orbit gave recurring encounters with Earth every 13 months, and it would collide with either Earth or the Moon within 30–40 years [2]. Its diameter was given as between 500 and 1,000 ft, depending on its composition and reflectivity. A dark, large, but loosely structured object would probably explode above ground without forming a crater, like the Tunguska one. 1989-FC had no cometary head and *presumably* wasn't icy (but see Chap. 1). And if the object were of rock or metal, with ten times the mass of a similarly sized iceberg, then the crater size could be ten miles or more, equivalent to the impact of an average-sized comet nucleus.

There was no overt response to the threat described in the *Gazette* report, which ascribed the impact prediction simply to 'scientists,' and came from a correspondent in the armed forces who could not be contacted. Numerous other reports around the world made no such prediction. Through the editors, and through press contacts in Australia, the author exchanged letters with a number of experts, including Dr. Duncan Steel, none of whom regarded the prediction as valid. It's now known that there will be no more close encounters with this object until after 2127, beyond which the asteroid's path is uncertain.

Nevertheless, Asclepius had passed straight through Earth's position in space 6 h earlier. The American Institute of Aeronautics & Astronautics held a conference on countering impact

D. Lunan, *Incoming Asteroid!: What Could We Do About It?*,
Astronomers' Universe, DOI 10.1007/978-1-4614-8749-4_4,
© Springer Science+Business Media, LLC 2014

threats, only 2 months after the *Gazette* article appeared, and in 1990, the U. S. government directed NASA to set up two committees, with international participation, to study the issues of detecting possible hazards and deflecting them. The detection group took the name of the Spaceguard Committee, after the proposal in Arthur C. Clarke's novel *Rendezvous with Rama* [3].

Ten years earlier, in June 1980, NASA's Advisory Council had suggested a "Spacewatch" to compile a catalog of all objects passing close to Earth [4], and a small search program was set up under Tom Gehrels at Kitt Peak Observatory in Arizona. Other watches for Earth-grazing objects were at Mt. Palomar under Elinor Helin, then by Gene and Carolyn Shoemaker of the U. S. Geological Survey, and in 1990 at the Anglo-Australian Telescope under Duncan Steel and Robert McNaught, at Siding Spring in Australia. In January 1992, NASA recommended that Congress set up a 'Spaceguard Survey' of six telescopes, to identify 90% of Earth-grazing hazards within a decade [5]. Towards the end of the year, NASA doubled its study funding for the project, which still left only a dozen people in the world working on the problem [6].

The favored answer at that time was that the incoming object could be diverted by exploding nuclear warheads beside it. The vaporized material would act like a crude rocket to thrust the comet or asteroid into a new trajectory [7]. But even passing objects are detected normally by the trails they make on photographic plates, not developed until after the event is over. Something coming towards Earth would be still harder to detect, and if it came out of the Sun like the Tunguska object it could arrive unannounced.

An asteroid coming straight at us wouldn't make a trail on a plate, even if it was developed in time—and even from the distance of the Moon, at average Earth-grazer velocities there would be less than 4 h to impact. It wouldn't show on Ballistic Missile Early Warning radar even then, because the computers were programmed to ignore objects at such distances (ever since the time in the 1960s, when they picked up the rising Moon and announced World War Three); and if by chance it passed through the USAF radar Space Fence, recording satellites traveling over the spine of North America, that would only be immediately before the impact. The space debris tracking program is scheduled to expand from 23,000 objects to hundreds of thousands in 2014, with a new additional site in Australia, but it still isn't geared to asteroid warning [8].

The chances that large numbers of missiles could be reprogrammed and launched with the required accuracy seemed very low. There would after all be no capability for mid-course corrections. The other problem was that if the object were shattered the consequences of multiple impacts could be worse than the original prospect—as Jay Tate put it, "you turn a cannonball into a cluster bomb." To protect Earth adequately, longer-term provision was necessary, and to neutralize a threat its physical properties must be known accurately, requiring visitation by spacecraft.

With a comprehensive catalog, dangerous objects could perhaps be intercepted at aphelion, their furthest point from the Sun, and deflected using neutron bombs [9]. This scenario was pushed by Dr. Edward Teller, the so-called father of the H-bomb, and by the 'Star Wars' SDI researchers at the Lawrence Livermore National Laboratory, whom Dr. Teller supported. In a time of decreasing tensions between the great powers, and with funding for the Strategic Defense Initiative under threat, critics such as Louis Friedman of the Planetary Society suggested that concern over impacts was just a cover for getting nuclear weapons into space [10].

A still more drastic nuclear proposal, not to deflect dangers but to detect them, was put forward by Arthur C. Clarke in early 1993. Project Excalibur would explode a 1,000-megaton device on the far side of the Sun for Earth's sake, and flood the Solar System briefly with microwaves, allowing the identification by radar of everything more than 3 ft across within the orbit of Jupiter—except objects on the Earth-Sun line at the time [11]. If the Test Ban Treaty was to be broken at all, for the sake of Earth's survival, perhaps there was no point in half measures.

Still the main Spaceguard program remained unfunded, and in 1995 a committee chaired by Gene Shoemaker proposed a more modest effort, followed in 1996 by proposals from the International Astronomical Union NEO Working Group and what was now the Spaceguard Foundation. Six 2-m telescopes would be used with centralized data collection, costing $50 million to set up and $10 million per year to run. Jay Tate noted for comparison that an F-16 aircraft cost $34 million and a single laser-guided smart bomb cost $4.5 million. The proposal gained verbal support from the Council of Europe, but still no funding. Worse still, in 1996 the Australian government withdrew funding for the southern hemisphere search at the AAT, which was still the only one of its kind south of the equator [3].

When Tate lectured in Scotland in 1998, still not a great deal was being done, and only the United States was taking the matter seriously in any way. NEAT, the U. S. Air Force's Near-Earth Asteroid Tracker, had been in use since 1995, with the more advanced LINEAR program to follow. LONEOS, a NASA program at the Lowell Observatory, began in 1993 and ran until 2008, using a 0.6-m Schmidt telescope; the estimated cost for a similar program in the UK had been put at £2.7 million to set up and £450,000 per year to run, with no hope then of raising it. David Asher was running a part-time search in Japan; Uruguay, Brazil, Argentina, Sweden, Finland and Italy all had plans but no funding. Part of the problem, as Tate saw it, was that this was a scientific-style program with a military-type objective. The scientific community would be satisfied with knowing the characteristics of a good statistical sample of asteroids, but to protect Earth, every asteroid had to be traced and classified as hazardous, harmless or a longer-term problem.

By 2001 the euphemism for asteroid deflection was mitigation, presumably in an attempt to sidestep the issue of putting nuclear weapons into space, and was the only issue on which the UK government of the day was not then fully backing the *Report of the Task Force on Potentially Hazardous Near Earth Objects*, published the previous year (Fig. 4.1). At that year's Charterhouse Conference Colin Hicks of the British National Space Centre criticized the word, which risked confusion with the detection strategy (and with alleviating the consequence if an impact does occur—see Chap. 7). Tate was even more forthright, stating "Detection and follow-up without mitigation equals scientific masturbation," and calling for public awareness and confidence to be raised in support of more funding. "Big explosions, dinosaurs and spaceships — we can't lose."

He was now in a position to speak more freely than he had been in 1998, when still a major in the Royal Artillery. He had noted the potential threat to the UK at the time of the 1994 impacts on Jupiter, and found that government departments

Fig. 4.1 Report of the UK government task force, 2001

knew nothing about the impact issue, let alone had any policy regarding it. In 1996 he submitted a paper to the Ministry of Defense and the Department of Trade and Industry, which the MoD dismissed. After the conference at Versailles in which the Spaceguard Foundation was established, he submitted another to the Ministry of Science and Technology, suggesting a three-person study. The Ministry of Defense responded that we had no rockets big enough to deflect asteroids, so nothing at all was done. Tate's first public lecture on Spaceguard was to the Society for Popular Astronomy in London, in April 1997, and the growing publicity required him to make a choice between Spaceguard and his army career. By the time of the Charterhouse Conference he was a free agent, and now runs Spaceguard UK (Fig. 4.2) from the

FIG. 4.2 Spaceguard UK logo

FIG. 4.3 The Spaceguard Centre, Knighton, Powys, Wales (© Spaceguard UK, 2013)

Spaceguard Centre in Powys, Wales, opened by Sir Patrick Moore in September 2001 (Fig. 4.3).

At this book project's two seminars in 2003 and 2012 (see Preface), Tate reported in detail on current search strategies. In 1998 Congress had directed NASA to locate 90% of all asteroids over 1 km in diameter within 10 years, and in 2003 that was thought to be half-way to completion, with five dedicated search programs running in the United States. Spacewatch and LONEOS had been joined by the Catalina Sky Survey, an NEO search set up by NASA, with a 1.7-m telescope north of Tucson, and was in process of upgrading the UK Schmidt telescope at Siding Spring for southern hemisphere scanning, to begin in 2004. Elinor Helin had retired, but the NEAT program was continuing with a 1.2-m USAF telescope. MIT's LINEAR, at the Lincoln Laboratories, Socorro, New Mexico, using a 1-m Space Command telescope, had been finding an asteroid per night by 2001 [12] and was responsible for 70% of current discoveries (surpassed by the Catalina survey in 2005). The German-French ODAS program in Europe had been succeeded by the German-Italian ADAS; Japan had its NISEI program, occasional work was being done in China, and there were 'useful ex-Soviet assets' that might become available (see below).

The new U. S. initiative was to extend 90% coverage to asteroids down to 150 m in diameter, in 7–20 years. It would require telescopes of 3–4 m in diameter, according to a NASA study published in September 2003. It might take 6–10 years to complete the initial catalog, but to determine the orbits of all those tracked could take 40–50 years. In response to an Announcement of Opportunity, the University of Hawaii and the USAF had proposed a whole-sky survey with a very large camera, the size of a refrigerator-freezer, and a 7° field to view. It would cost $3.5 million in the design phase and was to be called Pan-STARRS.

The first area of weakness was still follow-up observation and long-term tracking, both of which were largely left to amateur observers, reporting to the Minor Planet Center in Massachusetts. Further tracking was almost always necessary to determine the orbit with enough accuracy to know whether or not there was a threat from an asteroid, but there was no guarantee that it would be done. Months, or years, could be needed to identify a threat that was real.

A single radar observation was worth a thousand optical ones, pinpointing a target's position in 3-D and determining its velocity by Doppler shift. For example, Steve Ostrow had identified asteroid 1950 DA as having a one in 300 chance of hitting Earth on March 16, 2880, from one radar fix. Such radar fixes were accurate at distances of up to half an astronomical unit, but generally only to 0.1 au. There was only one two-component radar system capable of doing it, with transmissions from the Deep Space Network at Goldstone and reception at Arecibo in Puerto Rico, requiring both radio telescopes to be available at the same time (Fig. 4.4a, b, c). There was a threat that the Arecibo facility might be closed in 2011 [13], but hitherto that has been averted, and NASA has requested funding for it as part of its increased budget for asteroid searches in 2014. At Kennedy Space Center, NASA is testing 'KaBOOM' (Ka-Band Objects Observation and Monitoring Project—Fig. 4.5), an advanced radar that promises ranging to still greater distances, 0.5–1.0 au [14].

With most of the searching done from North America, small targets moving rapidly against the stars could easily slip by unnoticed, or not be found in time for follow-up observations. But the real gap in coverage was still the southern hemisphere, though the upgraded UK Schmidt telescope was to be joined by Foulkes South, one of two 2-m robotic telescopes primarily for school use but also available for astrometry. Foulkes North would be in Hawaii, widening the northern hemisphere coverage as well.

In 2010, there was interest in locating an observatory in the Falkland Islands. High humidity and almost permanent wind would compromise most professional astronomy, but with clear, dark skies, lower rainfall than the UK, a moderate range of seasonal temperatures, and full views of the area of sky not visible from Europe or the United States, there was a strong case for an amateur facility. The latitude of the Falkland Islands is shared only by the southern tip of South America and by other islands without observatories, and their longitude is shared only by Antarctica and Greenland. The similar northern latitude of the British Isles would allow paired observations with Spaceguard UK, and simultaneous or near-simultaneous observations would allow stereophotography of NEOs, to determine their distances and orbital paths without using radar. Among experts approached for comment, Sir Patrick

FIG. 4.4 (a) Deep Space Network radio telescope at Goldstone, California (NASA). (b) Arecibo radio telescope, Puerto Rico (NAIC/Arecibo Observatory). (c) Steerable Arecibo waveguide (NAIC/Arecibo Observatory)

FIG. 4.4 (continued)

FIG. 4.5 Experimental three-dish Ka-Boom radar at Kennedy Space Center (© Ken Kremer, *Universe Today*, 2013)

Moore was in favor of it [15]. But although John Braithwaite held and was prepared to donate much of the equipment to endow such an observatory, the authorities in the Islands said only a professional facility would meet their requirements, and unfortunately that view was backed by the British Antarctic Survey, so nothing happened.

In 2012 LINEAR remained the major finder of NEOs; NEAT, Spacewatch and LONEOS had all ended by early 2008. The Catalina Sky Survey was still going, and Pan-STARRS was now a reality, with the first telescope operational, the second soon to be joining it, but the third one was 'not going to happen' and the fourth was still not funded. Pan-STARRS was about to make its name by discovering a comet that was well seen in the southern hemisphere in March 2013, fading as it reached northern skies (Fig. 4.6).

Less welcome news was that NASA's Spacewatch 2, to extend the search down to objects 140 m in diameter, had been mandated since 2005 but was still not funded. After a good start, with David Asher's participation (see above), the Japanese BISEI Spaceguard

Fɪɢ. 4.6 Comet Pan-STARRS from Stirling Observatory, March 15, 2013 (© Alan Cayless, 2013)

project had turned its attention to space debris in Earth orbit; the DLR program in Germany and the Beijing one had been without funding for 3–4 years. The Australian government's component of the Sky Survey funding was about to end, once again leaving no southern hemisphere coverage—except by spacecraft, the major new development (see below). The South African government had been approached about closing the southern hemisphere gap in search programs, but would not be interested unless a full international program was set up.

Still less welcome news was that because of its unique observational qualities, Pan-STARRS was increasingly in demand for other uses, taking up 60% of the available observing time. It had been announced that the second Pan-STARRS telescope will be more focused on the asteroid search, but a privately funded Discovery Channel Telescope at the Lowell Observatory had been

Fig. 4.7 Jay Tate with Sir Martin Rees, then Astronomer-Royal, now Lord Rees, at the Spaceguard Centre (© Robert Law, 2002)

subverted from its intended NEO role within months of becoming operational, 75% of its time going to other uses.

Meanwhile, the Spaceguard Centre's own asteroid search had begun with a 13-in. refracting telescope and continued with a 14-in. Schmidt camera named Blofeld, accumulating 3,000 observations in 4 years, but with a small field of view. Now the center had been given the use of a 24-in. Grubb Parsons Schmidt from Cambridge, with a 5° field. Unused for many years, and made available at the suggestion of Lord Rees, the former Astronomer-Royal (Fig. 4.7), the telescope had been removed to Wales for refurbishment by a team of volunteers (Fig. 4.8), an intriguing task when there was no documentation of changes that had been made to the fork base, the equatorial drive and other essential features. By April 2013 more than half of the necessary funding had been raised, including a vital donation earmarked for converting the camera from twentieth-century glass plates to state-of-the-art CCD sensors. The instrument had been named Drax, continuing the tradition of naming Spaceguard Centre telescopes after James Bond villains. The total funding required is roughly the price of a new Landrover

Fig. 4.8 The future Drax telescope being removed from Cambridge (© Spaceguard UK, 2012)

Discovery, and a percentage of this book's proceeds will go towards the project (See Acknowledgements).

When operational Drax will be the only dedicated asteroid search program outside the United States, and one of the most important tasks will be to follow up discoveries made by the programs above. Pan-STARRS does some follow-up observation, but that too takes time away from detection. New discoveries are announced daily on the Minor Planet Center's NEO Confirmation Page, almost all of them calling for follow-up, but most of it doesn't happen. Almost all the follow-up is still done by volunteers, for instance using the Foulkes telescope. For example, PHA 2008 SE85 was recovered by an amateur observer, Erwin Schwab from Germany, in September 2012, after being found in September 2008 by the Catalina Sky Survey, tracked by a few observatories to October 2008 and then lost [16]. Its next Earth pass will be at 6 million km, twice the previously estimated distance, so there's no longer a danger from it, but it is only one case among many.

Alan Cayless of the Stirling Observatory suggested that with more coordination, amateurs could meet the need, but Tate replied that serendipitous follow-up was not adequate or practical. Pan-STARRS can detect objects down to magnitude 21–24, while the best amateur equipment can reach around magnitude 18.5. Erwin Schwab conducted his asteroid hunt during a regular observation slot at ESA's Optical Ground Station in Tenerife, Spain, sponsored by the Agency's Space Situational Awareness program. That program is 'probably' funded to build the first of four to six telescopes, each with a 6° field, for a complete survey, but a dedicated international network of 2- to 2.5-m telescopes is still needed, unless other means can be found.

With a telescope on the Moon measuring parallaxes with a baseline of a quarter of a million miles [17], optical determination of an NEO's orbit could be as good as radar's. Yuki Takahashi has made an observatory on Mt. Malapert, near the south polar crater Shackleton, a personal goal. His Master's degree thesis was on establishing a Very Low Frequency radio observatory in the peak's radio shadow [18], but it overlapped with a NASA study on a Lunar South Polar far-infrared telescope—just what's needed for the asteroid search, better than an optical one. With that, the size, composition and other parameters of the asteroid can be obtained from its albedo and spectrum much more accurately than in optical wavelengths, and it has to be done off-planet because the far infrared doesn't penetrate Earth's atmosphere.

Spaceborne telescopes may have a lot to offer, without going to such drastic lengths as Clarke's 'Excalibur' proposal. The WISE space telescope (Wide-Field Infrared Survey Explorer) was launched in December 2009 to conduct an all-sky survey of faint infrared sources, but after its coolant was exhausted in September 2010 it was turned to the NEOWISE survey of Solar System objects, using two of its scanners in shadow. For an estimated cost of $400,000, 100,000 main belt asteroids were tracked, and 33,500 new asteroids and comets were detected. Though the survey wasn't total, it was estimated that there are approximately 981 NEO asteroids, of which 911 have been detected, 93% of the likely total, none of them currently hazardous (Fig. 4.9). But all of their orbits will evolve over time and WISE was shut down in February 2011 [19] though it will be reactivated in September 2013 for three more years.

In the size range from 100 to 1,000 m, the WISE statistics indicate a population of around 19,500—fewer than the 35,000

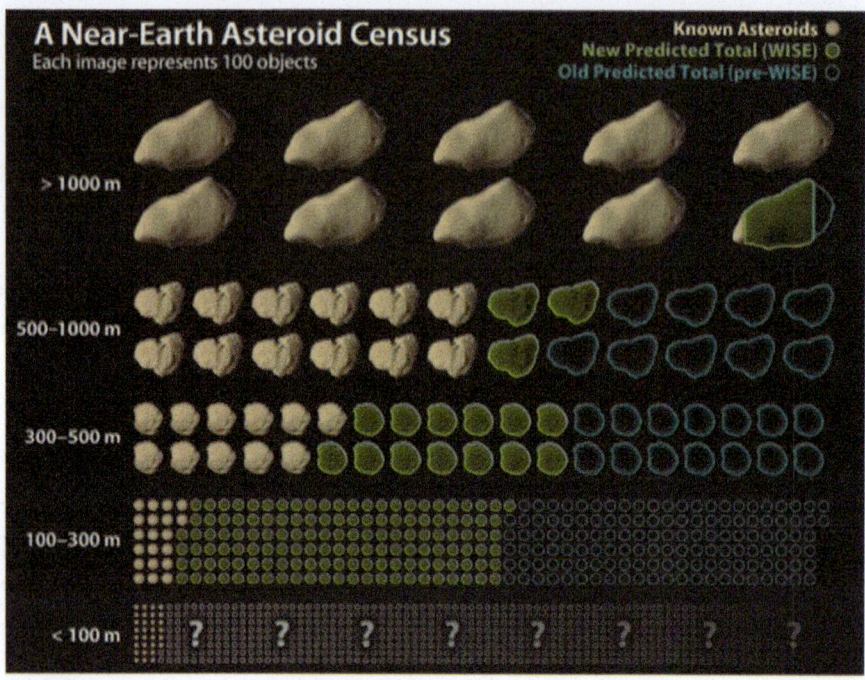

A Near-Earth Asteroid Census
Each image represents 100 objects

Known Asteroids ●
New Predicted Total (WISE) ⦾
Old Predicted Total (pre-WISE) ○

> 1000 m

500–1000 m

300–500 m

100–300 m

< 100 m

FIG. 4.9 Summary of NEOWISE survey outcome (NASA)

previously estimated, but only 5,200 detected so far. Although they're smaller than Goldilocks, these are still of a size to inflict what Tate called 'regional catastrophes,' devastating South America, for example; and below that, there may be millions of lesser ones ranging down to Asimov's city-busters, moderately regional or local threats in Benny Peiser's terminology [20]. Even by July 2013, the number of known NEOs had risen to 10,000 [21].

As for PHAs of all sizes, the ones that might actually hit us, WISE plotted 170 of a possible 4,700 at present, and combining those results with other surveys, it's thought that perhaps 20–30% of those have now been detected. More of them were in low-inclination orbits than expected, and they were brighter on average than those in high-inclination orbits, suggesting that they come from collisions of larger bodies in the Asteroid Belt. That's bad news because it means there's a continuing source of supply, rather than a thinning out of hazards as the Solar System ages.

NEOSSAT, a Canadian project drawing on the success of MOST, the Microvariability and Oscillation of Stars 'Humble

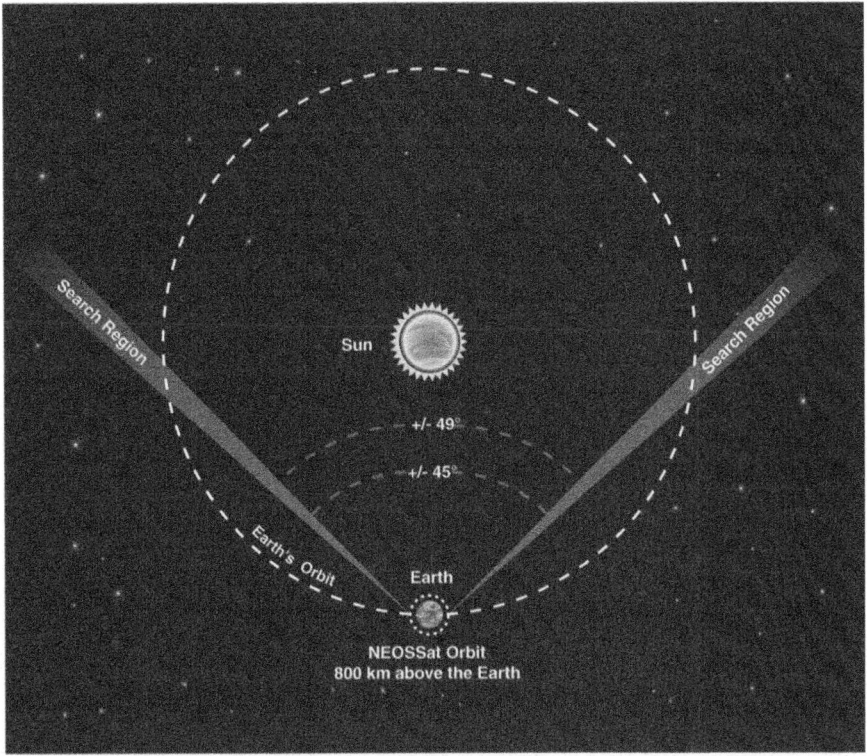

Fɪɢ. **4.10** NEOSSAT viewing angles (© University of Calgary, 2013)

Space Telescope' project, was launched in February 2013, and it will also track manmade space debris. NEOSSAT's field of view is limited, but importantly it can detect some objects coming from within Earth's orbit (Fig. 4.10), and in Sun-synchronous, near-polar orbit, it will operate 24/7 both north and south of the equator. It will be particularly useful for follow-up studies, but unfortunately it will be operational for only a year. A similarly named NASA project has been drafted, but is not funded so far.

Similar considerations could make a proposed WFIRST successor to WISE more useful than a ground-based Large Synoptic Survey Telescope, although costing $1.5 billion as opposed to $465 million [19]. But governments may not have to find the money, at least for this purpose, because private industry may now enter the picture. NASA has issued a Grand Challenge, under President Obama's Strategy for American Innovation, calling for that [21].

Fig. **4.11** Former astronaut Rusty Schweickart (© B612 Foundation, 2012)

The potential usefulness of asteroid resources has been appreciated for a very long time, going back at least to Konstantin Tsiolkovsky at the turn of the nineteenth century. In 1998, Tate said that the spinoffs to the space industry might provide sufficient incentive for privately sponsored missions; in 2003, he said that dual-purpose missions, with a view to industrializing the asteroids, might be the key. The big difference was that by then, for the first time, a commercial enterprise had been formed.

At a press conference in Seattle in April 2012, a group of entrepreneurs announced the foundation of a new company called Planetary Resources. Founded by Peter Diamandis and Eric Anderson, of the X-Prize and Space Adventures fame, the company also has aligned with major names such as film director James Cameron (not long returned from the Marianas Trench), Ross Perot Jr. and Larry Pegg, the co-founder of Google. Their declared aim is to access and utilize the resources of the asteroids, and as a first step, to establish a series of orbiting 9-in. telescopes, dubbed Arkyd-100s, to identify potential hazards and potential objectives [22].

Before then, a still more ambitious detection plan had been announced by a non-profit group called B612, named after the asteroid in Antoine de Saint-Exupéry's novella *The Little Prince*. Founded in 2001 at the Johnson Space Center, the group became a Foundation in 2002 with input from Rusty Schweickart (Fig. 4.11) and from Clark Chapman of the Planetary Society. In 2010, working

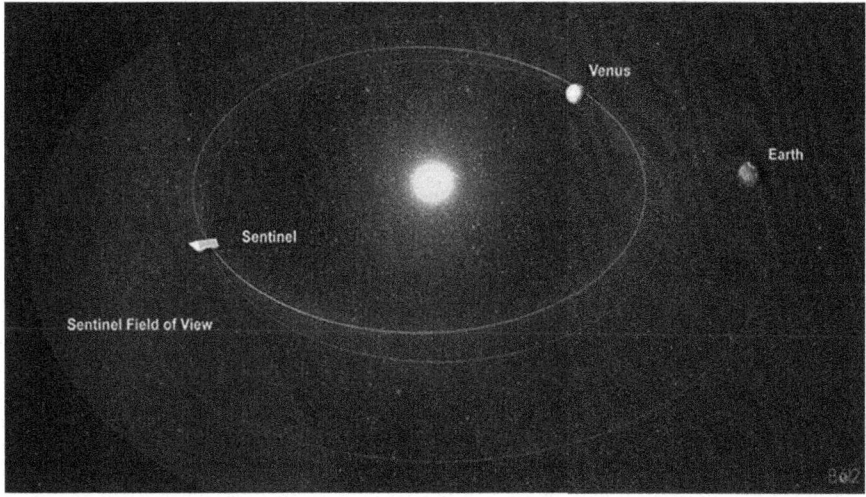

FIG. 4.12 Sentinel telescope field of view (© B612 Foundation, 2012)

with Ball Aerospace Corporation, the group unveiled plans for an infrared telescope, based on the Kepler exoplanet finder to be located near the orbit of Venus (Fig. 4.12), which would complete the cataloging of all asteroids down to 50 m in diameter within 7 years, meeting the objectives of Spaceguard 2. The development cost was estimated at $638 million [23]. Nevertheless, by 2012 development funding had been raised, and all that remained was to cover the cost of a launch by a Falcon 9 rocket in 2017 or 2018.

The Scenario Crystallizes

Find 'em early, find 'em early, find 'em early.

—Don Yeomans, Jet Propulsion Laboratory [24]

In our 2003 seminar, Tate was asked to cover how the Goldilocks threat would be recognized. Most probably the initial detection would be made by LINEAR. It would be reported to the IAU Minor Planet Center, where the Working Group on NEOs, a panel of six people, would assess the finding and make a judgment within 72 h on its reliability. The initial discoverer would then publish his or her finding, and an assessment of the threat would be made,

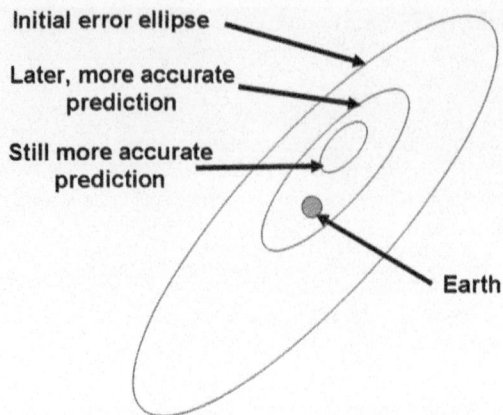

FIG. **4.13** Impact Probability Ellipse (Diagram by Lou Scheffer. Wikipedia Commons)

using the Torino Scale, which multiplies the size of the object by the probability of impact. The Torino Scale has been criticized for generating false alarms, and it's been suggested that a Torino rating of 1 should be reclassified as 'normal.' So a second assessment would be made using the Palermo Scale, which assesses the danger relative to the overall background hazard. Here, a rating close to zero would cause alarm.

It would be picked up by the NASA sentry website and would appear on the recursive, daily priority table of NEODYS, the NEO Dynamic Site, with a call for follow-up studies. At that point there wouldn't be a predicted impact but an 'error ellipse' within which the asteroid was predicted to pass on its return, and if Earth lay within that ellipse there would be initial cause for concern. Further astrometry, measuring the asteroid's position against the stars, would allow the orbit to be recalculated with greater accuracy, either shrinking the ellipse or moving it outwards until the probability of impact drops to zero (Fig. 4.13). The process could take weeks, or months.

In the aftermath of the February 2013 events and the WISE results, Keith Cooper wrote in *Astronomy Now*, "Suppose it did happen – what would be the process of disseminating this news?... If it is a NASA discovery, it will be sent up through the U. S. government before being announced, hopefully with enough time for us to do something about it" [19]. But it could be pushed off priority by objects discovered afterwards, which would soon be lost to observation: "NEODYS focuses on newly discovered objects," said Tate. If that happened to the Goldilocks sighting, recovery and follow-up might not happen for years, as in the 4-year case of 2008 SE85, above, and losing 4 years out of 10 could be fatal.

As Tate said above, this is a surveillance issue, not a scientific one. He maintains that it ought to be military, but the military say it's not theirs. The U. S. Space Command has been tasked with it, or part of it, but the UK Ministry of Defense has no such remit and refused to participate. The National Environment Research Council is tasked with assessing extreme natural events, but as of 2001, considered that impacts didn't qualify. The British National Space Centre had a watching brief, at least, but that disappeared when it became the UK Space Agency. On both sides of the Atlantic, the warning and response chain is not joined up.

A similar point was made in a panel at the 2013 Planetary Defense Conference in April 2013, in Flagstaff, Arizona, under the auspices of the International Academy of Astronautics. In a panel discussion, the chairman said that after 2 days of presentations on individual pieces of research, there was no sense of an overall strategy or even consensus on what needs to be done. Some original ideas emerged towards the end of the discussion that followed, but what was most striking was that it lacked the coherence which our project had from the outset, 10 years earlier. That was because the delegates were trying to discuss every possible kind of impact, all within the 50 min available.

One speaker said that part of the problem was that there was only one word for asteroid. How difficult it would be to talk about the different measures needed for hurricanes, typhoons, tornadoes and blizzards if we had only one word for storm. Actually we do have city-buster, continent-buster, Tunguska object, Chicxulub event and other such terms, including *Deep Impact's* ELE, pronounced 'Ellie,' for extinction level events. In introducing Goldilocks, our strategy was to define an object that conceivably could be tackled, so as to formulate a strategy that might be expanded to other cases. Should we have a more threatening name, like Kali, Shiva, the Hammer of God or Lucifer's Hammer, all used in fiction? 'Lia Fail,' the Stone of Destiny, came up as a suggestion (and the original one may well have been meteoritic), but would fully apply only if it can't be deflected.

(There's almost a tradition in spaceflight history of picking names or using place-names that seem like a gift to the opposition. NASA chose to run ground tests of its nuclear rocket for Mars flights at a place called Jackass Flats [see Chaps. 5 and 6]. Not to be outdone, Britain ran ground tests of its Blue Streak booster at Spadeadam Waste. But the U. N. Working Group on Near-Earth Objects (see Chap. 9) excelled both in February 2013 by proposing an International Asteroid Warning Network—"Whenever I think of the asteroid threat, it makes me want a IAWN.")

Updating his presentation in 2012, Tate told us that the reporting, assessment and follow-up procedures were essentially unchanged from 10 years earlier. It became clear that to bring about the situation we wanted, Goldilocks had to be defined still more tightly. It was a question of emphasis, starting from, "What if there was going to be an *impact*?" Bill Ramsay had asked, "What if we knew there was going to be an impact *in ten years' time*?", but now it had to become, "What if we *knew* there was going to be an impact in ten years' time?" The error ellipse had to be within the diameter of Earth, which meant that the discovery flyby had to have been close enough to have gone through the keyhole, and had to be tracked by radar, which meant that it must have been discovered on approach, possibly by spacecraft, far enough out for that level of importance to be assigned and the necessary arrangements to be made.

For that to happen now, before we have all-around coverage, almost certainly Goldilocks had to be approaching Earth's orbit from the outside, not coming from up-Sun. The orbital dynamics then have a major effect on the three-hit scenario. At the first window of opportunity for rendezvous at perihelion, 3.33 years after discovery, Earth will be 120° ahead of the asteroid when it crosses Earth's orbit coming inwards (see Fig. 3.4). For a minimum-energy Hohmann transfer, cotangential to both orbits and using chemical propellants, the launch from Earth then has to be 4–5 months before perihelion, so the time in flight to rendezvous would be comparable to a Venus mission, although the distance of the aphelion from the Sun makes the energy requirement comparable to a Mars flyby.

For the second perihelion, 6.66 years after discovery, the position is less favorable. As the asteroid crosses Earth's orbit coming sunwards the planet will be 240° ahead of it, and because a Hohmann transfer is the slowest possible transfer between two orbits, there is no Hohmann launch window for that configuration. A higher-energy propulsion system will be needed, and the rendez-vous will be on the asteroid's outbound leg from the Sun, giving more time to prepare the mission but less working time at the asteroid itself.

On the third time around, with impact at 9.99 years after discovery, the asteroid is coming almost straight at us, and any kind of rendezvous is virtually impossible with existing technology. That has a big impact both literally and figuratively on the final options (see Chap. 7), at the point when pretty well anything becomes worth trying.

The Political Reaction

We're here to defend democracy, not to practise it.

—Jay Tate, SAAMS Conference, Spaceguard UK, 2003

At that event Tate quoted a U. S. military study that had con-cluded that the threshold at which action should be taken is a political decision. The only UK politician to take a serious interest in the impact issue was Lembit Öpik, grandson of the astrono-mer Ernst Öpik and Member of Parliament for Montgomeryshire in Wales, 1997–2010, who had set up a parliamentary debate on impacts in 1999 and is a strong supporter of Spaceguard UK (Fig. 4.14). Öpik was unable to take part in the 2003 seminar due to a clashing constituency engagement, but kept his promise to come specially to Glasgow for a supplementary event in 2008.

To make the most of the opportunity he was briefed before-hand on a short list of questions for discussion. The first was, at what point in the detection process would governments take the threat seriously? Tate had predicted that the 'tipping point' (see below) would be when the error ellipse was twice the diameter of Earth, and Öpik added, quoting Tate, that a 10% probability of an impact would be enough to trigger action. That's beyond

Fig. **4.14** Lembit Öpik lecture at the UK Spaceguard Centre, 2002; the author at right (© Robert Law, 2002)

the point at which the actuarial factors quoted in Chap. 2 come into play. According to Craig Binns, 5% risk is something insurance companies take very seriously.

"When would it become a Security Council issue?" Ramsay had asked, but at that time there was no planning within the United Nations for such a contingency. The IAU's Working Group on NEOs would issue an alert, but not directly to the U. N. Delegates at the Unispace 3 conference in 1998 had agreed that there should be provision. An Action Team was set up, and in 2013 its efforts finally bore fruit (see Chap. 9).

At the 2013 Planetary Defense Conference, Dr. Alan Harris said that the United Nations was the only route through which agreement on the actions required could be reached, and a speaker from the audience added that the United Nations would have to be involved because of the existing treaties. It was noted that only two delegates attending the conference came from below the 20° north parallel; did that suggest that two-thirds of Earth was unconcerned? The chairman replied that many southern nations were represented on the U. N. Action Team, suggesting that for once opinion in the political field was better informed and balanced than in the scientific one.

Ramsay had suggested that because the impact threat was a gray area, like genetic engineering or climate change, where the public perception was different from the scientific one, politicians would wait to assess public opinion before acting. But had there been any poll to check on public awareness, or to raise it? Not as of 2003, nor in 2008, nor in 2012, as far as any of us knew. At the 2013 event, another contributor said that if fire broke out then and there in the conference center, everyone knew what would happen and what the response from outside would be, but the public has no such knowledge about asteroids.

Obviously Öpik could only answer on the basis of his own parliamentary experience, and it was one politician's view at that. Before a specific threat appeared, he felt, one would have to sensationalize the issue to make things happen. As Minister for Science, Lord Sainsbury had commissioned a White Paper in response to newspaper scare stories about weather. At the start of the meeting, Öpik had estimated that an all-sky search strategy might cost £12 million per year; John Braithwaite countered with a plan to use 24 pairs of matched 16-in. telescopes, north and south, in amateur hands, with a set-up cost of £528,000. A 5,000-mile baseline and synchronized sweeping would allow 3-D mapping, with sensitivity comparable to radar and therefore with accurate orbit determinations. But even if he carried his points on costs, and on entrusting the search to amateur observers, there would be no political will to take action, said Öpik, for fear of ridicule. The fear factor is a very big one. Partly anticipating the remarks at the 2013 Planetary Defense Conference, Braithwaite suggested that we should find a scarier name because 'asteroids' is too neutral.

This was before we realized that for our scenario to work, we had to have near-certainty, as above. So what would the political reaction to that be? In Öpik's view the reaction would be swift, and the first response would be anger, with particular recriminations in the UK because no prior action had been taken. After the Chelyabinsk event, Tate told *Astronomy Now*, "When we explain to people that the UK government is doing absolutely nothing, we're met by surprise and anger. People really can't understand why something's not being done, given that it would be relatively cheap to sort out, but the Government's view has always been that we shouldn't bother because the Americans are doing it all, which is not the case" [25].

Continuing, Öpik said he now realized that he had never given adequate thought to the huge problems that would arise if there *was* significant warning time before an impact. An early sign of trouble would be a collapse of house prices—and before the sub-prime mortgage issue, not all of us realized how serious that alone would be. At the same time there would a flight of capital, not just from the target country but from the target hemisphere, as illus-trated by the cartoon quoted at the end of Chap. 2. And that would be followed by massive migration of people out of the target area, as H. G. Wells foresaw (see Chap. 7), and any attempt to conceal the location would worsen the collapse everywhere. Very shortly there would be public panic, civil disobedience and martial law almost everywhere.

Worldwide, the threat would have center stage in politics up to the day of the expected impact. In his 1998 and 2001 lectures, Tate had characterized the available strategies as (1) the Ostrich Option—ignore it and hope it will go away; (2) the Bruce Willis option—blow it to bits, now increasingly coming to be seen as fantasy; and (3) change the velocity of the coming impactor.

With impact near certain, said Öpik, finally, governments worldwide would listen to the experts. "The lonely furrow in poli-tics that I plough now would be gone." They would demand solu-tions from the experts, but there would be a free hand to achieve them. Worldwide, astronautics would leap forward with a sud-denness not seen since the Second World War. The G8 countries would have to put up the money, but money itself would be no object, as Ramsay had been insisting since 1998. Tate's reply to him that cost-effectiveness rules all would no longer apply at the technical level.

That would stop short of large-scale mitigation on Earth, in Öpik's view. It had been estimated that it would cost £250 tril-lion to evacuate Europe, for example, and insurance companies wouldn't pay. Evacuations are very uncommon in warfare, Craig Binns pointed out, and when planned for, have involved a variety of improbable scenarios. In the 1930s and 1940s, the French government's line on evacuation was that people should stay put to keep the roads clear.

Likewise, asked to comment on Dr. Arthur Hodkin's plans to protect populations in situ (see Chap. 7), Öpik's reply was that governments won't do it. The rationing and other measures required would be reminiscent of Stalin, too unpopular to undertake and too big a distraction from deflection measures. "We have to deflect: if we can't, we're screwed, and if we can't convince the public that it'll work, we're screwed anyway." By no means everyone agreed with this view, and we'll come back to that in Chap. 7.

However, if deflection by our favored methods doesn't work, then at about a year before impact, decisions will have to be taken on more violent methods—nuclear warheads, kinetic impactors and beam weapons (see Chap. 7). Each of these carries a risk of fragmenting the asteroid. Not all the pieces would be on impact trajectories, but a million Barringer-level events, say, spread over a hemisphere, could bring the world closer to mass extinction than one Nordlinger-scale one. Would it be attempted? Yes, said Öpik. Those who realized the consequences could be worse would object, but most people would be pragmatic and want it to be tried. Public demand would force it to happen, but whatever the result, it would be a big issue at the next election, if there was another election…

Should the threat be kept secret? Obviously we're not talking about the Hollywood scenario where an extra star appears in the Plough (the Big Dipper), but only children notice it and adults pay no attention. In a situation like that, every amateur astronomer would make the same discovery very quickly. But a 1-km asteroid approaching Earth may never attain naked-eye visibility, even at its closest. Could the discovery be suppressed, say at the point where its Torino scale rating included a 10% chance of impact? If the Torino scares of the early 2000s had continued, public alarm might be muted because this was one scare too many, but in the detection scenario above, too many people would already know about the real one for it to be kept quiet. The radar determination of the precise orbit by Goldstone and Arecibo might be secret for a time, but people around the world who'd been asked to make optical observations would be expressing disquiet and asking what the radar results were. Those 'useful ex-Soviet assets' that Jay Tate mentioned earlier could include military radar near Ussuriysk in the Far East [26], and if the Russians determined a hazard that the United States was trying to keep secret, they might not be willing to help in setting up international cooperation to deal with it.

Even if it could be kept quiet, the massive shift in space policy throughout the developed nations could not be hidden for long, and the nature of the intended missions would soon become apparent. At the Planetary Defense Conference, Alan Harris suggested that as there were only so many spacefaring nations in the world, the others would probably trust them to deal with the problem.

Deflection: The Available Methods

At the end of the day, even the best strategies have to degenerate into action.

—Dr. Uno Öpik (1926–2005), son of Ernst Öpik, father of Lembit

What Tate called the Bruce Willis scenario (blow it away) has been around in Hollywood productions for a long time. One early example was the 1950s TV series *Men into Space*, starring William Lundigan, in which manned spacecraft were used to place conventional explosives on the surface of an asteroid called Skyra, which simply disappeared when they were detonated. More than 20 years later, the same thing happened in *Meteor* when nuclear missiles reached the potential impactor. There have been many imitators since, most of them assuming that it would be easy to reprogram existing ballistic missiles to hit a fast-moving target in space, and that they can be steered in flight or if necessary destroyed, as in Bond movies.

In reality, destruct systems are only fitted to missiles under testing. Designed to destroy the vehicle when it's out of control, they're very vulnerable to outside interference, as in a famous case in the early 1960s involving an Atlas launch and a radio taxi. The Giotto mission to Halley's Comet shows that near head-on interception can be done, but that took years of planning, months of flight time, precise two-way telemetry, input from other spacecraft for the final guidance, and sufficient onboard propellant, so much so that the reserves were later used to set up an encounter with a different comet.

However, the biggest mistake is the belief that solid rock or metal simply disappears in the vicinity of a nuclear explosion. Although the energy density of a nuclear explosion is huge, a great deal of it is normally unused. There was no crater at Ground Zero in Hiroshima, and indeed the shells of concrete buildings remained standing. In John Baxter's novel *The Hermes Fall* one of the characters says of the asteroid, taken to be blade-shaped and a mile in length, "To an H-bomb, that thing's just a pebble" [27], but it wouldn't be true (Hermes is now known to be dumbbell-shaped, 0.45×0.3 km, but the point is unchanged). In space, the only blast comes from vaporization of the warhead; most of that is dissipated into space, but even so it generates 20% of the estimated deflection effect, which suggests the whole method is highly inefficient [28]. A surface explosion, as in *Men into Space*, would be still less effective. On the other hand burying the device, as in *Deep Impact* or *Armageddon*, might succeed all too well. Prof. Colin McInnes pointed out that with a yield of 104 kilotons, the 1962 SEDAN underground nuclear test had formed a $1,200 \times 230$ ft crater in the Nevada desert [29]. As the author pointed out in a 1973 short story, doing that to a 1-km asteroid could raise some very undesirable results [30].

The serious advocates of using nuclear warheads against asteroids or comets propose that with sufficient notice, preferably decades, a rendezvous could be achieved at aphelion, and the detonation could be positioned off-surface for maximum effect, as well as making sure that effect was in the right direction. David Asher presented the case at our 2003 seminar. Material vaporized from the surface of the asteroid would be blasted off into space, in effect creating a brief rocket thrust. Much of the energy released by a normal nuclear explosion in vacuum would be X-rays and gamma rays, producing a definite heating effect, but more penetrating radiation would be more effective, and some variety of neutron bomb would probably be best, expelling material at up to 12 km/s from up to 100 m below the surface. For each type and yield of device there would be an optimum detonation distance for the maximum capture of energy.

However, we don't know what would actually happen next, and the result may vary according to the composition of the asteroid. 'To every action there is an equal and opposite reaction,' indeed, but what if the object's outer layers are so loosely compacted that they blast off into space without imparting a major kinetic impulse to the rest of it? What if the inner layers are structured such that they are simply compacted by the blow, heating the interior of the asteroid but not significantly changing its path? That was the predicted result from studies of a carbonaceous chondrite that fell at Lake Tagish in Canada in 2000 [31].

And on the gripping hand, to quote Niven and Pournelle [32]—what if they are so rigid or brittle that the blow shatters the asteroid, creating thousands or millions of PHAs where only one was before? Braithwaite suggested that many of them might then miss Earth, but some studies suggest that on a long timescale they'd tend to converge, if not coalesce, once more [33]. Even if they mostly missed Earth the first time, they'd still be in similar orbits and sooner or later they would all be back.

The seminar also noted a worrying level of uncertainty about what would be required. Asher had with him two papers by Thomas J. Ahrens and Alan Harris, highly respected in the field, the first of which gave the yield needed to deflect a 1-km asteroid as 0.01–0.1 megatons [34], and the second gave a full megaton [35], ten times as powerful, with no explanation for the increase.

Dr. Nigel Holloway, contributing by letter, was highly confident. He could see no need to discuss other scenarios when he calculated that a single 1-megaton warhead, massing 1 metric ton, could deflect a 1-km asteroid by more than a full Earth diameter, even with only a year's notice. "In practice, several smaller devices would probably be better—reducing the chance of a total failure, and allowing each explosion to be rather nearer the asteroid without risking its fragmentation. So—this results in a fairly neat 'baseline' for deflection assessments, i.e., to deflect a 1-km object with 1 year to go you need a 1 t device" [36]. With decades in which to experiment, repeated experiments could be tried; but given the uncertainties above, with our 10-year timescale and limited launch windows we felt very reluctant to put all humanity's eggs in that one nuclear basket. Even trying it with a succession of small devices seemed too risky. What if you sent one bomb too few? If the characters of

Where Eagles Dare had been carrying just one bomb fewer, or if just one of the 40 or so they did have had failed to detonate, the ending of the entire movie would have been different.

Certain advocates of the militarization of space continue to insist—some of them shrilly—that while gentle methods are perhaps very well in the long term, nuclear weapons are all that are really required. We know that they work, and in the short term such as our 10-year timescale, they are probably all that we have. Nevertheless there is a great deal of unhappiness about proposals to put nuclear weapons into space, even for the purpose of protecting Earth. It would spell the end of U. N. treaties that have successfully prevented the deployment of weapons in space, restricted development of space weapons to ground-based systems, and kept other weapon testing to a minimum. Calling the bomb 'a device' does not help—as Tom Lehrer said of that euphemism in another context. It makes one feel "like a Christian Scientist with appendicitis." It would be going too far to suggest that humanity should accept the global consequences of a 1-km impact in order to preserve the treaties, but for those who are *not* for the militarization of space, seeking other methods seems a high priority. Holloway's assurance did make us look again at the third hit, though, in which those options are to be seen as last resorts (see Chap. 7).

At the 2013 Planetary Defense Conference, discussing the nuclear option, one delegate asked how the technique could be tested in advance, to allow an informed decision to be taken? Alan Harris replied that the same question applied to all methods, each of which were bound to have surprises, and tests of all of them were needed for effective responses to be made when required. Some might argue that it would be better to test all the others first, and not to compromise the treaties on nuclear testing and the peaceful uses of outer space unless it had become clear that nothing else worked.

In the 1970s, in the ASTRA Interplanetary Project, the only viable alternative to the nuclear option seemed to be the industrialization of the inner Solar System, to the extent that a task force could be sent to any threatening asteroid and dismantle it, shipping back its usable resources as finished products or as raw materials, particularly to orbiting factories in Earth-Moon system, or delivering them to Earth in Waverider carriers manufactured in

FIG. 4.15 "Industrial action" (© Tom Campbell, 1985; personal collection of the author)

situ. Unwanted silicate residues could be dispersed harmlessly as dust, but *Man and the Planets* later suggested uses even for those [37]. The task force units might never be as large as the stylized one in Tom Campbell's painting "Industrial Action" (Fig. 4.15), and Gavin Roberts designed a more realistic Standard Interplanetary Operations Spacecraft (SIOS) that could spend years or decades on long-term missions—note the centrifuge cabins for artificial gravity (Fig. 4.16a, b).

The idea seemed the best option and still is, in the really long term. The final paragraph of the asteroid chapter in *New Worlds for Old* reads: "At a high enough level, such technology could more than handle any dangers presented by wandering small fry like Eros or Icarus. But, whereas most of the development of the Solar System can be presented on a take-it-or-leave-it basis (with a grim future for Earth if we leave it), the reach for the asteroids is ultimately a matter of unadorned survival. If we do not wipe ourselves out, no other outside forces intervene, and yet we do not claim the little worlds, then eventually one of them will claim us" [38].

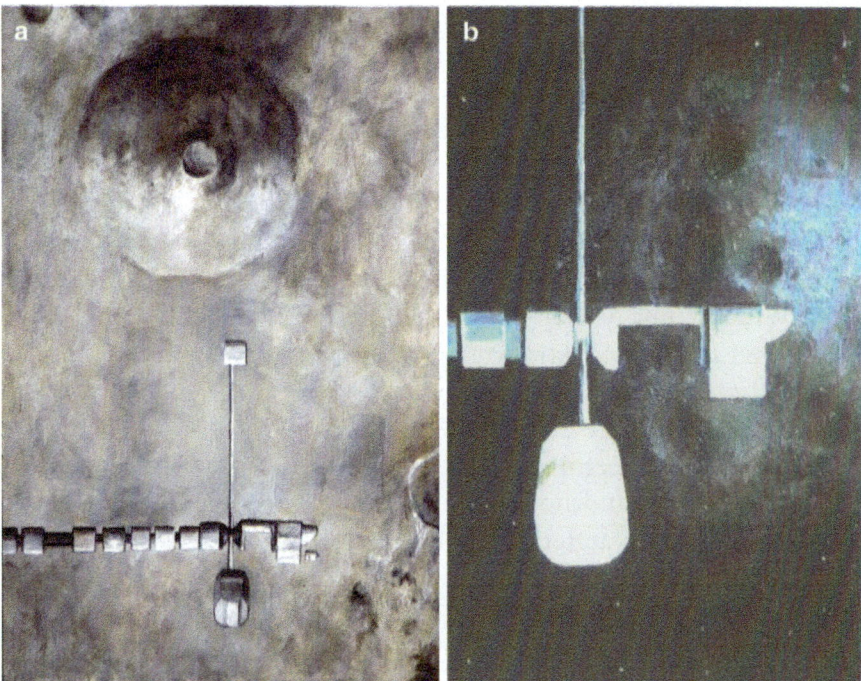

Fig. 4.16 (a) SIOS over middle spot (Pavonis Mons) (© Gavin Roberts, 1975, based on Mariner 9 imagery for *Man and the Planets*). (b) SIOS mission to an asteroid (© Gavin Roberts, 1979)

In 1986, however, Gordon Dick (now Gordon Ross) came up with a design for a 'comet-chaser,' based on his 'Solaris' design for a parabolic solar sail (see Chap. 5) [39]. They could be mass-produced and stationed at comparatively low cost at Earth-Moon and Earth-Sun Lagrange points, providing a means to protect Earth much sooner, without involving nuclear weapons and without a huge preliminary investment in manned spaceflight. It had been recognized for a long time that sunlight exercises pressure in a vacuum, causing comet tails to point away from the Sun, affecting the orbits of the 1960s Echo satellites, and used for attitude control on the Mariner 4 and Mariner 10 missions. But Gordon was perhaps the first to realize that the light and heat reflected from the sail could be collected and channeled for other uses.

It took a surprising time to get the idea into print, and by then similar ones were being evolved elsewhere, particularly by Jay Melosh [28, 40]. In 1989 we had made a serious effort to push the

Solaris itself as a contender for an international solar sail race [41], and in 1995 it had seemed for a time as if a more ambitious version might find government funding as a project to commemorate the Millennium [42]. We were also in touch locally with experts such as Dr. Colin McInnes, Dr. John Simmonds and Prof. John Brown, who were bringing different kinds of expertise to solar sail applications [43]. So once Ramsay's project was launched in 2002, it was natural to begin with Ross' concept, as a first answer to the question 'What would we do?', and the numbers indicated that even in a worst-case orientation it would work as a viable alternative to the nuclear option, so we selected it for our 'first hit.' We will consider this in more detail in Chap. 5.

Immediately before the 2003 seminar, Prof. McInnes invited us to consider another violent but non-nuclear method of deflection, kinetic deflection [44], previously discussed by Melosh [40]. McInnes's version would use a solar sail to carry a large inert payload into an orbit that would be rotated over the poles of the Sun, so that it would be traveling in retrograde orbit when it met the threatening asteroid. With an asteroid in an orbit close to Earth's, such a solar sail could theoretically deliver a mass of 2.8 t at an impact velocity of 60 km/s, 7.5 years after Earth escape, with a deflection effect equivalent to a 0.1 megaton nuclear device. Foreseeable improvements in solar sail efficiency could reduce the in-flight time by at least a year [45].

For an asteroid such as ours, the closing velocity would be higher, making the projectile more effective but possibly increasing the risk of fragmentation. Goldilocks may well be in an inclined orbit, and Graham Dale pointed out that if a faster impact at a 30° angle could achieve the same effect, a year could be shaved off the time to interception. So if the solar sail/impactor is launched 3 years after the asteroid's discovery, it would take 6.5 years to reach it and the deflection achieved would be comparable to a nuclear device, at the lower end of the range quoted by Ahrens, and if the deflection is as stated by Holloway, then the impactor can just get the job done within our 10-year deadline. We are, again, looking a last-ditch option for Goldilocks, but one that requires a launch date around our first window at 3.33 years, even though it will travel to the asteroid by a very different route.

Unlike nuclear weapons, kinetic impactors can be tested in space without breaking any U. N. treaties. Space probe missions to investigate the internal compositions of asteroids have not so far taken flight. The Department of Defense Clementine 2 spacecraft was to have fired penetrators to embed them in asteroids including Eros, Toutatis and a possible dead comet. It was to have been launched in 1997, but because the Clementine series was also testing sensors and guidance systems for anti-missile systems, the Carter administration canceled it as overly provocative.

The 2005 Deep Impact mission treated us to an initial demonstration that impactor missions can work. The intention was to study the composition of the comet, rather than to change its orbit, and the European Space Agency proposed to follow it with a mission called Don Quixote. Two spacecraft would be launched together by a Soyuz-Fregat booster, possibly in 2015 [46], and separated on different trajectories by a subsequent Earth flyby. Sancho Panza, named after the squire who looked on while his master tilted at windmills, would reach the asteroid first and deploy penetrators around it to set up a seismic network, while Hidalgo (Cervantes' original name for Don Quixote) would gain speed in a Venus flyby and impact the asteroid with a mass of nearly 400 kg at a minimum 10 km/s (Fig. 4.17) [47]. One particular aim would be to see how the impact energy was absorbed by the asteroid and what the effect on its orbit would be. The reference mission considered an asteroid 500 m across to be suitable, and Apophis was one of the two possible targets in the final selection, only for the mission to be canceled when Apophis was no longer considered dangerous [33].

Another impactor mission in preparation is Hayabusa 2, scheduled to launch in July 2014 to reach asteroid 1999 JU3 in June 2018, returning to Earth with samples in December 2020 [48]. A payload slug will be fired at the asteroid using a shaped charge, in order to release samples for collection (Fig. 4.18a, b).

Meanwhile in Europe, the German DLR space agency's Institute of Planetary Research has launched the NEOshield initiative, with an impressive list of partners including the Russian Rovcosmos, the Paris Observatory, Astrium, Surrey Satellite Technology and Queen's University, Belfast. NEOshield is a 5-year engineering scrutiny of deflection methods, particularly the question of energy coupling—how good is the transfer of energy to the target? It has led to the joint U. S.-European Asteroid Impact and Deflection

FIG. 4.17 Sancho Panza watches Hidalgo impact (ESA)

mission (AIDA), which will send two small spacecraft to study the binary asteroid Didymos at its closest to Earth, 11 million km, in 2022, and impact into its moon in 2022 [49]. Didymos is only 800 m across, its unnamed moon just 250 m (Fig. 4.19), so the impact should have a marked effect on its orbit and provide a lot of information on the effect of collisions. The U. S. collider, DART (Double Asteroid Redirection Test) will strike at about 6.25 km per second, watched by ESA's Asteroid Impact Monitor (AIM), which will carry out a detailed science survey of Didymos before and after the collision.

Hayabusa 2, AIDA and OSIRIS-REX will all greatly add to our knowledge of Earth-grazing asteroids and our abilities to deflect

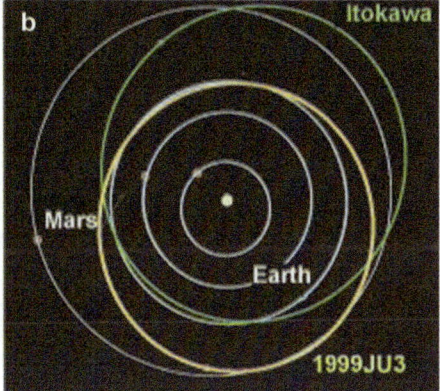

FIG. 4.18 (a) Hayabusa 2 impactor (© JAXA, 2012). (b) Itokawa and 1993 JU3 orbits (© JAXA, 2012)

them. If no major threat develops in the meantime, by 2022 the ongoing surveys may also have been completed and should tell us whether we have a problem, and how much time we have. In our scenario, Hayabusa 2 with its ion drive (Fig. 4.18b) might be redirected to Goldilocks in time to provide useful data ahead of the 3.33-year mission, but the probe probably wouldn't be fast enough to catch it. AIDA might be able to achieve a flyby, if its timescale can be accelerated.

Fig. 4.19 Didymos and moon (NEOShield)

Meanwhile more gentle, non-nuclear deflection methods have entered the discussion. In 2005 Drs. Edward Lu and Stanley Love, both experienced astronauts, advanced the concept of the gravity tractor—a massive spacecraft, powered by an ion drive, which would hover ahead of the asteroid in the desired direction, holding station by low-thrust, continuous propulsion and 'tow' the asteroid by gravity alone (Fig. 4.20a). The propulsion would have to be angled so that the exhaust didn't strike the asteroid (Fig. 4.20b), otherwise the tractor effect would be canceled out, but even so, currently proposed versions of the ion drive could power a 20-t spacecraft that could deflect a 200-m rock asteroid in 20 years. Solar-electric propulsion would suffice; the composition or fragility of the asteroid is no longer a factor, because the tractor tows the asteroid as a body instead of impacting a part of it [50].

The method has been enthusiastically taken up by Rusty Schweickart and the B612 Foundation, and has everything going for it except timescale. Applied to the Goldilocks mission, the tractor would need a mass of 750 t for the 6.6-year window, or twice that at 3.3 years. That couldn't be done with currently available

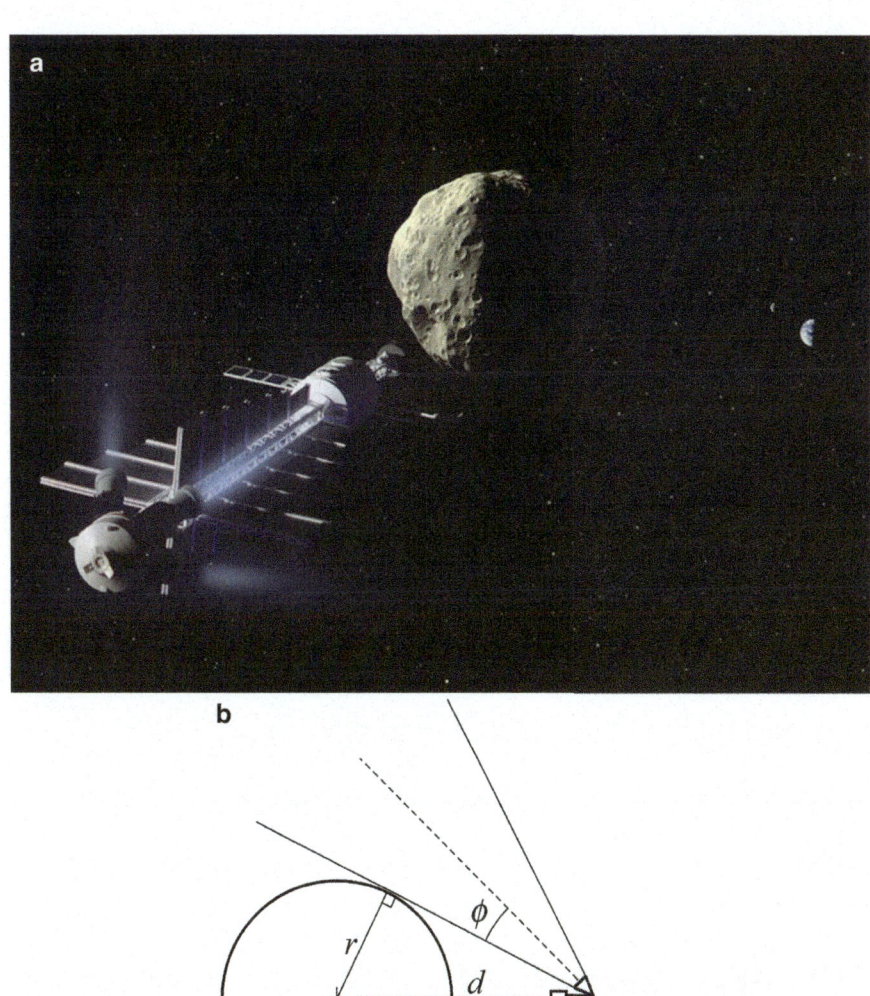

FIG. 4.20 (a) Gravity tractor (© Dan Durda, B612 Foundation, 2008). (b) Gravity tractor geometry (After Lu and Love)

technology (see below), but it generates some interesting options for the second hit that we propose—see Chap. 6.

Instead of angling the ion drive away from the asteroid, another option is to aim the ion beam at the asteroid and push it very gently in the desired direction. A similar thruster has to be operating simultaneously in the opposite direction, to hold the spacecraft at the required distance. Twice the diameter of the asteroid is best, and significantly reduces the risk of collision if the rock is irregular in shape. The spacecraft mass required is much lower than for the gravity tractor, but the advantage diminishes for larger asteroids and disappears at a diameter of 2 km [51], so in the Goldilocks case it might be marginal. The timescale for deflection is equally long (20 years for a 200-m asteroid, in the example discussed) [52]; there's the same uncertainty as with solar collectors and mass drivers about how the thrust is transmitted from the asteroid surface, which is assumed to be inelastic.

A similar 1950s proposal for hovering flight with nozzles pointed in opposite directions was thought likely to produce 'a horrible grinding noise and then crash,' but while the forces involved here are too small for that to be a problem, there could be difficulties in balancing the thrust, particularly over such long periods. Continuous low-thrust propulsion is not always wholly predictable: ESA's SMART-1 ion-drive mission to the Moon surprised its builders by arriving two months early (see Chap. 6).

Until fairly recently, discussion of the effects of sunlight on small bodies in space focused on the Poynting-Robertson effect, which causes dust grains to migrate inwards towards the Sun. Due to their velocity in orbit, sunlight is subject to aberration, and seems to be coming from slightly ahead of the Sun's true position. Some of the radiation absorbed by a particle would exert a retrograde thrust on it, causing it to slow down and spiral sunwards.

Around 1900, however, the Russian engineer Yarkovsky realized that the heat absorbed and re-emitted by an asteroid would also have a significant effect, because the greatest amount would be absorbed at noon, when the Sun was overhead, and re-radiated around '3 p.m.', generating a net thrust on the body. Ernst Öpik is credited with reintroducing the idea in 1951. The Yarkovsky effect may explain one major statistical anomaly concerning NEOs: spectroscopic evidence indicates that 66% of them are Type LL chondrites, similar to the

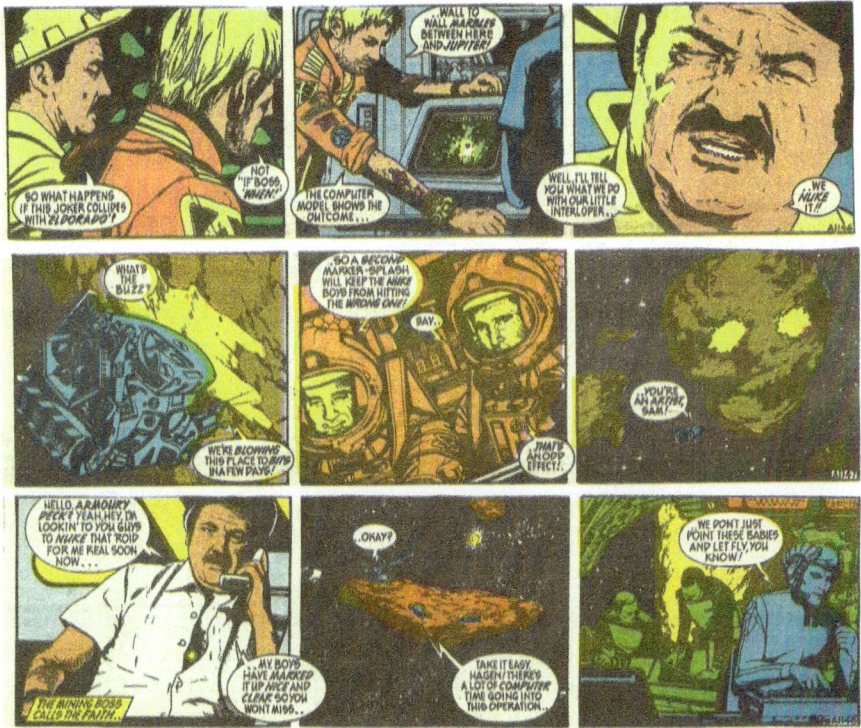

Fig. 4.21 Painting parts of an asteroid, though not for Yarkovsky effect in this case (© Sydney Jordan, 1984, What Dreams May Come. Story by Marise Morland, Lance McLane comic strip, *The Daily Record*, 1984–1985)

Chelyabinsk object (and to Goldilocks), but they make up only 8 % of known meteorite falls. The Yarkovsky effect would have been operating throughout the Asteroid Belt throughout the history of the Solar System and could explain why the population of smaller meteoroid-sized bodies is more diverse [53].

On a body the size of an asteroid the thrust would be very small, and its power and direction could vary widely, depending on the asteroid's rotation and the orientation of the axis, on its albedo (the percentage of light that it reflects) and on any variations of it across the surface, not to mention irregularities of shape. Nevertheless the Yarkovsky effect on such objects is measurable and has been deduced from the motions of Bennu, for example. It could be accentuated by painting parts of the asteroid [54] (Fig. 4.21), or by partly sheathing it in reflective plastic [55]. Note that it's the heat emitted from the unpainted or uncovered areas that generates the thrust; the purpose of the reflective paint or sheeting is to reduce absorption and re-radiation from the rest of the surface. While such methods could speed up the changes in an asteroid's orbit due to the Yarkovsky effect, decades would be needed to move one out of a collision path with Earth.

Ramsay's three-hit scenario allows interception of the Goldilocks asteroid 3.33 years after discovery and again after 6.66 years. The latter period roughly equals the time between John F. Kennedy's "this nation should commit itself" speech and the first human mission to the Moon, Apollo 8, so a second, crewed mission is conceivable and may be necessary. Below and in Chap. 5, we'll look at the possibilities for unmanned missions in the first launch window. But they may not succeed, and for political as well as practical reasons, the planners are likely to opt for human supervision on the spot during the second attempt.

On the timescale we have, the one suitable method in the latter phase is the one we haven't yet discussed—electromagnetic launch, specifically the version known as the mass driver. Mass drivers had featured quite prominently in *Man and the Planets*, as systems for payload launch and capture, and even as propulsion for spaceborne manufacturing units (Fig. 4.1, Chaps. 7 and 8). We hadn't thought of them as propelling whole asteroids, but as with the solar collectors discussed above, the technique harnesses solar energy, does not break U. N. treaties, delivers thrust in controlled increments instead of one mighty blow and can achieve the results that we want within what remains of the timescale (see Chap. 6) [40].

So we concluded that our 'first hit' would be with solar collectors, the second would be a manned mission with mass drivers, and the nuclear and kinetic options would be held in reserve, with some significant modifications, for potential third hit measures a year or so before impact, if the other methods failed.

After we reached those decisions, working from first principles, it was gratifying to learn that a later, detailed mathematical study at the University of Strathclyde had come to the same conclusions [28]—except that they had limited their remit to payloads that could be launched with existing boosters, where we had assumed that with money no object, a crash program would bring bigger ones into service.

The Tools to Hand

Governments are like seesaws – when they tip, things happen.

—Jay Tate, Glasgow, 1998

Before the creation of NASA in 1958, the U. S. space program was in the hands of its three armed services, and there was a lot of criticism of the supposed wastefulness of the competition between them. The U. S. Navy had its Viking and Vanguard programs, leading to the Titan booster; the U. S. Army had Redstone and Jupiter, produced by Wernher von Braun's Huntsville Arsenal, whose technology led directly to the Saturn I and thence to Saturn V. The U. S. Air Force had the Thor and Atlas, and was particularly criticized for competing against itself when it acquired the Titan program in its missile phase. But Titan and Atlas utilized different technologies and from the 1960s to the 1990s were turned to different uses as space boosters.

Behind the military programs and their merger into NASA stood a wide range of contractors, among them Boeing, Convair, Douglas, Grumman, Lockheed, Martin (later Martin Marietta, later Lockheed Martin) and North American, later North American Rockwell and still later Rockwell International, all competing for the contracts and subcontracts as they became available. Again there was criticism of the supposed waste and the risks of cost-cutting. Asked what goes through one's mind in the final moments of the countdown, John Glenn is supposed to have replied, "You think that you're sitting on top of six million parts, each one made by the guy who submitted the lowest possible tender." But in the Moon race, the choice of technologies and manufacturers proved ultimately superior to the monolithic Soviet system.

When this project began those days seemed to be past, with mergers on both sides of the Atlantic forming fewer, larger companies and the range of launch vehicles constantly shrinking. But 10 years on, the position is rather different and more encouraging.

Admittedly the heavy-lift capability is absent, at least for now. With continuing development the Saturn V was to have been the workforce for the U. S. program into the twenty-first century, but it was discarded by the Nixon administration. There are still those who advocate resurrecting it, but it's frozen in time and could no longer be duplicated. For example, it's said that the end caps of the first stage tanks were the largest cold-hammered castings ever made, and even then, the Huntsville recruiters had to scour Europe for retired craftsmen who knew how to do it.

When the ASTRA Interplanetary Project that led to *New Worlds for Old* and *Man and the Planets* began in 1973, we assumed that the lack would be made good either by heavy-lift boosters using the Space Transportation System (space shuttle) technology, or by larger, single-stage-to-orbit mixed-mode successors to the space shuttle, and that what the late Krafft Ehricke called 'the strategic approach to the Solar System' [56] would be based upon in-space uses for the shuttle's external tank. Forty years on, such developments are no nearer, but that's not to say that under pressure, some of them—particularly a heavy lifter using the external tank—couldn't be brought forward in the 10-year timescale of this project.

The Soviet Lenin booster for the Moon race never flew successfully, but at the end of the Soviet era the Energia booster emerged as a substitute, flying successfully in both the missions it performed. With added upper stages and clip-on boosters it could have lifted 250 t to low Earth orbit, 100 t more than Saturn V. After years of inaction the assembly building housing the remaining Energias and the Buran shuttle collapsed during inspection, with loss of life, and hopes of a resurrection were dashed. The George W. Bush administration initiated the Constellation program for a return to the Moon and for Mars missions with the Aries V booster, but that was canceled under Obama and has been replaced by the Space Launch System (Fig. 4.22). With upper stages, it will be capable of launching with a payload to low Earth orbit of 70 metric tons, up to 140 t with upper stages, equivalent to Saturn V. The main engines of the first stage are probably to be an upgraded version of the Saturn V's J-2 second stage engines, designated J2-X, and although the current plan is for solid strap-on boosters, a liquid variant has been proposed using a simpler version of the Saturn's F-1 first stage engines [57]. Encouragingly, both are performing well in initial tests, and liquid-fuel boosters would give the vehicle a significantly higher payload.

At the moment the principal vehicles supplying the International Space Station are the Russian Soyuz and the unmanned

Fɪɢ. **4.22** Space Launch System (NASA)

Progress ferry, both launched by the booster whose venerable first stage launched Sputnik I, the first lunar probes, and Yuri Gagarin. With the retirement of the space shuttle, the field has become open for players both old and new to compete for contracts to supply the ISS, and almost as importantly, to return payloads to Earth. First with successful deliveries to the ISS were Europe's Automated Transfer Vehicle and Japan's HTV, soon followed by the Space-X Dragon capsule launched by their own Falcon 9 booster. ATV and HTV currently have no return-to-Earth capability, but their Ariane V and H-II boosters are both man-rated, and a crewed version of the ATV is possible. Space-X is using a prototype called Grasshopper to test systems for flying back the first stage to the launch site and recovering a crewed Dragon on land. CEO Elon Musk has said his goal is to make the whole system reusable.

Taurus is offering the Cygnus capsule, with more cargo carrying and longer on-orbit stay time than the Dragon, to be launched by its own Antares booster. Boeing has the CST-100 capsule, to be launched by any of the available boosters, and Blue Origin's capsule is another contender, to be launched by Atlas V. Orbital Sciences Corporation are bringing forward a lifting body Orbital Space Plane particularly for crew transfers to the ISS; the Sierra Nevada

Dreamchaser is another, for Atlas V; and among the outsiders is the Excalibur Almaz, a Russian alternative to Soyuz originally developed for the Soviet military and now intended for space tourism. Most if not all of these could be launched by the ATK Liberty, a booster combining a European upper stage with the solid booster developed for the Constellation program's Aries I.

The Liberty booster was not one of NASA's three prime contenders selected for ISS delivery contracts, but under pressure to deal with Goldilocks we could expect all of these programs to be funded, at least until the best ones for the missions were chosen, and they might all have a role to play. Perhaps prophetically, Sydney Jordan's "Jeff Hawke" strip in the 1950s portrayed Chinese participation in an international space effort to meet a global threat; China's space program is independent of the rest at present, but the capabilities of the Shenzou spacecraft make clear that it's intended for the Moon, and there's no question that the ultimate goal is Mars.

Shenzou is launched by the Long March 2 F, and the Tiangong space station units by Long March 3; but Long March 5 is in preparation, and a still bigger booster is scheduled for manned lunar missions, though not yet begun. The indications are that it will have a payload to low Earth orbit of about 100 metric tons, so Earth orbit rendezvous would be required to put people on the Moon [58]. Meanwhile the idea of Chinese participation in the ISS comes up repeatedly, although human rights issues remain an obstacle to international cooperation.

Innovative technologies for the manned mission may also come from suborbital spaceplane projects such as the Virgin Galactic Spaceship 2 and the X-Cor Lynx, both of which already have contracts for short-duration science missions as well as their space tourism plans. Still more innovative is the Reaction Engines Skylon, which could fly off and land on conventional runways with its radical Sabre engine, liquefying oxidizer in flight for rocket boost into orbit. Skylon has recently passed major tests of engine components and (at last) has backing from the UK Space Agency. It's intended primarily for automated deliveries to low Earth orbit, but its designer Alan Bond has put together a manned Mars mission concept based on it to show what Skylon is capable of [59].

For our unmanned mission at 3.33 years the United States has the Delta IV Heavy, Atlas V and the various private launch vehicles above. Europe has Ariane V, scheduled for upgrading, and with Ariane VI coming behind, as well as the smaller Vega booster and, very importantly, a new launch complex at the Kourou spaceport in Guyana for the Russian Soyuz booster—something the author and Bill Ramsay jointly suggested as long ago as 1991 [60]. Russia itself has the Proton booster, supposedly never used for manned flight, though it was rumored that it launched the first Soyuz and excessive vibration caused the damage that killed Vladimir Komarov after re-entry. Proton was repeatedly used in the 1960s for circumlunar flights of the Zond, a lighter version of the Soyuz, and a manned mission was scheduled for December 1968. Delayed by an electrical problem, it was pre-empted by Apollo 8. But Proton was the launcher for the Salyut and Mir space stations, and the Russian components of the ISS, so it might well launch components of the asteroid mission even if it remains unmanned.

As mentioned earlier, Japan has the H-II, which has been used for space probes to the Moon, Mars and Venus, and two successful missions to Halley's Comet were previously launched with smaller rockets. The frequency of launches is constrained by the powerful fishermen's lobby, and efforts to find an equatorial launch site have so far been unsuccessful, due to unhappy memories in Oceania. Ukraine has the Zenit, which is launched from Russia's Baikonur complex and commercially from the Sea Launch platform, and can be positioned in latitude to take maximum advantage of Earth's rotation for orbits inclined to the equator. Israel has launched satellites, but has the major disadvantage of having to launch against Earth's rotation, along the Mediterranean, to avoid overflying unfriendly neighbors. Brazil has the Alcântara launch site, the nearest to the equator in the world, but has yet to launch a satellite with its VLS booster; still, it is developing larger ones jointly with Russia. Iran has launched a satellite with its Safir booster and is developing larger ones, and North and South Korea have both launched satellites.

Outside the United States and Russia, only Europe and Japan have so far developed vehicles for manned spaceflight. Europe built the Spacelab modules repeatedly flown in the cargo bay of the space shuttle (Britain's contribution was the pallets for external payloads); Italy built the cupola for the ISS, now returning spectacular views from orbit, and also three pressurized reusable cargo modules, one of which, Leonardo, is now permanently attached to the station. Both Europe's Columbus module and Japan's larger JEM were originally intended to be autonomous, free-flying space stations, to be crewed and serviced using the Hermes and Hope mini-shuttles, respectively. After their cancellation, Europe was to contribute 40% of the airframe to the X-38 crew return vehicle, before the United States canceled that. As a member of the European Space Agency Canada has no launch capability of its own, other than the highly successful Black Brant high-altitude sounding rocket, but has carved its own niche by providing the remote manipulator Canadarms for the shuttle and the ISS, followed by the still more versatile Dextre remote arm installed in 2008. Manned missions to the Moon and Mars are the long-term objectives of ESA's Aurora program, and as steps to that end, Europe has a Lunar Polar Lander in preparation and has conducted simulated Mars missions with the Russian space agency.

China's achievements have made it clear that manned missions to the Moon are intended, so providing the basic architecture for Mars, and Japanese ambitions are no less. India, too, plans manned lunar missions, raising the prospect of a three-way race. It's by no means certain that the flags raised on Mars will be Stars and Stripes—after all, it is the Red Planet.

The U. S. Constellation/Aries program had the same premise, as its name implies, and in July 2007 NASA and the Bush administration defeated proposals in Congress and the Senate to have manned missions to Mars deleted from the program's objectives. At the same time NASA began discussing asteroid missions as an initial step, extending mission times to 120 days.

The suggestion that NASA should try for asteroid missions rather than Moon landings goes back to the mid-1960s [61], at least, but although it's easier to get to some asteroids in energy terms, the life-support requirement is much greater. As first imagined in 2007, an Orion mission to an asteroid passing near Earth could have been launched with upgraded versions of the Aries 1 booster (now canceled), Atlas V or Delta IV Heavy, using Earth orbit rendezvous for assembly with a space station module as living quarters and a liquid hydrogen upper stage derived from the long and

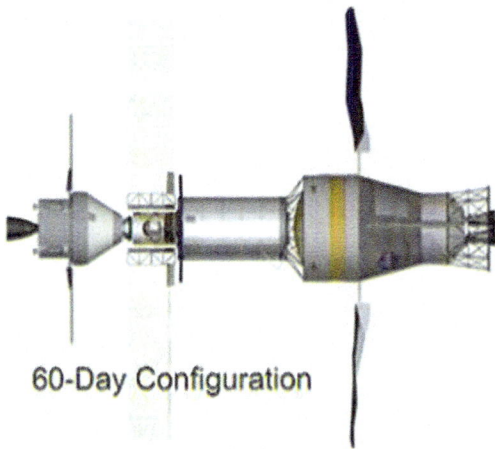

60-Day Configuration

FIG. 4.23 NASA Deep Space Habitat, Orion spacecraft with living quarters and Centaur-based propulsion stage (NASA)

successful Centaur series (Fig. 4.23). Going for a target in the range of 25–60 m diameter, and a near-circular orbit around the Sun with eccentricity less than 0.026, there were 14 candidate asteroids in the 2015–2024 launch window: eight Atens, five Apollos and one Amor (Amor itself), all approaching to within 0.2 au (18 million miles approximately) but none of them currently hazardous [62]. For longer missions to a larger list of targets, more advanced propulsion such as a plasma drive would be needed (Figs. 4.24 and 4.25) and Boeing outlined ambitious proposals based on its CST-100 module and ion drive. Lockheed Martin developed detailed proposals for a 'Plymouth Rock' mission, with paired Orion modules (Fig. 4.26) [63], and went so far as to build a simulator to practice approaches to it (Fig. 4.27).

The Constellation did not survive the end of the Bush administration, but President Obama replaced the Moon and Mars with a new objective for NASA, to prepare for a crewed mission to a near-Earth asteroid. No target date was set, and the objective was widely criticized as mere window-dressing to keep NASA in the manned spaceflight business. But the new interest aroused by the events in February 2013 has allowed both NASA and the government to say they're taking the impact threat seriously. With the first flight of the Orion capsule scheduled for launch on a Delta IV booster in 2014, from 2017 the service module for ESA's

2019 Opportunity: 2008 EA9

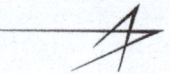

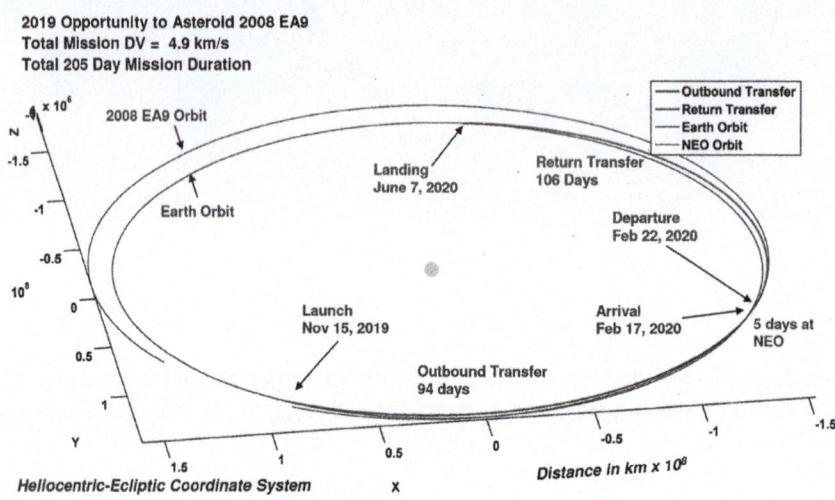

FIG. **4.24** Lockheed Martin NEO mission example. Note the long flight times (© Lockheed Martin Inc., 2009)

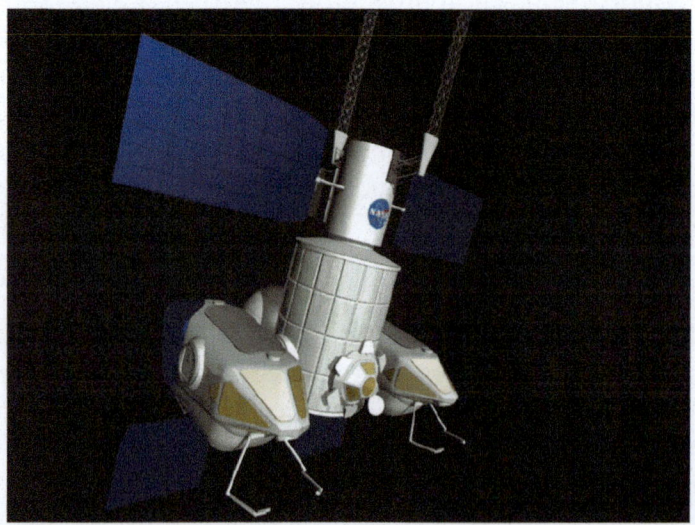

FIG. **4.25** The 500-day mission Deep Space Habitat with Tranquillity Node 3 space station module, Cupola and custom-built MMSEV landers (NASA)

Fig. 4.26 Lockheed Martin twin-Orion Plymouth Rock mission (NASA)

Fig. 4.27 Lockheed Martin Space Ops Simulation Center, Colorado (© Lockheed Martin Inc., 2009)

FIG. **4.28** (**a**) Asteroid capture (NASA). (**b**) Rendezvous with the captured asteroid at Earth-Moon L1 point (NASA)

Automated Transfer Vehicle, used for space station supply, will be fitted to the Orion for missions to the ISS and to lunar distances; and in April 2013 it was announced that NASA is looking seriously at a proposal to capture a 7-m NEO with a mass of 500 t within an inflated bag (Fig. 4.28a, b), bringing it to Earth orbit at Earth-Moon L1 point, for an Orion rendezvous with it in 2021 [64].

Figures 4.28a, b are from a NASA simulation whose first version made the capture look more simple than it's likely to be in real life. The asteroid didn't appear to have any axial rotation, which is impossible. It might be Sun-locked or star-locked, by very unlikely coincidence, but it can't be both. A later version shows it with one axis of rotation. The spacecraft approaches along it and matches the spin rate, like the space station docking in *2001, a Space Odyssey*. In his ongoing argument with Stanley Kubrick about that sequence, Arthur C. Clarke claimed to have been told by an astronaut that it couldn't be done. That seems unlikely. Practicing for it hardly seems more difficult than the 'flying bedstead' lunar landing trainer that so nearly killed Neil Armstrong, or the modified Gulfstream in which the astronauts practiced dead-stick shuttle landings.

Unless the axis of rotation happens to be in the required direction of thrust, the rotation has to be canceled before propulsion is applied for the transit to L1. Otherwise unwanted gyroscopic effects will occur—as pilots new to rotary engines often learned to their cost in World War 1. If the rotation hasn't been canceled, the upcoming Orion will have to match the spin to dock with it, and even if that poses no problems, it could make the proposed EVA activity to gather samples very difficult (Fig. 4.29a, b). Just imagine that pole and those sample boxes persistently trying to drift outwards, to say nothing of the astronauts themselves. But in the later version of the simulation, the spin is canceled.

Missions to NEOs in orbits similar to Earth's are very easy in energy terms, but usually they would be of short duration, a maximum of 4–6 months. Beyond that, as either Earth or the asteroid draws ahead in its orbit around the Sun, the requirements in time and fuel for the return trip become excessive; and in most cases, there will be a synodic period on the order of 19 years before the next launch window occurs. Growing experience with the ECLSS life-support system on the International Space Station has relaxed those constraints somewhat, and Lockheed Martin's example was a longer 205-day mission to asteroid 2008 EA9 (Fig. 4.24). Missions to the Earth-Moon L2 point have been considered as rehearsals [65] (Fig. 4.30), and bases there may be available in future years. Our mission to Goldilocks will have bigger propulsion requirements than those above, and will need bigger boosters or more launches to be put together in low Earth orbit. L1 or L2 might be preferable

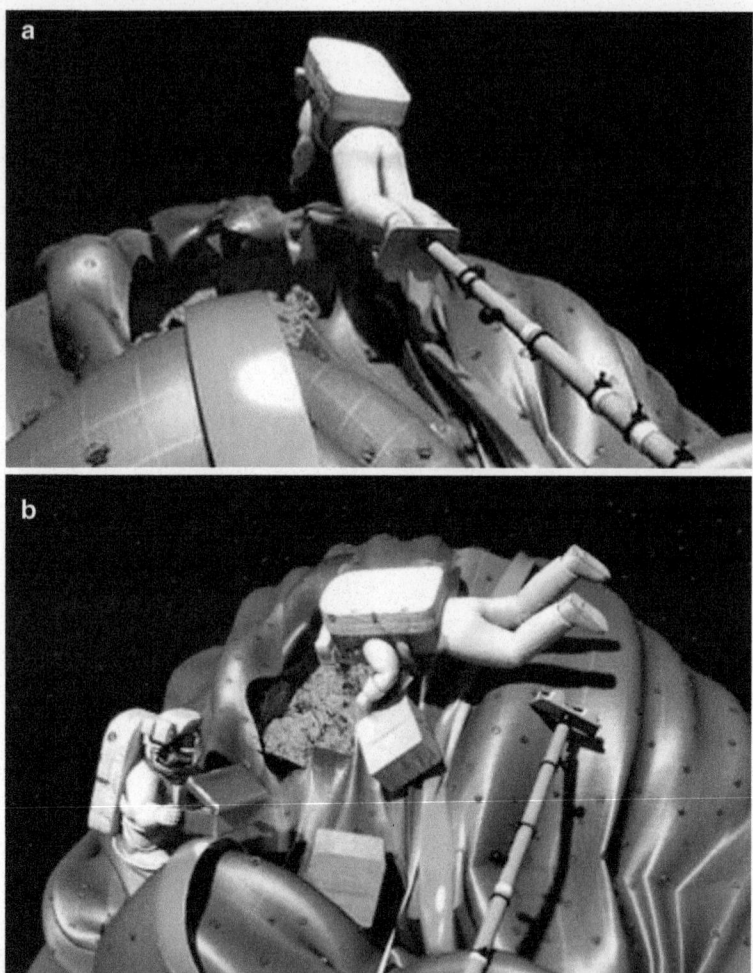

FIG. 4.29 (a, b) Asteroid sampling at L1 (NASA)

jumping-off points to leave the Earth-Moon system, particularly for the low-thrust propulsion systems discussed below, but it's not likely that such a complex program could be put together in the time we have in hand.

Still, whatever agency emerges as tasked with the prime responsibility for deflecting Goldilocks would already have a very broad suite of resources to call upon, and still more in just a few years' time, but decisions would have to be taken quickly. As Braithwaite summed it up, "government decisions have to have the academic imprimatur for credibility—but for action, government has

FIG. 4.30 Orion mission to Earth-Moon L2 point (NASA)

to hand over to a contractor." As Ramsay stressed, money will be no object, cooperation will be total, and human space capabilities worldwide will be greatly extended by the time it's over. In 1998 Jay Tate predicted that there might be a two-track program, with the United States as lead nation, and the international cooperation under the IAU and COSPAR, the Committee for Space Research established in 1958, partnered with the International Astronautical Federation. Technologies might be radically different, but ultimately they would be subsumed (he said 'syphoned') into a U. S. program under an international banner. Under threat, the feasibility study might take a year, and to build whatever is launched could take 2 years—which ties in nicely with a first rendezvous at perihelion 3.33 years after detection.

By contrast it was slightly shocking to hear Rusty Schweickart declare, at the 2013 Planetary Defense Conference, that if faced

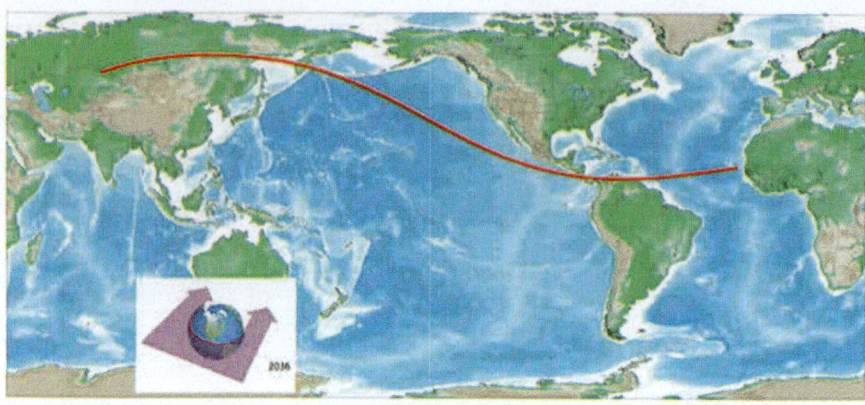

FIG. **4.31** The previously calculated *'redline'* for Apophis in 2036 (© B612 Foundation, 2006)

with an actual threat, the spacefaring nations were likely to take unilateral and possibly conflicting action. There would be a 'corridor of threat,' sometimes called 'the redline' (Fig. 4.31) extending across the face of Earth from the most likely impact point. The impact will come on the evening hemisphere of Earth, possibly soon after sunset. If the asteroid is slowed in its path, but not enough to clear Earth, it will strike further west; if advanced, it will fall further east. In pushing it off the rim of the planet, the United States might try to push it over Russia, and Russia might try to push it over the United States. Other delegates feared that any action might be paralyzed because if it wasn't wholly successful, under existing treaties the nation(s) doing the pushing would be required to compensate the nation(s) now in the impact zone. One can't help thinking that after a 1-km impact, once again few claims if any are likely to be filed.

After years discussing what could be accomplished with total international cooperation, it was depressing to hear such a negative view of what international relations might descend to in such a crisis. One of the other delegates, obviously sharing that disappointment, said he hoped the international response might at least be comparable to the campaign to eradicate smallpox (Lesley Wright had made the same point in 2001) [66].

While hoping for the best, though, the redline concept does raise other issues, to which we'll return below.

The Deflection Dilemma

I am but mad north-north-west: when the wind is southerly I
know a hawk from a handsaw.

—*Hamlet*, Act II, Scene ii

Once incoming objects can be detected, should defenses be
prepared? When that question first came up, the only proposed
solutions were for specialized nuclear weapons to be deployed
in space—something forbidden in any form under U. N. treaties.
Given that the defenses might not be needed for centuries, that they
might prove not to work in practice, and the unstable situation of
the Cold War, Carl Sagan thought that the risks of undermining
the treaties, opening the door for advocates of space militarization
such as Teller, probably outweighed the benefits [67].

Sagan was even more concerned that the spaceborne weapons
might be misused:

Can we humans be trusted with civilization-threatening technologies? If the chance is almost one in a thousand that much
of the human population will be killed by an impact in the
next century, isn't it more likely that asteroid deflection technology will get into the wrong hands in another century—
some misanthropic sociopath like a Hitler or a Stalin eager to
kill everybody, a megalomaniac lusting after 'greatness' and
'glory,' a victim of ethnic violence bent on revenge, someone
in the grip of unusually severe testosterone poisoning, some
religious fanatic hastening the Day of Judgment, or just technicians incompetent or insufficiently vigilant in handling the
controls and safeguards? Such people exist. The risks seem far
worse than the benefits, the cure worse than the disease. The
cloud of near-Earth asteroids through which Earth ploughs
may constitute a modern Camarine marsh.

One answer to Sagan was given with comparable eloquence
by science writer James Oberg in a submission to a conference
organized by the U. S. Space Command in 1998:

In discussing what can be done about such threats, we come
across a number of conceptual problems with the purely academic approach. It turns out that the operational skills possessed

by NASA and the U. S. Space Command are much more easily and reliably applied to this asteroid problem than are the theoretical skills of the scientific and intellectual community.

For example, Dr. Carl Sagan and some of his colleagues suggested that it was statistically more dangerous to build an asteroid defense system than simply to wait for the impacts. Their argument was based on the thesis that a system-in-being could under some circumstances be abused by 'a madman' to deliberately divert otherwise-harmless objects towards an Earth impact. Although admittedly unlikely, the manmade danger was deemed MORE likely than the original natural threat of asteroid impact.

But this view is erroneous. The concern fails to account for operational issues in navigation, targeting, guidance and control, issues which real-world spaceflight operators deal with on a daily basis. By assuming that a space rendezvous—bringing two objects into contact—is merely an inverse process of avoidance—guaranteeing that two objects do NOT come into contact, this concern is unrealistic.

The 'avoidance' maneuver is already in the repertory of spaceflight operators today, in low Earth orbit. If the predicted path of a piece of debris comes 'close enough' (defined in the dimensions of the avoidance zone around the shuttle), the shuttle makes a small orbital adjustment to take it (and the zone centered on it) away from the predicted path of the candidate impactor.

Rendezvous is also routine in low Earth orbit, but it is a far different process than merely reversing the avoidance maneuver. As the active vehicle nears the target it receives more and more precise relative position data (navigation), which it converts into desired course corrections (targeting), which converts into required rocket burns (guidance), and which it then performs—to the required level of precision—using onboard rockets (control).

As the range and time-to-contact drops, so does the size of the uncertainty zone around the target, where the chaser is aiming. At the same time, the effect of rocket maneuvers on miss distance can easily drift outside the 'uncertainty zone' to such a great distance that the active vehicle's rockets simply cannot bring the aim point back onto the target fast enough. In other words, there is not enough 'control authority' in the system. And the active vehicle flies past the target. The rendezvous fails.

For the proposed asteroid-deflection schemes of the next several decades at least, their control accuracy is far too poor to perform a 'rendezvous,' a deliberate collision, with Earth. Such systems would be fully effective in diverting dangerous asteroids, but would be physically unable to do the opposite, bringing them into contact with Earth. As a threat for misuse, they would panic only those who don't understand real space operations [68].

The point was underlined in a study conducted by David Asher and Nigel Holloway, who pointed out that while the path of an impactor has to be moved by a maximum 10,000 km to deflect it from Earth, to make an average Earth-grazer strike the planet the distance will probably be nearer a million km. They set themselves the task of devastating the UK with a strike on the town of Telford, while making it look like an accident. Telford wasn't a whimsical choice of target: the asteroids Hathor and 1997 XF-11 were discovered from there, though both were too big for the intended purpose of their scenario—both about a kilometer in diameter. A strike by either would have global consequences, probably doing too much harm to the senders. 1997 XF-11 had seemed a possible natural hazard, with a close approach in 2028 and a possibility of later impact, but more observations showed it to be no danger.

The chosen asteroid was 1998 HH-49, expected to pass Earth at 60 times the planet's diameter on October 16, 2023 (later revised to 78 diameters, which would upgrade the deflection requirements below by nearly a third). Near the geographical centre of England, a hit by it would devastate the country. Scotland would have some protection due to the shadow effects of the Southern Uplands, so some of the UK's military assets would survive, if still based there; but the supposed object was disruption, not preparation for an invasion.

Deflection would be attempted in July 2022, one quarter of an orbit away from Earth, near the Earth-Sun line. At that range, to hit Britain would require an accuracy of 0.01%—much greater to hit a specific target within the island. The asteroid would come out of the Sun, hitting at midday, and would never be brighter than 23rd magnitude, so the impact should be entirely without warning. Comparing the operation to putting ('Golf War'), if everything went perfectly, a minimum of ten shots would be needed of 1–2 megatons each, probably 15 at least, at 1-day intervals; even then in simulations only 30 hits were attained out of 40, with five falling elsewhere in the UK and one missing Earth altogether [69].

Examined in detail, however, the plan had many weaknesses. The detonations would have to be within one diameter of the asteroid to be effective. Each was intended to change the asteroid's velocity by up to 5 m per second, adding up to a total deflection of 33 m per second, so the first five explosions would have to be still closer or have larger yield. The bus vehicle would have to be far enough back to escape the worst of the blast and debris—ideally, it should be on the far side of the asteroid from it—and after each detonation the placing of the next would have to be recalculated, after the bus had matched the change with its own thrusters. Whatever asteroid was used, there would be no flexibility in the arrival date unless the orbital plane was changed, and because of 1998 HH49's relatively high inclination to the ecliptic, any plane change would cause it to miss Earth altogether. Pushing it by the relatively crude method of exploding bombs next to it could all too easily introduce an off-axis vector. Fine guidance would be no small task, as Oberg had pointed out, and would have to be done by onboard computers. Otherwise, two-way communication with Earth would probably be detected by radio astronomers.

The asteroid, 150 m in diameter with an estimated mass of 50 million tons, would strike with an energy equivalent to 15 hydrogen bombs. The thesis was presented at a special day in the 2001 British Rocketry Oral History Conference at Charterhouse (see above), and Roy Dommett, a veteran of Britain's independent deterrent program, asked, "If you're going to all that trouble to attack Britain with the equivalent of 15 H-bombs, and use 15 H-bombs to do it, why not just use 15 H-bombs on Britain itself?" The only reason for using an asteroid would be to keep the operation secret, but secrecy seems far from certain. Such a large payload, equivalent to at least five intercontinental ballistic missile warheads plus a much larger 'bus' vehicle, could not be launched in secret and would have to be disguised as a civilian space probe. Even so, its outbound trajectory would be known and suspicions would surely be aroused in other nations after the catastrophe, even if Britain was no longer in a condition to investigate—and especially if it was done with a known object like 1998 HH-49, which 'turned up missing' at the same time as the spacecraft.

At deflection time in 2022, the asteroid would be within 10° of the Sun for a month, and 30° for 3 months, by which time any visible debris around it would have dispersed. The explosions would be out of SOHO's field of view, even from the different perspective of the L1 point, but now other solar-observing probes such as STEREO, with all around views of the Earth-Sun line, would see them for sure.

So rather than being secret, it's more likely that the attack would be an overt display of power, which could easily backfire due to retaliation or pre-emptive strikes by those whom it was meant to 'shock and awe.' Tate considered that the whole Deflection Dilemma would be negated if there was sufficient warning of incoming threats. If revenge was the motivation, one of the possibilities Sagan suggested, then since planning, target selection and waiting for a launch window would take between 20 and 40 years, it would definitely be a dish served cold. Altogether, the whole scenario seems too farfetched.

In our Goldilocks scenario, Oberg's point about comprehension of the threat could be more of a problem than the Deflection Dilemma. Andy Nimmo thought that the space nations would get together to deflect an impactor whether others liked it or not, but as noted above, Rusty Schweickart feared that spacefaring powers who didn't understand the issues might try to push the impact into one another's territory. Craig Binns thought people not in the impact zone would not want deflection in case it put them in danger, not realizing that, as Lembit Öpik put it, "A continent-level event will destroy *all* the global economy." And quoting his late father again, he added, "Common sense is not common."

At Tate's 1998 lecture in Glasgow, it was suggested that only a lunatic fringe could possibly oppose deflecting an incoming impactor. Having discussed the matter with many people outside our group, particularly in the 10 years of this project, the author is not so sure. It has been both surprising and disappointing to find a widespread response that nothing should be done: 'Let it wipe us out, and a good thing too.'

Environmentalists seem to be particularly inclined to that view, even if it means the extinction of many other species.

A great deal of that response has come from older people, and there seems to be a definite echo of the attitude to nuclear war, which psychologist Michael Bradley characterized as 'The Cronos Complex' [70]. (Cronos destroyed his own children for fear that they would replace him.) But the older people now are the generation whom the survivors of World War 2 seemed to fear so much, and having rebelled against the Cronos attitude in my youth, as part of the process leading to the Politics of Survival, it is depressing to find it reborn in our own generation, with much less reason.

A somewhat more reasoned philosophical attitude is expressed by Prof. Ted Nield, editor of the journal *Geoscience,* in his book, *Incoming! Or, Why We Should Stop Worrying and Learn to Love the Asteroid* [71]. In regard to the Chicxulub impact, having followed and reviewed the issues in detail, Nield is not convinced that it is the proverbial 'smoking gun.' Scientists who say the Chicxulub impact didn't kill the dinosaurs are now finding it hard to get a hearing, by their own account, and while Nield points out that nevertheless their views *are* known, he feels a degree of sympathy. In his view, attributing the extinction of the dinosaurs and other life forms at that time to a single cause is too simplistic, and at odds with what is known about other such events. He believes it has a great deal more to do with the near-simultaneous but prolonged and complex effects of the volcanic plume that formed the Deccan Traps in India over a period of 30,000 years. The K-T impact may have contributed, may even have provided the *coup de grace,* but wasn't the major cause.

One reason for his skepticism is that no other mass extinctions have been correlated with impacts, but all of them coincide with the formation of Large Igneous Provinces by volcanic plumes. But there are known impact events that have still to be dated, particularly a very large crater under the Antarctic ice that was discovered by the Soviets in the 1960s [72]. As mentioned above, Sagan suggested that only large impacts at sea could explain the apparent connection between reversals of Earth's magnetic field and 'megadeaths' of marine organisms. Of the various books attempting to describe the destructiveness of the K-T impact, one of the most graphic is *The Great Extinction* by Michael Allaby and James Lovelock [73]. But the horrors they describe are largely overkill, when Enever's 'Giant Meteor Impact' provides enough damage to create mass extinctions every time something of that mass hits the sea [74].

So we might argue that the Chicxulub impact *could* have caused the mass extinction on its own, even if as it happens, it didn't. Here the biggest revelation of Nield's book comes from Sweden, where geological conditions have preserved the record of a sustained bombardment of Earth, lasting up to 15 million years but with a peak of one to 2 million years, during the Ordovician Period (488–444 million years before present). The falling bodies stemmed from the Gefion family, a group of asteroids that were formed when a body 100–150 km in diameter was a shattered by a collision, roughly 485 million years ago. The micrometeorite flux was about 150 times the present one, and evidence for that is now turning up all over the world [75].

There were at least four events big enough to form craters in Sweden alone, and statistically it's to be expected that at least one such event elsewhere would be in the range of the Chicxulub impact. Yet far from being associated with a Great Dying, the period represents one of the most prolific phases in the evolution of life on Earth—some of the changes probably triggered when ecological niches were freed by locally destructive events, a hypothesis credited to Ernst Öpik. (There was a mass extinction at the end of the Ordovician, but that was due to global cooling as the Gondwana supercontinent drifted across the South Pole).

From all of that, Dr. Nield concludes that impacts or even bombardments are not the threat to life on Earth that has been supposed. Then again, if there was only one Ordovician impact in the Chicxulub range, but it happened on land, things could have been very different had it fallen into the sea.

Even if he's right, though, his subtitle takes a very long view of what's good for life on Earth and what's not. It recalls Dougal Dixon's *After Man, A Zoology of the Future*, which imagined that we take ourselves out and most of the higher species with us, but after 50 million years all the evolutionary niches (except ours) have filled up again, so that's all right. Our descendants in 65 million years might think our extinction was a good thing, but only if they *can* think. The best candidates for that in Dixon's future were the giant, fully aquatic penguins who had filled the niche of the cetaceans [76].

Incoming! mentions proposals to deflect impactors only twice, and then only to say that the Bruce Willis scenario with nuclear weapons is probably a bad idea. But while Ted Nield was in Glasgow, the author put the question to him, "What would you want done if we knew there was going to be an impact?" He conceded that if he had to share in its consequences, if it happened, he would just as soon that some efforts were made to prevent it. At the 2003 Powys seminar and at the subsequent meeting with Lembit Öpik in 2008, the majority view was that most people who want us all dead would change their minds very rapidly if they were in line for it themselves. Many were reminded of Spike Milligan's "Carrington Briggs," who

> ...care not two figs
> Whether he lived or died.
> But when he was dead, he lay on his bed
> And he cried and he cried and he cried [77].

The prospect of religious opposition should not be taken so lightly, though Öpik said that it wouldn't rule the day. Within Christianity, in the first year one might expect a lot of talk about bowing to the will of the Almighty, and in the novel *Shiva Descending* by Gregory Benford and David Brin, the United States largely succumbs to a sect called the Gabriels, who are determined to enforce that perceived will by violence [78]. But as Vince Docherty commented, more probably as the impact day drew closer "half of the Christians would be in the churches, the other half would be putting in the windows of the big stores"—and that was before the summer riots of 2011.

Undoubtedly, sects would arise claiming revelations about the coming end, and the tragedy of the Heaven's Gate followers shows just how bad that could get [79]. Their cue for mass suicide was the approach of Comet Hale-Bopp, and a long succession of similar prophecies have been made in relation to comets before and since [80]. With an asteroid actually on an impact course with Earth, there would undoubtedly be many more. In *The Hammer of God*, which is given its name by 'Chrislamic' fundamentalists, Arthur C. Clarke wrote, "From now on, their antics were no concern of a planet which had more serious matters to worry about.

It was an understandable mistake—and a disastrous one" [11]. We need to think about what form such a disaster could take.

In *Broca's Brain*, Sagan states, "In a more or less spontaneous gesture, the Apollo 8 astronauts read from the first verse of the Book of Genesis [presumably he meant to say 'chapter'], in part, I believe, to reassure the taxpayers back in the United States that there were no real inconsistencies between conventional religious outlooks and a manned flight to the Moon. Orthodox Muslims, on the other hand, were outraged after the Apollo 11 astronauts accomplished the first manned lunar landing, because the Moon has a special and sacred significance in Islam" [81].

According to astronaut James Lovell, the text was suggested by a U. S. Information Agency representative, and was chosen because it's shared by the Christian, Jewish and several other mid-Eastern religions [82]. Nevertheless, apparently in many Muslim countries it is still widely believed that the Moon landings never happened, and it's unclear whether that disbelief extends to other events in space. As other commentators have remarked, it is odd that such an attitude has developed, when Muslim astronomers played so great a part in the history of astronomy that most of the stars still bear Arabic names. A combination of religious and political fervor could result in some nations refusing to believe in the Goldilocks threat, at least outwardly, just because the United States endorsed it—though one feels that some leaders might still invest in the deflection effort for their own protection. Ironically, if their nations didn't undergo the civil disobedience and other disruption that Lembit Öpik foresaw, they might emerge economically and socially stronger than many western countries when it was over.

A great deal may depend, then, on where the impact point and the redline fall when they're first calculated. Early in our discussions, it was suggested that it might best to keep the result secret, or to falsify to reduce public fear. But when the truth became known, not only would there be mass panic, but if whole nations were left out of the truth, that would very likely lead to the ill-considered free-for-all of unilateral actions that Rusty Schweickart feared.

Also, to be blunt about it, if the predicted impact is within the continental United States, or off either coast, then there will be factions in the Muslim world who will acclaim it as divine judgment, and oppose any attempt to deflect it as defiance or blasphemy. Inflight sabotage is perhaps the least likely result, even though it's been a staple of fiction from *The Conquest of Space* to *Fantastic Voyage* to *The Hammer of God*. All those scenarios postulated a saboteur who had infiltrated the program long before, had been brought in as a late entry, or had changed his beliefs during the mission. In real life it's impossible to think of an example, in space, in military aviation, at sea or under it; either the attempt has been successful, or it's never happened.

Attacks on space centers are perhaps more likely, and form a major theme of *Shiva Descending*. Mass invasions by mobs unafraid of death seem improbable, but there are few launch sites, and even fewer sites for manned spacecraft launches. Isolated on the steppes of Kazakhstan, Baikonur might be easiest to defend with a cordon of troops thrown around it. Kourou, on the coast of French Guyana and surrounded by jungle, would pose more problems; Kennedy Space Center, in a nature reserve on an island on the coast, would be almost impossible to protect against determined infiltration, and suicide attacks from the air cannot be ruled out.

Once the missions are in flight, the targets for terrorists may change, especially in response to the issues discussed below. Hostage-taking seems all too likely. The families of the mission controllers and the astronauts can be taken into protective custody, but with radiological weapons and 'suitcase' bombs, entire cities in the spacefaring nations might be held to ransom. These are grim possibilities, but if our scenario becomes reality they will need to be faced.

They might come to a head if the asteroid can be moved a little, but not enough to clear Earth. Where would it do least harm? As Tate pointed out, in deep enough water, deeper than 5 km perhaps, a 1-km impactor wouldn't penetrate Earth's crust. There would still be global effects on climate, but more water vapor and less dust would be thrown into the upper atmosphere than there would be if it hit a continent. It might perhaps be compared to evaporating the content of Loch Ness, which would produce a global deluge, as Fred and Geoffrey Hoyle pointed out [83], but not the devastating quantity of water that would be injected by evaporation above a Chicxulub-type crater penetrating Earth's crust on the seafloor. The atmosphere might even clear more rapidly than it would from a comparable injection of dust, which would produce darkness worldwide for 6 months to a year, and would precipitate increased rainfall in any case.

The real problem is that coastlines everywhere around the ocean concerned will be swept by tsunamis, on a scale far beyond previous human experience. The waves from the Eltanin impact off the tip of South America didn't just scour the Pacific rim, starting at 60–80 m high and still up to 10 m high when they reached Alaska and South Australia, but they diffracted around Tierra del Fuego to hit the west coast of Africa from the equator southward. New Zealand and Indonesia shielded China and the Indian Ocean, and the Antarctic peninsulas protected western Australia, but at the price of higher waves at the edges of the pattern [3].

The movie *Deep Impact* was rightly criticized for its focus on just one aspect of the disaster. As the asteroid strikes, we see the blast wave radiating outwards, and we see the ejecta plume burst out of the top of the fireball and start to fall back, but neither reaches the nearby continental United States. There's no ground shock, no air blast preceding the tsunami, and no rain of molten rock (The ejecta plume wasn't shown on the promotional poster, and looks like an afterthought on screen). But since the *Deep Impact* object is about a mile in diameter, the film's scenario for a strike off New York might just work if it was far enough out. Because of the contours of the seafloor, 500 miles out would be needed—very little more or less, on a narrow spur of the North-western Atlantic Basin. Nares Deep, southeast of Bermuda, would be preferable and a bigger target. But there are only a dozen or so targets like that, most of them small, and over most of the oceans some penetration of the seafloor seems inevitable.

Still, the dust plume would be a lot less than on land, and evacuating coastal regions all around the ocean rim might be easier to accomplish than evacuating one entire continent. The responsibility for moving part of their populations to higher ground would fall on many coastal nations, but it could be preferable to the extinction of inland ones, or of larger ones such as the United States. In 2003 Arthur Hodkin said that recent experience in North Carolina showed that the United States itself was becoming better able to handle evacuations from hurricane landfalls. Unfortunately Katrina 2 years later was to show that the lessons weren't being learned elsewhere, and as of February 2012, there were no plans anywhere for evacuating impact zones [84]. When it is done the displaced populations would have to be fed and housed, but so will people everywhere, unless governments refuse to face up to that as Lembit Öpik predicted. We'll come back to that question in Chap. 7.

As the world squares up to those difficult choices, there will be panic and counsels of despair among the leaders of nations, as there will be in the world at large. But we may hope there will also be Kennedys, urging their nations to commit themselves for a decade, and Churchills, urging them to aspire to their finest hour for a thousand years.

Jay Tate predicted that it would take a year to evaluate the options for deflection and decide which ones to adopt. If by chance the scenario does become reality, perhaps we can spare the new agency's leaders some of the uncertainty and delay. With existing technology, solar energy is the best option for the first hit; mass drivers, with human supervision, for the second; and the nuclear and kinetic options should be held in reserve for the third. With cost not a factor, all of those will be pursued, as well as newer techniques that might be ready in time.

Early in this book we quoted H. G. Wells and his vision of destruction. Looking ahead, we'll end it with Conan Doyle and a moment of reflection. Between the two comes the need for action. Come, then—the game's afoot!

Part III
What Would We Do?

5. The First Scenario: Deflection

> Turn again, turn again, turn again I bid ye...
>
> —"The Burning of Auchindoun" (traditional)

As noted above, at the time of the author's *New Worlds for Old* in 1979, the best answer to any form of impact threat seemed to be the industrialization of the inner Solar System, and that idea was developed in *Man and the Planets* (1983). A fast-response system would still be needed for dangerous comets coming in from the Kuiper Belt or the Oort Cloud, but as the book envisaged there would be courier links between the planets maintained by pulsed fusion spaceships, there should be time enough to intercept such comets, even if they were on retrograde orbits [1].

Space industry on this scale remains a long-term prospect. But a technology 'annexed' by the Strategic Defense Initiative in the mid-1980s may provide the answer—a quick, relatively inexpensive, non-violent answer that breaks no international treaties and creates fewer political issues. The comet-chaser designed by Gordon Ross of Glasgow School of Art, to counter the expected threat from Comet Swift-Tuttle in the late twentieth century (Chap. 1), could probably be made effective soon enough to deal with the Goldilocks hazard.

Ross is an aerodynamicist and former sail-maker, winner of the Duke of Edinburgh's Award for the double-surface sail design now in competition use worldwide, designer of the standard competition hang-glider. In the early 1980s he formulated the concept and design of Solaris, a parabolic solar sail with a unique method of pleated storage and deployment by inflating pressurized ribs. The details are still confidential; obviously, the sail is not to scale with the rest of the system and the structural details are schematic.

The comet-chaser he envisages marries the Solaris, his own design of parabolic solar sail, with the flexible parabolic mirror

D. Lunan, *Incoming Asteroid!: What Could We Do About It?*,
Astronomers' Universe, DOI 10.1007/978-1-4614-8749-4_5,
© Springer Science+Business Media, LLC 2014

FIG. 5.1 The Strathclyde University flexible mirror (© Gordon Ross, 2012)

created at Strathclyde University by a team led by Dr. Peter Waddell (Fig. 5.1) [2].

At the start of UK National Astronomy Week in 1985, the author was treated to a demonstration in Dr. Waddell's laboratory by John Braithwaite, who was a consultant to the Strathclyde team (Fig. 5.2). Ironically, the return of Halley's Comet that year was to provide a major boost to John's own business as a telescope maker, later to become the only one in Scotland and a continuing success until his death in 2012.

The author was placed before a 26-in. mirror (Fig. 5.1), which appeared simply to be a plain circular mirror such as one might find in a bar or hotel lobby. Braithwaite then turned on a pump that withdrew air from behind the mirror, and the Mylar reflecting surface began to bow away (Fig. 5.3). As it did so the author's reflection, originally at three-quarters length, grew more magnified until his face filled the frame. But there was none of the distortion of a fairground mirror, and even at the limit the image was marred only at the edge where the Mylar was pinned

Fig. 5.2 John Braithwaite in Braithwaite Telescopes workshop, Dalserf, Lanarkshire (© Jared Earle, 2012)

to the frame. Braithwaite then placed an optical target where the author had been standing, and a plain secondary mirror in front of the primary, handing him a card on which to catch the image. Wherever the card was moved, Braithwaite would adjust the pressure behind the mirror to bring the image back into focus each time, proving that the mirror was changing instantaneously from one parabolic figure to another—a feat that had been claimed early in the twentieth century but denounced by generations of optical experts as unattainable in practice.

A few days later, in the foyer of the Theatre Royal, Glasgow at the opening of Brecht's *Life of Galileo*, Braithwaite provided a further demonstration with an outwardly conventional Newtonian 4-in. reflecting telescope. Pointing it at a cornice on the ceiling, he had the author adjust the focus to suit his own eyesight. He then

① VACUUM DEFORMABLE MIRROR

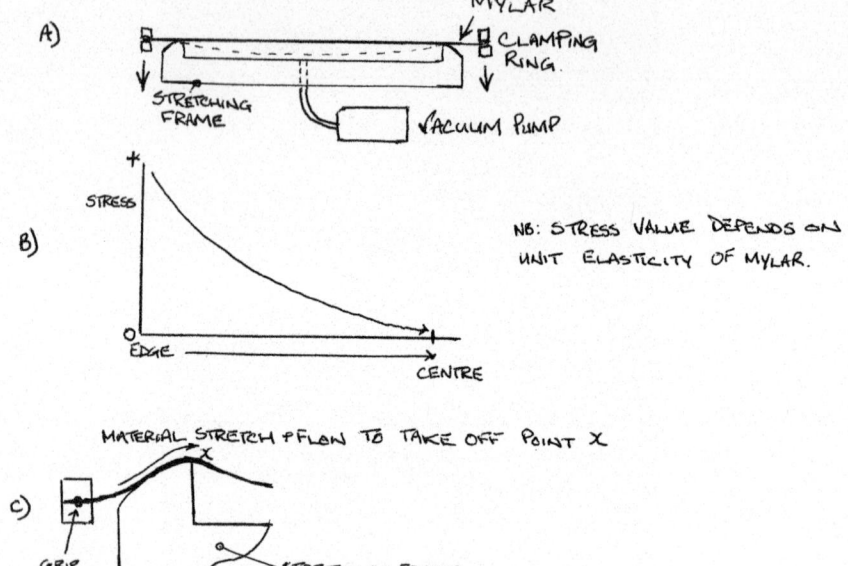

A)

MYLAR
CLAMPING RING.
STRETCHING FRAME
VACUUM PUMP

B)

STRESS

NB: STRESS VALUE DEPENDS ON UNIT ELASTICITY OF MYLAR.

EDGE

CENTRE

MATERIAL STRETCH + FLOW TO TAKE OFF POINT X

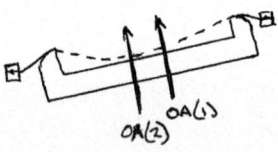

C)

GRIP SEAL

STRETCHING FRAME.

CLAMPING RING IN OPERATIONAL POSITION

D) SHIFT IN OPTICAL AXIS   WITH DIFFERENTIAL TENSIONING

OA(2)
OA(1)

FIG. 5.3 Flexible mirror diagrams (© John Braithwaite, 2008)

slid the assembly of focusing mount and secondary mirror down a slit in the tube until it was hard up against the primary mirror, sucked air from behind the primary mirror through a rubber tube, and waved the author back to the eyepiece. The image was still in focus. Despite the skepticism of professional astronomers (one particularly eminent one saying, "They will never get it to focus"), the classical problem of the flexible mirror had been solved.

Nevertheless, the university declined to invest even the most basic funding in what they had, and finally sold the patents to a U. S.

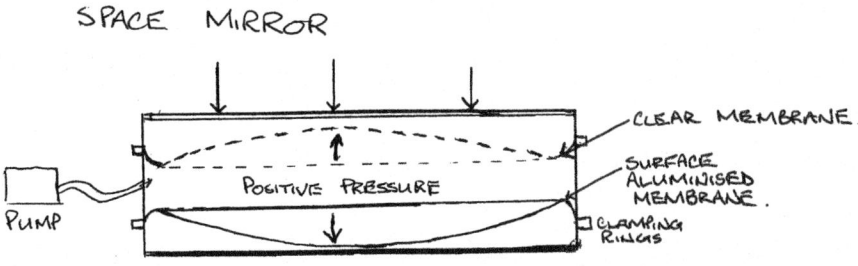

FIG. 5.4 Space mirror diagram (© John Braithwaite, 2008)

aerospace company for defense weapons' applications. (Having written it up for amateur publications in the United States and the UK in 1986 [3], the author was asked to speak about the mirror to a lunchtime astronomy club at NASA's Johnson Space Center during a lecture tour, and then asked to miss part of the afternoon's tour in order to brief a senior air force officer on the matter.) Braithwaite was invited to go to the United States to continue working on the mirror but politely declined. When pressed with the comment, "Does that mean you want the Russians to have it?" he replied, "No, when I see a friend of mine embarking on a costly and potentially fatal error, if I try to dissuade him it does not mean that I wish another friend to make the same mistake." Little has been heard of the flexible mirror since, but it may provide the way out of the impasse over nuclear weapons and impactor deflection.

To function the Strathclyde mirror requires a pressure differential, which in space would be maintained by a layer of gas contained by a transparent, flexible membrane (Fig. 5.4) [4]. Other adaptive optic systems, developed meantime, include mirrors made of multiple segments, individually mounted and computer controlled [5], and very thin mirrors, mechanically adjusted from the rear [6].

Either of these technologies might in theory be used in the comet-chaser instead of the Strathclyde mirror, perhaps more effectively for directing the beam off-axis. They are now established technology, because they became available when major breakthroughs had been made in understanding the distortions in 'seeing' that Earth's atmosphere inflicts on optical telescopes. Catastrophe theory provided the first account of how a pressure wave, passing the observer, makes the light from a point source like

FIG. 5.5 Solaris comet-chaser (© Sydney Jordan, 1994)

a star flicker as the wave comes and goes; the poetic description, 'Twinkle, Twinkle Little Star' turned out to be correct [7]. But the wave would typically be traveling at about 60 miles per hour, so if it was measured by an 'artificial' star generated by a laser beam, a computer could distort a telescope mirror to correct for it—giving ground-based telescopes performance equivalent to space ones. That's why the James Webb Space Telescope, as a successor to the Hubble, has been designed to work primarily at infrared wavelengths, which don't penetrate the atmosphere.

Comet-chasers could be stationed at the five Lagrange points of the Earth-Moon system, equidistant from Earth and the Moon, maintained in place by sunlight pressure against the perturbing pulls of the Sun, Earth's equatorial bulge and the other planets [8]. Even in an emergency at least one should be able to intercept an incoming comet by the time it reaches the orbit of the Moon (Fig. 5.5), and burn a hole in its crust to release a gas jet that perhaps could deflect the comet past Earth [9]. At the distance of the Moon, the deflection required would be only 1°, or 4,000 miles in space. For asteroids, more time would be required for deflection,

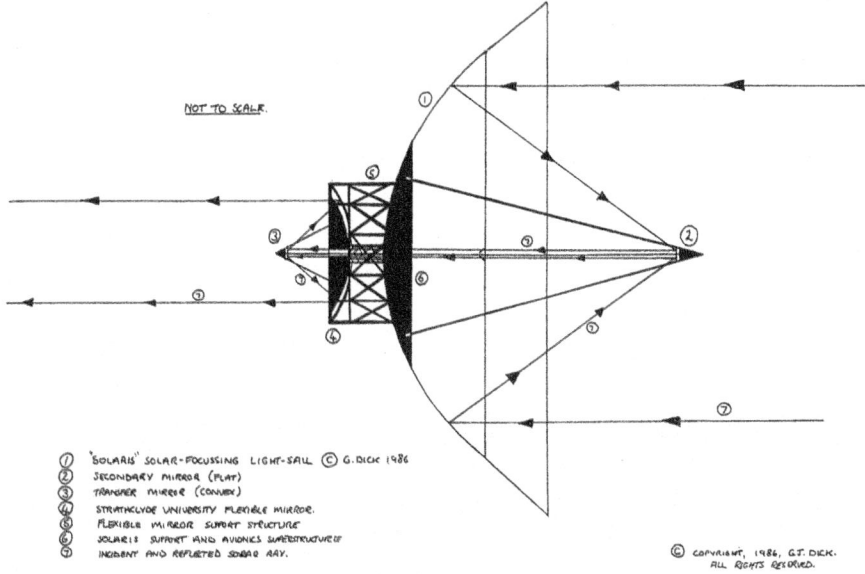

NOT TO SCALE.

① 'SOLARIS' SOLAR-FOCUSSING LIGHT-SAIL © G. DICK 1986
② SECONDARY MIRROR (FLAT)
③ TRANSFER MIRROR (CONVEX)
④ STRATHCLYDE UNIVERSITY FLEXIBLE MIRROR.
⑤ FLEXIBLE MIRROR SUPPORT STRUCTURE
⑥ SOLARIS SUPPORT AND AVIONICS SUPERSTRUCTURE
⑦ INCIDENT AND REFLECTED SOLAR RAY.

FIG. 5.6 Comet-chaser diagram (© Gordon Ross [then Gordon Dick], 1986)

but an interplanetary version of the sail would be able to cope with that, given enough notice. The more comprehensive the catalog of Earth-grazers, the more likely that the comet-chasers can get to a threatening object in time to turn it aside. And if on further research the burn technique isn't going to work, if nuclear warheads have to be used, then at least the comet-chaser could use passive propulsion to place them in exactly the right place.

The optical system of the comet-chaser combines four concentric parabolic mirrors in what might be called a double Cassegrain configuration (Fig. 5.6). Once rendezvous has been achieved, the vehicle concentrates all the energy gathered by the sail on the secondary mirror, from there to a tertiary one behind the sail, and from there via a quaternary, adjustable mirror to the optimum spot on the crust of the comet. While the beam goes down-Sun, net thrust on the vehicle will be zero while it is in burn mode.

At first we envisioned a fifth, free-flying mirror to keep the beam focused on the target spot as the nucleus rotated; if there was time, the first objective might be to slow the rotation to make it Sun-synchronous, so that the orbit-changing jet would be fully effective. The free-flyer would probably be needed even then, unless the hot spot was right at the sub-solar point on the stabilized nucleus. By the time we achieved detailed publication [10], a similar concept was being studied independently by Dr. H. J. Melosh of the University of Arizona's Lunar and Planetary Laboratory. Dr. Melosh calculated that at Earth's distance from the Sun, a 1-km parabolic solar sail with a free-flying secondary mirror could illuminate a 10-m spot on an asteroid at a range of 1 km, hot enough to vaporize rock or ice but probably not iron. Asteroids up to 3.4 km in diameter could be deflected from impact in a year [11].

In theory, deflection in any direction would take the asteroid or comet out of its collision course with Earth, so the obvious place to focus the beam was at one of the asteroid's or comet's poles. For a comet, that might be the only effective strategy because once started, a jet would probably continue under the action of sunlight and if it wasn't at one of the poles, rotation could cancel its effect, as seems to be the case with Comet Swift-Tuttle's jets (at present).

In a BBC Horizon documentary of January 2003, Melosh made a similar proposal [12]. Even then off-axis focusing would probably be required, and as the comet's orbit began to change, the optical system would have to be de-focused intermittently, to allow the sail to follow, unless there was enough energy available for a beam-splitter system to balance up and re-adjust the forces.

Three questions arose, however. The first was the sheer size of the 1-km sail. This was within the payload capability of the space shuttle to lift, but ten times the size Gordon had pictured. A 100-m comet-chaser would require a minimum of 4 years to move a 1-km asteroid, but that was within our allowable timespan of 6.6 years. Well before then it would be obvious whether the technique was working and whether a further mission, manned or unmanned, was required.

Melosh had originally calculated that there was no need to direct the sunlight at a fixed point on the asteroid's surface, because vaporization would be sufficiently rapid for the jet to be continuous even if the asteroid rotated under it. Figures supplied to us by Dr. Jim Randolph of JPL, from tests at the solar furnace at Mont-Louis in the Pyrenees (Fig. 5.7), with a collecting area of 1,830 m² and equivalent to a solar sail of 24 m diameter, suggested that a 100-m collector in space at 1 au from the Sun could raise the target spot to 7,000°C. The tests were conducted on pumice, as the nearest readily available material equivalent to chondrite.

Fɪɢ. **5.7** Solar furnace at Mont-Louis, Pyrenees, France (Wikipedia Commons)

Unless there's compelling reason to do otherwise, the hot spot should preferably be in the plane of the asteroid's orbit, at the center of the advancing or retreating hemisphere. To quote James Oberg again:

> Here's another example where today's experienced space operators already have a better grasp on required mid-century environmental modification operations. The obvious response to an approaching asteroid is to 'deflect' it sideways to miss Earth. This is the mode for an approaching tomahawk in The Last of the Mohicans, sure. But this 'common sense' idea fails to appreciate the unearthly nature of out-of-plane dynamics in space.
>
> Assuming a long enough lead time, the last kind of impulse you would ever want to impart to an asteroid is perpendicular to its motion. This would merely make it wobble in its orbit, but it would for the most part still arrive at future points close to the original predictions.
>
> Instead, the impulse should be directed along its flight path—slowing it down (from in front) or speeding it up (from behind) would work equally well. This would alter the energy of the object and cause it to arrive at predicted future intersection points at a different time. When it got there, the fast-moving Earth wouldn't be there—and the impact would be avoided.
>
> But try and explain this to someone unfamiliar with orbital operations. Tell them that in order to make it miss Earth, you want to speed up the approaching asteroid, and see the reaction! [13]

Nevertheless, it does move it.

The qualification has to be added that the 'wobble' is in the orbit of the asteroid, not Earth's, so the return to the intersection point isn't a hazard unless the asteroid's orbit is commensurate (resonant) with Earth's. More importantly, as Asher explained at Powys in 2003, the energy required to alter the asteroid's orbital period, making it cross Earth's orbit behind or ahead of the planet, is less than that needed for a right-angled deflection, by a factor of three—and the effect is multiplied with each subsequent revolution around the Sun.

However, the shape of the asteroid, its internal structure, the orientation of its axis of rotation, and whether it has more than one of them, will all have a bearing, literally, on the preferred direction of thrust. It would be nice if the optimum thrust direction, as determined by orbital dynamics, coincided with the one that the asteroid's properties dictated, but it's probably too much to hope for. Likewise it would be good if there was just one axis of rotation, and its pole lay in the orbital plane, or its equator crossed it, but neither of those can be taken for granted. If the preferred thrust vector for the asteroid is none of the above, it would be good if—considering Earth as a circular target—we only had to move it a short way to clear the edge, but again that may be asking too much.

Even based on the most pessimistic assumptions we could make—that the asteroid is going to hit near one rim of Earth's trailing hemisphere, but can *only* be moved at right angles in the direction of the other rim—that we need a safety margin of an additional 2,000 km—and giving the asteroid an average density of 2.75 g/c.c., a bit higher than Eros's—the thrust required came out to 10^3 N. The space shuttle's main engine has a thrust of 13×10^6 N, so to deflect the asteroid we would need 0.00008 of the SSME's power. That doesn't sound like very much, so perhaps a more useful comparison is with the thrust of a wartime V2. The V2's takeoff thrust was 25 t, 2.49×10^5 N, so whatever drive we apply to the asteroid has to achieve 0.004 of that—an average thrust of 224 lb, maintained over 6.6 years. Put that way it's still quite substantial, but remember it's the worst imaginable case. The actual thrust required will probably be 10–30 times lower. The acceleration of the asteroid would be about 7.5×10^{-11} g, while the surface gravity would be about 10^{-4} g (see Chap. 6), so there's no danger of pulling it apart.

The third issue that concerned us was the possibility that the ejecta plume would damage the spacecraft. The Mont-Louis tests suggested that for safety the mirror should be at least 10 km from the target, rather than 1 km. The Solaris would be operating from up-Sun, so its main collecting surface would be facing away from the plume, and since the structure would be almost at right angles to it, the flexible mirror could be shielded from damage. As the transparent membrane over it would be pressure-braced, that would provide added protection against any debris than did impinge on it. The secondary mirror would be much more exposed, but even that wouldn't have to be right in the firing line; it could be made a lot more robust, and multiple replacements could be carried.

Nevertheless the concern remained, and another communication from Dr. Jim Randolph suggested another possible answer. Formerly the head of Advanced Mission Planning at JPL, by the early 1990s Randolph was director of NASA's Solar Physics division. Our connection with him stemmed from ASTRA's research into the Waverider atmospheric entry vehicle, which was devised by Prof. Terence Nonweiler as the man-carrying vehicle for the British space program based on the Blue Streak launcher, canceled by the Macmillan government in 1960. Waverider has major potential in the future exploration and development of the Solar System [1], and ASTRA had been promoting those since adopting Waverider as its flagship in 1977, undertaking low-speed research on it in the 1980s and 1990s, claiming the first free flight of a Waverider, first rocket launch to free flight and first radio-controlled flight [14]. We had begun with Nonweiler's original concept of a V-shaped caret wing, although it had become increasingly sophisticated with modifications such as laminar flow and flexible airfoils. Meanwhile the University of Maryland and several aerospace companies had pursued an alternative design approach, resulting in a much chunkier vehicle, which has now been tested successfully in the U. S. Air Force X-51 program [15].

In the visit to JPL mentioned in Chap. 1, the author was asked to lecture on Waverider. It took a certain amount of nerve for an Arts graduate to talk aerodynamics to the world's foremost experts, when the only response to the hard questions had to be, "You'll have to ask Terence Nonweiler (or Gordon Dick) about that." Dr. Randolph was interested in Waverider as a carrier for the Starprobe project, to send an instrumented vehicle to within four solar radii

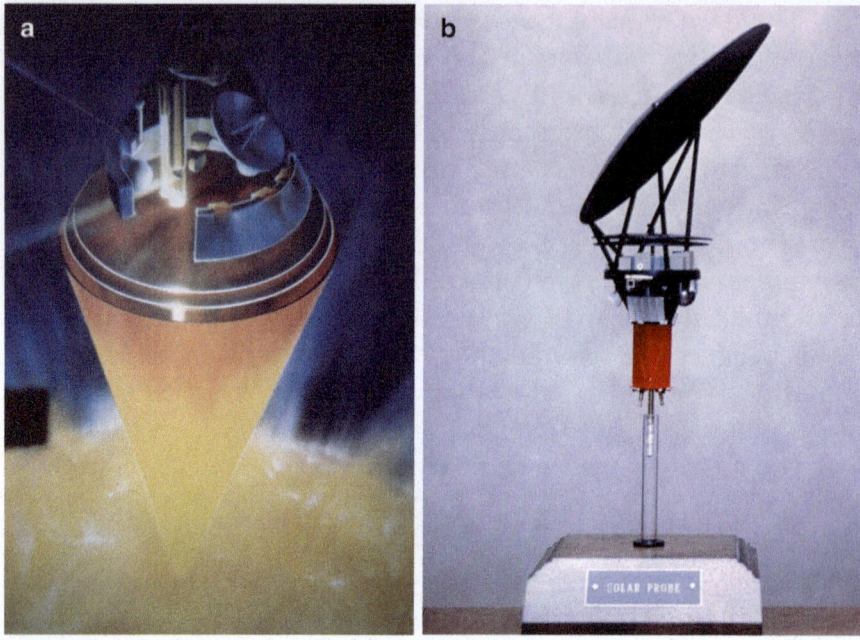

FIG. 5.8 (a) Original Starprobe concept (JPL). (b) 1990s Solar probe redesign (NASA)

(3 million km) of the surface of the Sun. It could be accomplished by a very close Jupiter slingshot, but the radiation hazards and very long flight time made that unpromising. Aerogravity maneuvers in the atmospheres of the inner planets could give the probe a trajectory with solar encounters every 2–3 months, but required a carrier with a very high lift-to-drag ratio at high Mach numbers—for which Waverider was the best candidate. For a time, extraordinarily, the enabling technology for 'the physics mission of the century' was in the hands of an unfunded amateur group in the west of Scotland.

The Starprobe instrumentation was to be mounted on the base of a heavy ceramic cone, gravity stabilized so that it pointed at the Sun throughout the flyby and the instruments remained in shadow (Fig. 5.8a). But with advances in material technology in the 1990s, the same effect could be achieved with an inclined, elliptical carbon-fiber disc that would also function as the radio antenna of what was now called Solar Probe (Fig. 5.8b), making the spacecraft so much lighter that it could be launched with a Delta II booster and achieve the solar pass by Venus slingshot. It restricted the timing of the mission, because Earth had to be at right angles to the probe during the solar flyby (to the right in the illustration), but the larger antenna allowed a much higher bit rate for returning data, and resonances between Venus's orbit and Earth's created recurring launch windows.

Still the mission didn't get the go-ahead, and with continuing improvements in instrumentation one of its major objectives—to measure the frame dragging that relativity theory predicted in the vicinity of a massive rotating body [16]—has been measured, though to less accuracy, by Gravity Probe B in Earth orbit [17]. For a time the mission was called Fire, to be paired with a Pluto mission called Ice (not to be confused with International Cometary Explorer in Chap. 1), but eventually Pluto Express flew solo, as New Horizons, and Solar Probe Plus is now scheduled for multiple passes within 8.5 solar radii, after launch in 2018.

The Solar Probe redesign ended the potential role for Waverider as a carrier vehicle, at least for the time being, but the off-axis elliptical reflector suggested a possible way to beat the solar collector's problems with the ejecta plume. Gordon's new design, Archimedes, would position four of them down-Sun of the asteroid (Fig. 5.9a), with pressure bracing on the reflectors again supplied by a transparent membrane that would protect them from debris (Fig. 5.9b), while the main body of the plume passed between them. Unlike the comet-chaser, this one was an 'asteroid burner.'

Ross wrote:

> The new design is a far simpler and more reliable system, designed to divert comets or asteroids. It uses more sophisticated, pressurized membrane reflectors, which are elliptical in shape, to concentrate enormous amounts of sunlight onto a spot about 10 meters across, on the target surface. This vaporizes surface material, which forms a jet or plume. The exhaust gases cause pressure against surface of the burn 'trench,' producing a net thrust of several tons magnitude. As the asteroid rotates about its centre of mass, the burn spot traverses the surface, burning a 10 m wide trench into the surface regolith. The walls of the formed trench help to direct and augment the thrust, by containment of the exhaust flow. The burned trench cools rapidly and the surface contacts to form a glass-like residue. The burned area is weakened by this huge heat gradient, so that during future passes through the 'Beam,' the surface material vaporizes more easily and more rapidly.

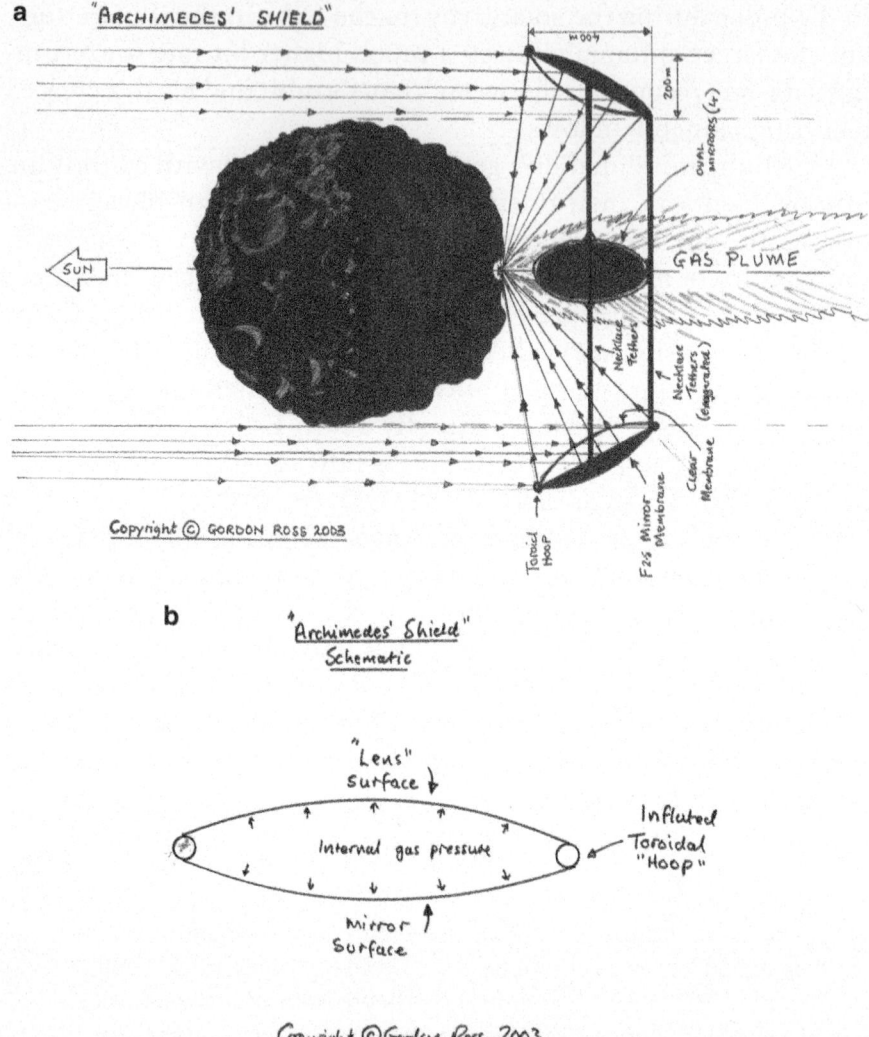

a "ARCHIMEDES' SHIELD"

b "Archimedes' Shield"
Schematic

FIG. 5.9 (a) Archimedes system (© Gordon Ross, 2003). (b) The "Archimedes' Shield" mirror (Designs and drawings) (© Gordon Ross, 2003)

The reflector array consists of four or more elliptical reflectors, around 30,000 square meters in area. A 'necklace' of lightweight Dyneema cords would connect these reflectors to each other. The reflectors' membrane mirrors would each be supported by a single toroidal tube, which would be inflated using inert gas, and solidified by a pressure sensitive polymer.

The deployed structure would resemble a large, extremely thin, double skinned drum. One skin would be made from alumi-nized polyester, the other from clear polyester. When pressur-ized, these two skins would bow outwards, forming a reason-ably accurate parabolic reflector, with a clear parabolic 'lens' covering it. The complete mirrors are then solidified, using UV activated polymers coating their inner surfaces. This means that the mirrors are not relying solely on maintaining a gas-tight envelope for their six-year operational lifespan. (A similar method of deployment using matrix resin, cured after inflation of the sail ribs, was described by Bernasconi and Zurbuchen in 1995, for use in a 4-lobed 'Cloverleaf' steerable sail for asteroid missions [18] – DL.)

By using the gas in the toroidal collar as reaction mass, the mirrors may be gently maneuvered into a diamond formation. They would then be nudged into a position directly down-sun of the target object. The Array would then be slowly spun up, to stabilize the formation and the necklace would then angle the mirrors, to converge all four foci on the center of the dark side of the asteroid. The reflectors, when positioned in this way, are angled radially outwards, like four large wing mirrors. By virtue of their formation, the sunlight from the reflectors would strike the asteroid surface regolith at an angle, but the exhaust gases from the vaporization would pass through the large gap in the center of the spinning array.

A small amount of station keeping may be necessary to counter the minute thrust from the sun's light pressure, but it may be possible to position the burner array close enough so that the asteroid's gravity field could balance this out.

The system could be mass-produced, giving multiple redun-dancies. The design is consistent with the use of current materials technology, and is very lightweight. Dozens of these arrays could be launched to the target object, using existing chemical rockets. Damaged reflector arrays could be removed and replaced by back-ups [19].

John Braithwaite's comment on the revised design was that using off-axis parabolas eliminated the reflection losses that would be incurred with the multiple mirrors of the original design. On the other hand, collimating the last mirror in that sequence would allow the spacecraft to operate at greater distance from the aster-oid, and that was to become an issue later.

Fɪɢ. **5.10** Dr. Max Vasile with solar power demonstrator (© University of Strathclyde, 2012)

In 2005, a study of asteroid deflection methods began at Glasgow University's Department of Aerospace Engineering, funded by the Energy and Physics Science Research Council and headed by Drs. Gianmarco Radice and Massiliano (Max) Vasile (Fig. 5.10). Previous studies at the University of Milan and ESA had suggested that most deflection methods were not economical, so Dr. Radice had proposed to find one that would specifically include reaching the target asteroid as well as deflecting it, taking the natures of the asteroids into account, focusing on unmanned missions and existing launch vehicles, and not assuming any advances in robotics. As Max Vasile said, "Robots to repair robots will be much more complicated" [20].

When we shared our concepts with the university team, initially they were in favor; indeed we were surprised to hear, from Jamie Bentley in Australia, that by 2007 they had solved the problem of protecting Earth from asteroids and the answer was mirrors [21]. More accurately, their initial study had concluded that with 10

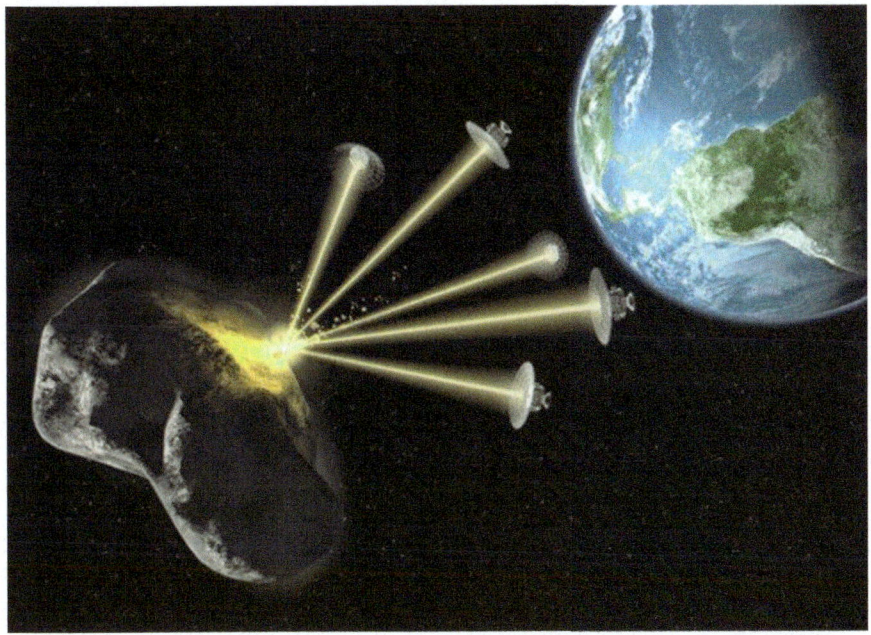

Fɪɢ. **5.11** Mirrorbees constellation (© The Planetary Society, 2007)

years' warning for an asteroid similar to Apophis, the best options were the nuclear and solar sublimation methods [22]. Mass drivers entered the argument later, on publication of the MADMEN paper (Chap. 6), and displaced nuclear weapons from their first-equal ranking [23]. With initial misunderstandings sorted out, however, differences between the two approaches became apparent.

The team concluded that a single mirror 1–10 km in diameter like Dr. Melosh's would be impractical to deploy and control, and lay perhaps a hundred years in the future. Instead, they were developing the concept of a swarm of solar collectors they termed fireflies or mirrorbees (Fig. 5.11). Two big advantages were the existing experience of flying satellite constellations (the Cluster satellites, for example), and that the system was scalable. For a bigger hazard, add more mirrors.

To deflect an Apophis-sized asteroid in 200 days from arrival would take a fleet of spacecraft stationed 12 km away, each with a mirror 20 m across and a dry mass of 200 k—an order of magnitude less than the Melosh single mirror. Their initial aim was to deflect the target by the full radius of the Earth-Moon system, and for a shift of only 30,000 km just ten spacecraft would be needed; for our less ambitious aim, four might be enough [25]. Forced orbit would be required, specifically 'funnel orbits' ahead of or behind the asteroid, depending on whether the asteroid was to be decelerated or accelerated, with 'cross orbits' inside the funnel for sample collection.

Studies of what were now called 'polesitters' or 'statites' at Glasgow University and by Dr. Robert Forward in the United States, showed that solar sails would be more than adequate for this form of station-keeping. Bigger mirrors would have equilibrium points closer to the asteroid, and in general they would move inwards and outwards with varying distance from the Sun. The operation would become more complicated if the asteroid wasn't spherical, but sideways thrusts on the order of 1 mN would be adequate to compensate for that. Generally, more work was needed on swarm control, beam control and deployment techniques, and considerable headway had been made by the following year; dual-mirror systems, possibly free-flying, off-axis circular mirrors and collimating lenses, were all under consideration to improve the focusing [26].

Even by then, however, doubts were being raised about whether the focusing requirements could be met. According to the optics department at Glasgow University, the biggest difficulty was that the Sun is not a point source of light. Dr. Melosh had considered that in 1994 and thought that his twin-mirror solution adequately addressed it [11]; but his starting point was the much larger 1–10 km diameter for the primary mirror, and the hot spot generated on the asteroid was to be 10 m across, whereas with their smaller mirrorbees Max and his team were trying for a much tighter focus of 1 m or less.

The other big question was the danger to the spacecraft posed by the plume of debris coming off from the asteroid. Again, Melosh had considered the issue and had suggested ways of dealing with it, including replacing secondary mirrors as they became damaged, a gas cloud in front of them to provide a measure of shielding, and a Cassegrain system like Ross's so that the main reflector could face away from the asteroid and the back could be armoured [11]. Nevertheless he noted that the last mirror in the system had to be close to the surface and was unavoidably at risk. A 2006 study by Kahle et al. had concluded that the lifetime of the focusing mirror

Fig. 5.12 NERVA nuclear rocket test, Jackass Flats, Nevada, 1960s (NASA)

would be measured in minutes, at most [27]; at the Association of Space Explorers meeting in Glasgow in 2007, the author attempted to raise the subject of solar deflection with Rusty Schweickart, but the astronaut dismissed it out of hand for that reason. A later study drafted in 2010 thought the survival time might be extended to an hour, but even so, up to 5,000 secondary mirrors might be required to complete the deflection of an Apophis-sized asteroid by 10,000 km in a year [28]. At a safer distance, up to 200 would be needed to do it in 20 years; the converse of that, Vasile calculated, was that at a safe distance his 20-m mirrors could only achieve a deflection of 500 km in 5 years [25].

However, a great deal depended on the properties of the plume. Dr. Melosh's tests with high-energy lasers fired at basalt samples in air had produced plumes remarkably like the atmospheric tests of the NERVA nuclear rocket in the 1960s (Fig. 5.12), penetrating and fracturing the surface so that the gas plume was slowed down by the rock fragments it carried away. Vasile and Ross disagreed on whether this would happen in space, Vasile thinking that a high-velocity stream of gas was more likely to be produced, while Ross believed that the heat would excavate a trench as the asteroid rotated, producing a focused plume of gas and rock fragments.

While appreciating why Vasile's team had decided to work with a maximum of two mirrors, rather than a train of four or possibly five as in the original Solaris concept, the author felt that removing the focusing system provided by the Strathclyde flexible mirror, central (both literally and figuratively) to the original Solaris concept, had created a great deal of unnecessary difficulty. A partial answer might be to use the concentrated sunlight to power a laser, rather than focusing it directly on the asteroid. But when challenged by Max Vasile to suggest an alternative solution, Braithwaite and Ross took little time to do so. This new version made the primary and secondary mirrors adaptable, with a fixed tertiary one, so that the focal length of the array could be adapted by varying the pressure within the transparent membrane that now covered the main mirror in full (Fig. 5.13). With that pressure bracing, they were sure that the system could resist impacts by slow-moving debris, especially if both surfaces were self-sealing. Subsequently Braithwaite suggested that the necessary focusing could be achieved by building a lens into the membrane, and varying its curvature by the same method. That might seem overly complex and vulnerable to damage, but has possible advantages when it comes to preventing contamination (see below).

Contamination was potentially the show-stopper. As Vasile said, focusing doesn't matter if you can't continue the operation. The big question then was, what would the nature and shape of the plume actually be? Ross's Archimedes drawing (Fig. 5.9a) suggests a spear-like shape, reminiscent of a chemical rocket flame in the atmosphere. We know from slow-motion films of the Saturn V and space shuttle engines igniting that the flames are, briefly, ragged, but as the motors reach full thrust the exhaust becomes much more directional. Seeing a Firestone Corporal launch from South Uist in 1959, and the Apollo-Soyuz rendezvous launch in 1975, the author was surprised at how narrow and bright the flames appeared to the eye. What we see in photographs is lens flare, because the flame has to be over-exposed to show the rocket.

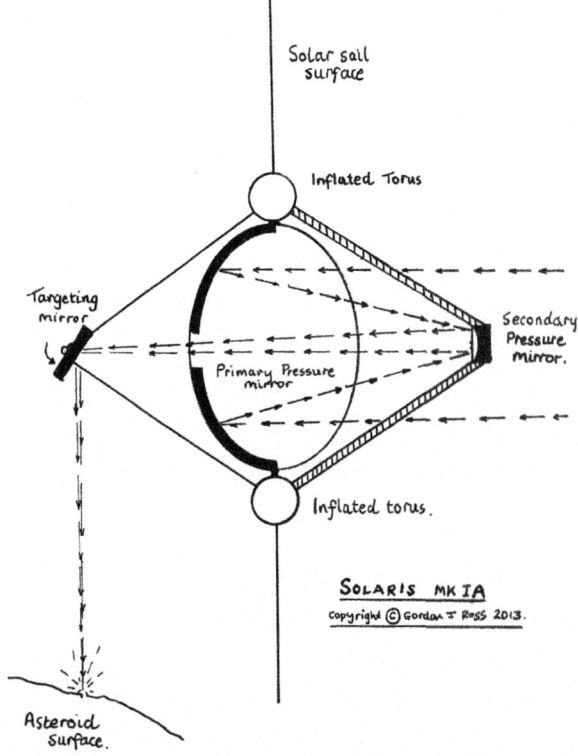

Solar sail surface

Inflated Torus

Targeting mirror

Secondary Pressure mirror.

Primary Pressure mirror

Inflated torus.

SOLARIS MK IA
copyright © Gordon J Ross 2013.

Asteroid Surface.

Fig. 5.13 Solaris Mark IA (Design by John Braithwaite and Gordon Ross, drawing by Gordon Ross)

Even as a rocket enters the upper atmosphere, the exhaust shape becomes noticeably different (Fig. 5.14a), and still more so in space (Fig. 5.14b). In sunlight, in a vacuum, rocket flames are invisible; a UK scientist pointed that fact out in the *Daily Express* in the mid-1950s, even before Sputnik, and was finally proven right nearly 20 years later when we watched the later Lunar Modules lift off from the Moon. Stanley Kubrick portrayed it accurately in *2001, a Space Odyssey*, and *Moon Zero Two* followed suit; *Battlestar Galactica* set the clock back, with flames like spear-blades, but George Lucas had it right again in the *Star Wars* trilogy (though the dogfights in space are impossible). But in darkness, the true situation is revealed.

When the Soviet Union was still keeping its military launch complex at Plesetsk secret, there were numerous reports

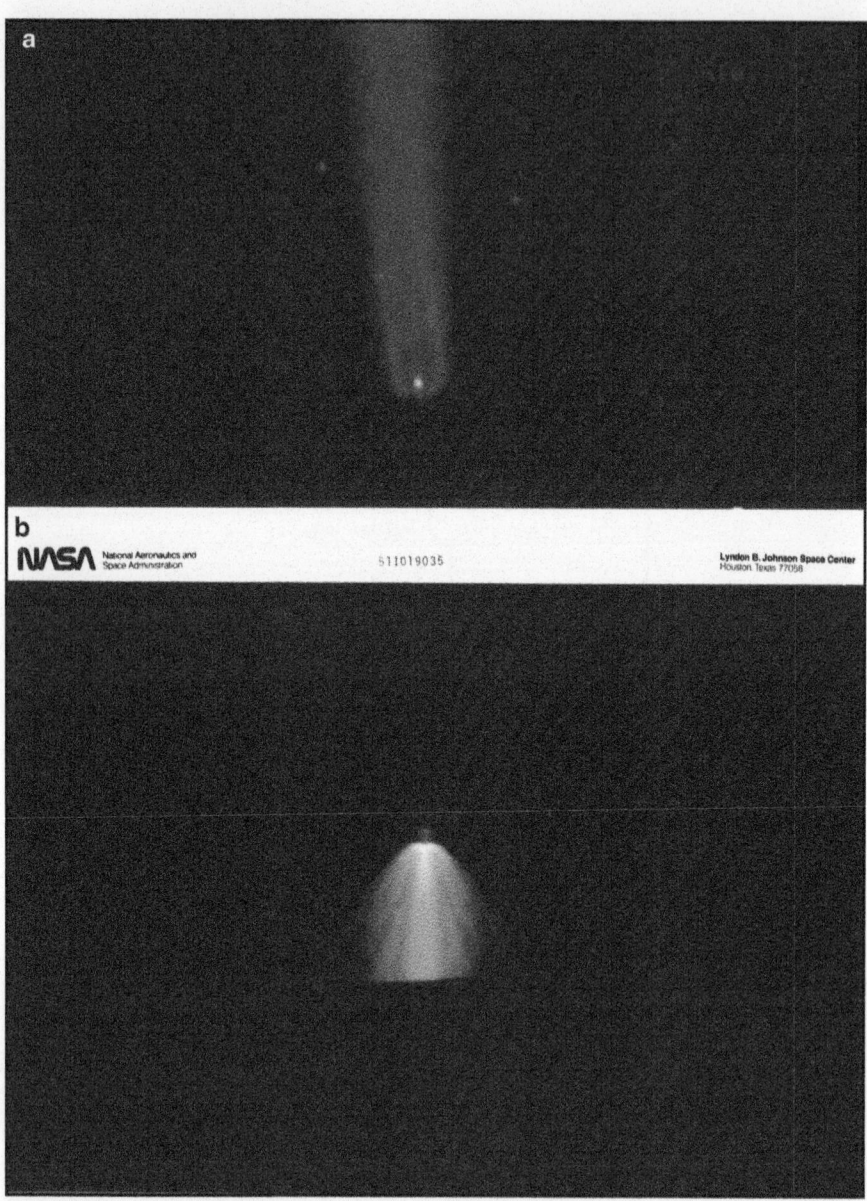

FIG. 5.14 (a) Plumes from Soyuz booster core engines at high-altitude separation of liquid-fuel boosters (NASA). (b) Ignition of the LEASAT solid-fuel booster in darkness, after hot-wiring on-orbit by the STS-51i astronauts, August 1995 (NASA). (c) NERVA manned mission leaves Mars, 1969—see Chap. 6 (NASA)

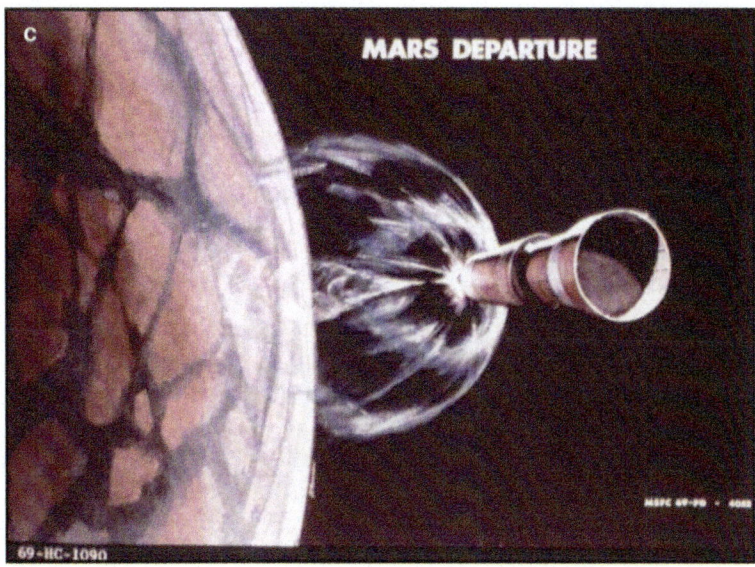

FIG. 5.14 (continued)

of 'Jellyfish UFOs' in the night sky, and officially encouraged societies were set up to study them. They were also seen in South America, and links were formed between the watchers on the two continents. But James Oberg, as an expert on the Soviet space program, noticed that they were only seen in opposite seasons—e.g., autumn in Russia, or autumn in Chile, never both at the same time. Correlating the sightings with NATO satellite tracking, he discovered that the Russian sightings were spy satellite boosters leaving the atmosphere, and the South American ones were the orbit circularization burns [29].

However, the jellyfish comparison captures the true shape of a rocket exhaust in a vacuum (Fig. 5.14c). The engine bells of the Apollo Service Module and Lunar Module were shaped to take maximum advantage of that, and that's why there's no crater below the Descent Stage engine on the lunar surface. The hoax claimants make a big deal of that, but ironically if there were one, *that* might be evidence of a hoax!

To stand off further from the asteroid and reduce the risk from the plume, Vasile and his team proposed to substitute lasers for the concentrated sunlight of the mirrors, increasing the range from 1 to 4 km [30]. On examining previous studies of the plume

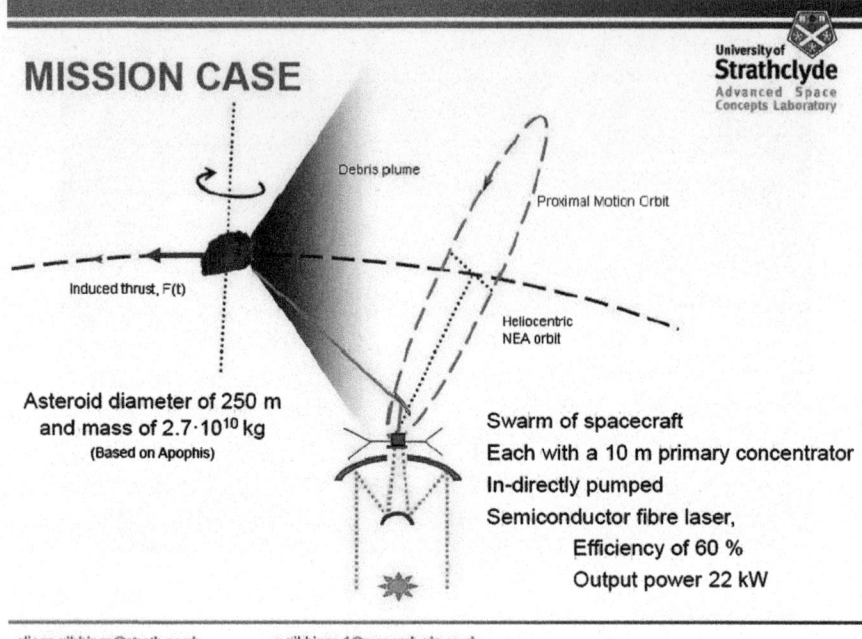

alison.gibbings@strath.ac.uk   a.gibbings.1@research.gla.ac.uk

FIG. 5.15 Assumed plume spread during asteroid deflection (Courtesy of Dr. Max Vasile)

issue, they found that some key assumptions had been made. The ejecta plume was assumed to be equivalent to a rocket exhaust, consisting of uniformly expanded gas with no ionization and no solid content. Dr. Melosh's experiments had indicated that might not be the case, and Gordon Ross was quite sure it would not be. The target asteroid was assumed to be dense, homogenous and spherical, with the mineral fosterite being taken to be typical. All were reasonable starting points, but a wider range of possibilities had to be considered.

Crucially, though, calculations of the degradation rate for the optical surfaces of the spacecraft, and the attenuation of the concentrated beam, took it for granted that the debris condensed on them immediately and stuck. If the debris was evenly distributed over a half sphere (Fig. 5.15), contamination would be immediate and would reduce the performance by 85%, cutting the deflection on the reference mission from 30,000 to 4,500 km—still enough to achieve deflection into deep water, worthwhile for a 1-km asteroid (Chaps. 4 and 7), or with luck, enough to clear the rim of Earth.

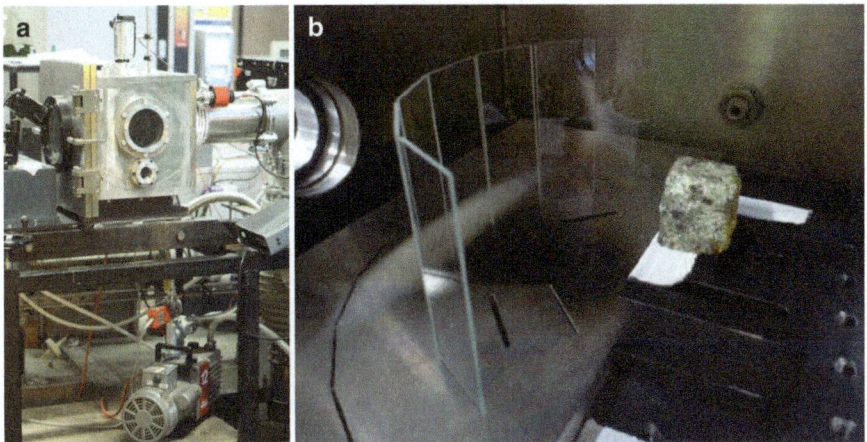

FIG. 5.16 (a) 90 W continuous-wave laser used in plume simulations (Courtesy of Dr. Max Vasile). (b) Rock sample for laser testing (Courtesy of Dr. Max Vasile)

Having moved to the University of Strathclyde, Vasile set up a series of tests in 2010–2011, financed by the Planetary Society. They were conducted using a 90 W continuous-wave laser (Fig. 5.16a), initially in a purged atmosphere of nitrogen and later in a vacuum, with targets of sandstone (to represent a rocky, dense asteroid), olivine (for a rocky, dense S-type asteroid) and a composite mixture to represent a rubble pile (Fig. 5.16b).

In the nitrogen atmosphere, the plume sputtered off was indeed a broad cone, containing multiple fragments (Fig. 5.17a), but in a vacuum the effect was much more like a gaseous rocket exhaust, and a focused one at that, though it began to expand a few centimeters out from the hot spot (Fig. 5.17b). Strikingly, the material deposited by the plume was a lot less than expected, and it didn't bond to the surface, being easily removed by a vibrating self-cleaning system. Even without that, when compared with the extreme degradation of the optical surfaces predicted by Kahle [27], the initial effect was 67% less and the achievable deflection distance for the asteroid was at least doubled [30].

The next move was to repeat the tests with the target suspended as a pendulum, and by February 2011 some dramatic movie sequences had been obtained. The laser did excavate a trench in the surface as had been predicted, with a gas jet traveling at 6 km/s. The lasers might have to be pulsed to keep the jet manageable [25].

a

NITROGEN PURGE

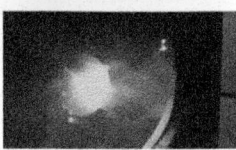

Small, and extended rocket plume
Similar mass flow rate, compared to the model

Variation in cone angle and ejecta distribution
    Ablation process included solid ejecta particles
    Subjected to the volumetric removal of material
    Resulted in the laser tunnelling into the subsurface
    Technique is sensitive to the focal point of the laser

Subjected to the
structure and
composition of the
target material

alison.gibbings@strath.ac.uk   a.gibbings.1@research.gla.ac.uk

b

VACUUM

Small & extended rocket plume. Little ejecta
At 3, 7 and 10 cm away from the spot:
    Measured the deposited mass/area, $(\Delta m/A)_{SLIDES}$
    Measured the height of the ejecta, Δh_{EXP}

From this the density of the deposited material can be calculated $\rho_{EXP}(r,\theta)$
Derive the expected collection rate of ejecta on each slide

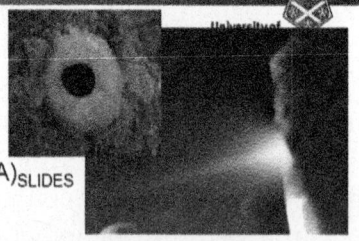

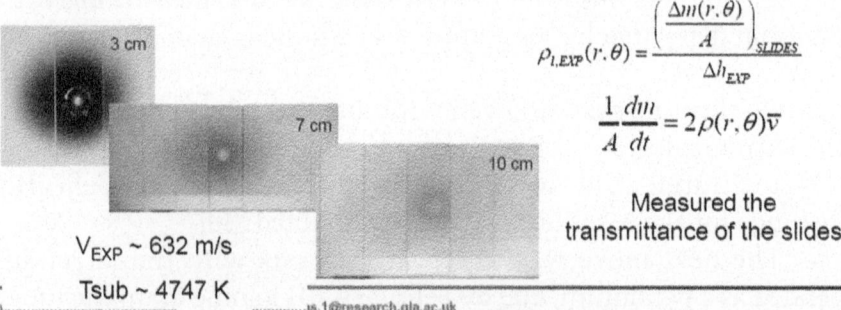

$$\rho_{l,EXP}(r,\theta) = \frac{\left(\frac{\Delta m(r,\theta)}{A}\right)_{SLIDES}}{\Delta h_{EXP}}$$

$$\frac{1}{A}\frac{dm}{dt} = 2\rho(r,\theta)\bar{v}$$

Measured the
transmittance of the slides

$V_{EXP} \sim 632$ m/s

Tsub ~ 4747 K

FIG. 5.17 (a) Laser-generated plume tests in nitrogen atmosphere (courtesy of Dr. Max Vasile). (b) Laser-generated plume tests in a vacuum (courtesy of Dr. Max Vasile)

Any of the three adaptive optic systems described at the beginning of this chapter might lend themselves to 'self-cleaning.' The multi-mirror system and the thin mirror deformable from the rear might seem particularly well suited to it, because the cleansing operation is much simpler than the phased distortion to correct for twinkling starlight. But the clear, pressure-based membrane has its advantages. It's more resistant to damage, especially if self-sealing. The contamination never reaches the reflecting surface, and the transparent one could be given an electrostatic charge to help it repel both dust and gas. The induced vibration could be equivalent to pulsing the laser, if that proves to be necessary. Over years of operation, a maintenance robot might even be able to fit new membranes as the original ones wore out.

Braithwaite and Ross undertook to build a free-flying demonstrator in the atmosphere in the summer of 2011; it could even have practical application, as a tethered platform for optical and infrared cameras over disaster sites. To begin development work Braithwaite borrowed the original 36-in. mirror from Strathclyde University, but other commitments intervened for all concerned, and the work was not begun. The author and his wife were in East Anglia for historical research in January 2012 when a call from John's wife told us he was in hospital for tests. By the time of our return to Scotland he had been diagnosed with untreatable liver cancer and had resolved to spend the next 3 months ensuring that his many projects would continue. But within days he was too ill to be visited, and by mid-February he was gone.

Ross joined forces with Chris O'Kane, inventor of the Vistamorph™ panoramic lens, whose spectacular images of ancient sites in Scotland and Egypt appeared in the author's *The Stones and the Stars*. O'Kane had been working to marry his invention to Braithwaite's 'Real-View' system for 3-D projection without glasses [31], and although the main problems of domestic television and cinema versions had been solved, these were all projects whose continuance John had been unable to ensure. The optics for the free-flying demonstrator were another. As O'Kane said at the asteroid seminar that October, it illustrates that if you have a great idea, you need to get on with it. You may not have next year.

Without Braithwaite's input, Ross and O'Kane found their ideas converging to some extent with Vasile's. The secondary

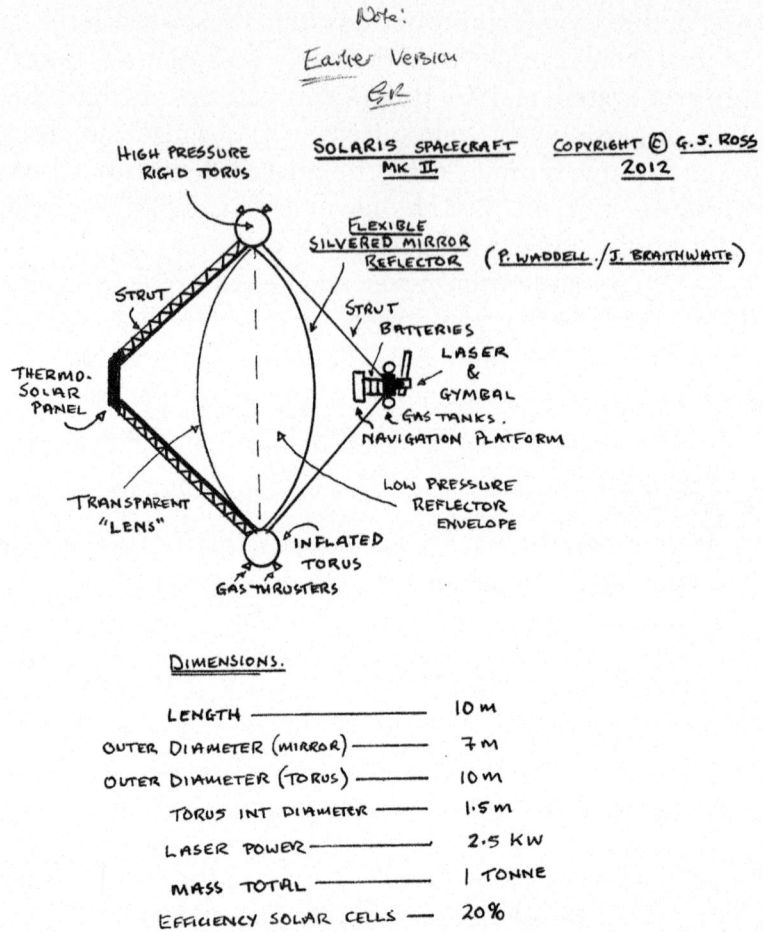

FIG. 5.18 Solaris 'Mark 1A,' intermediate between Figs. 5.13 and 5.20 (© Gordon Ross, 2012)

mirrors, lenses and other possible solutions to focusing were laid aside in favor of photovoltaic panels, or more probably Peltier thermal panels in view of the energy loading at the prime focus of the reflector. (Thermovoltaic panels are used in blast furnaces now, at operating temperatures up to 1,000°.) They would power a laser on a rotatable turret (Fig. 5.18), firing from a distance of at least 1 km; they could be placed up-Sun of the mirror (Fig. 5.19a) or down-Sun (Fig. 5.19b). For practicality, the model Ross and O'Kane brought to the 2012 seminar fired down-Sun; even with the heavy curtains of the Bridie Library closed to a chink (Figs. 5.20a), and even when

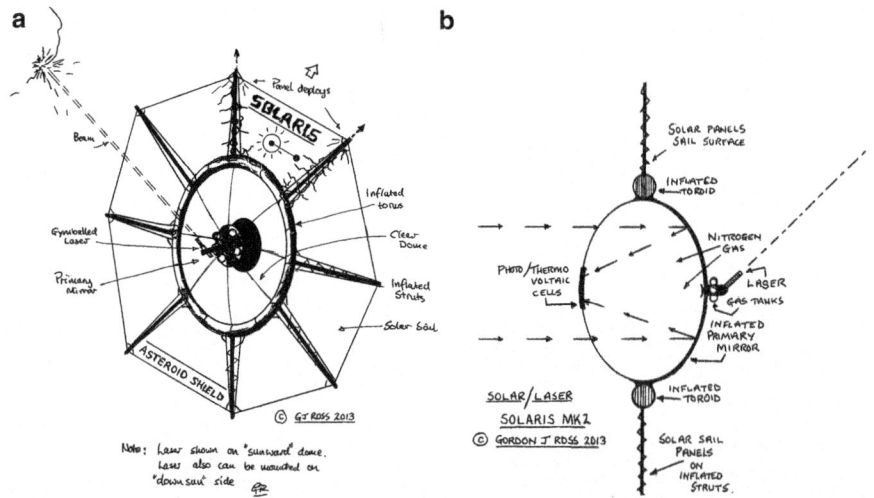

FIGS. 5.19 (a, b) Project Sunburn: Solaris Mark 2, front view and central section side view (© Gordon Ross, 2012)

the sky outside was overcast, its laser pointer would have ignited paper if allowed to remain steady (Fig. 5.20b).

Since the reflectors of Solaris Mark II are no longer primarily for propulsion, they can be made somewhat more robust, to withstand impacting fragments. Dishes 20 m in size are expected to be in use on communications satellites by the 2020s [25]. They will need initial flexibility for deployment, like the ones on the Tracking and Data Relay Satellites (Figs. 5.21a, b), because they're too big to be carried in rigid form. (Ariane V has a diameter of 5.4 m, Proton 7.4, and even Saturn V was only 10 m across.) Prof. McInnes calculated that to move a 1-km asteroid with 10 years' notice, a 100-m mirror would be needed, and it would take twenty-five 20-m collectors to match that. The standard payload fairing for SLS would be 8.4 m across, with an increase to 10 m under consideration, but with fixed mirrors of even that size, up to 100 would have to be launched.

Whatever size we adopt, if we want to use existing boosters for launch in the 3.3-year window, the mirrors will have to be folded. That should be no problem. Trial inflatable structures were launched on Aerobee rockets as long ago as the 1950s. The Echo satellites were orbited at the beginning of the 1960s and followed by PAGEOS in 1966, and both the United States and Russia have

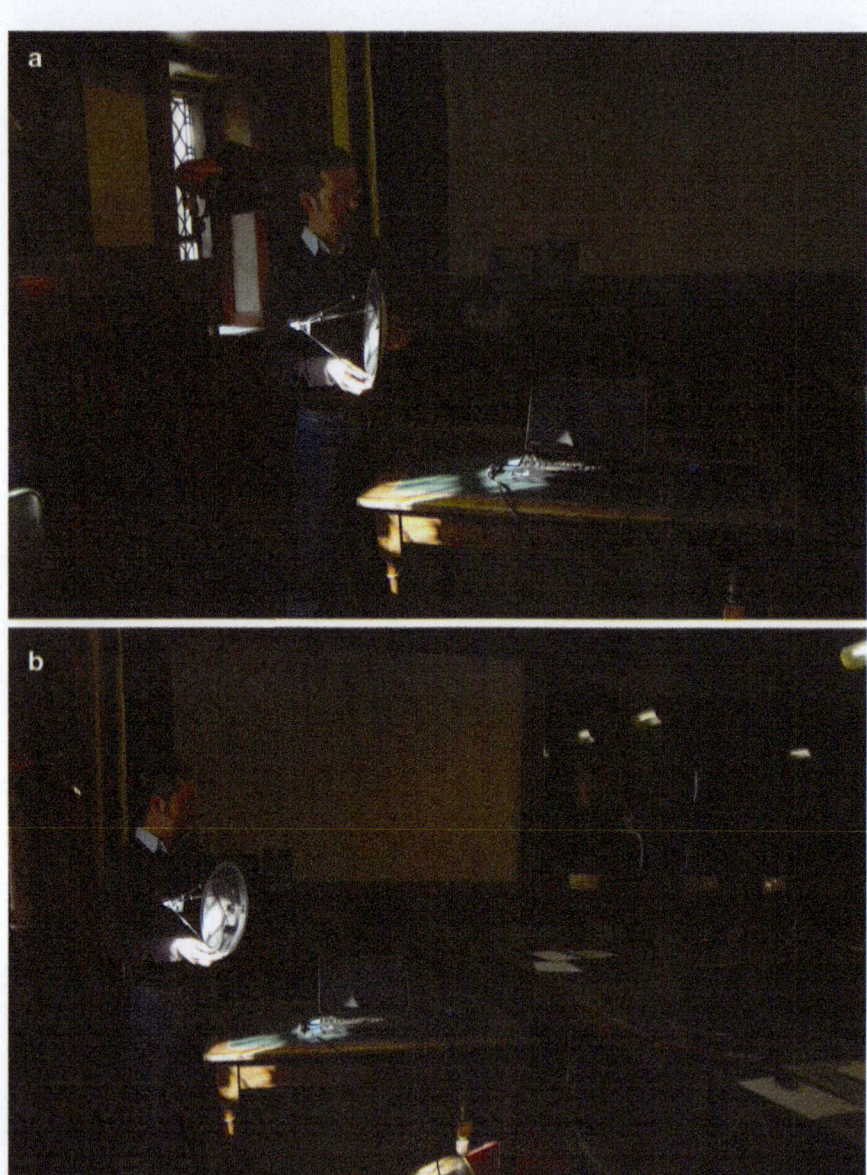

FIGS. 5.20 (a, b) Gordon Ross with Solaris Mark 2 'Sunburn' demonstrator, Bridie Library, Glasgow University Union, October 14, 2012 (© Chris O'Kane, 2012)

FIG. 5.21 (a) TDRS-1 release from Challenger, STS-6, April 1983 (NASA).
(b) TDRS deployed in geosynchronous orbit (NASA)

inflated trial mirrors in orbit. Finally, two solar sails, flimsier than
our mirrors, have been deployed in space—one a demonstrator in
low orbit, the other the Japanese Ikaros, which passed Venus in
December 2010.

Space testing for final selection between the different Ross,
Braithwaite and Vasile designs would be very desirable, and if
NASA does bring a small asteroid to the Earth-Moon L2 point that
will be very useful indeed. Ideally the perfected versions would be
parked at the Lagrange points, in packaged form, ready for deployment on warning from Sentinel systems.

Deploying one in space would be like inflating a torus-shaped,
half-silvered party balloon, with the solar collector then extending
around it like a frilled skirt. Prototypes may have been flown in
Earth orbit before then. They could be used to address the space
debris problem. As Gordon remarked, only the thermal tiles on
the space shuttle would be immune to them. Potential sponsors
have already been approached—a live TV demonstration would
be very good for credibility. Jay Tate stressed that the emphasis
throughout would have to be on peaceful uses such as deflection
and asteroid mining. Any hint of weaponry would attract a very
hostile reaction from the U. N. Committee for the Peaceful Uses
of Outer Space. When an actual deflection was a priority, that

situation would be different, said Bill Ramsay—particularly if the United Nations owns the spacecraft.

In his presentation Ross said that Max Vasile's group was now calculating on 12 'Laser Bees' in a halo orbit around Apophis, like a necklace, balancing its gravity with sunlight pressure (see Chap. 8). The incomplete figures illustrate the difficulty of comparing like with like. Are we talking about an Apophis-like object, i.e., a rocky Apollo asteroid, or Apophis itself—and if the latter, is it the small deflection required to miss the keyhole in 2026, or the much bigger deflection in 2035? What is the timescale, when so many assessments build in the assumed time from discovery to launch, and an average time from launch to rendezvous, whereas we have fixed launch windows and arrival dates but generally longer times in which to get results? What size is Apophis taken to be—the more recent measurement, or somewhere in the lower size range previously estimated? If it was 328 m, as one of the Strathclyde studies assumed [23], then Goldilocks would have 3 times its diameter and 27 times its mass. Depending on the values assigned to these uncertainties, then up to 350 10-m reflectors might be needed—putting them less than 10 m apart, rim to rim, if grouped around the asteroid in a single orbital plane.

The launch mass would be no problem. With its lightweight, pressure-braced construction, Solaris should be less massive than the mirror bees, which Max Vasile planned to launch on existing boosters. Almost all the structure except the thermovoltaic panels, the laser and the gyros of the inertial platform could be pressurized. A 10-m spacecraft might mass 50 k and have the volume of two suitcases. A single Proton could launch 400 of them to low Earth orbit, or 100 at least with individual transfer stages for onward launch to Goldilocks, more if each transfer stage carries several burners. With Delta IV, Atlas V, Ariane V, H-II, Long March and others to call on, adding more should not be difficult—see Chap. 4.

"Not long ago," wrote James Oberg, "astronomers thought of asteroids as rocks, perhaps rubble covered, but still mostly single bodies. But evidence has accumulated that asteroids are rubble piles all the way through, loosely bound together by what is generously called gravity (escape velocity is 11,000 m per second on Earth but less than 1 m per second on a typical small asteroid.)....

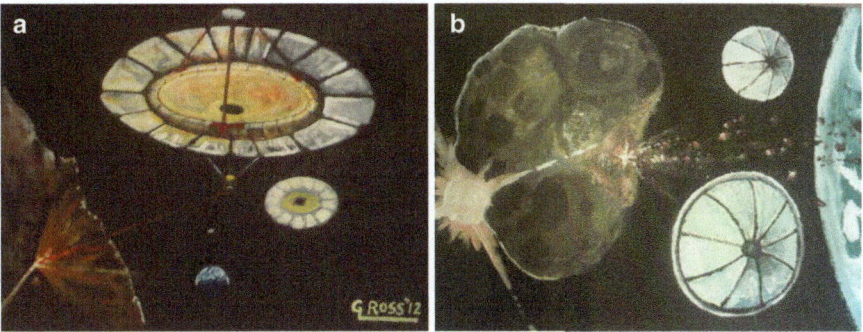

FIGS. 5.22 (a, b) Solaris Mark 2 fleet in action (© Gordon Ross, 2012 & 2013)

Pushing an asteroid has been likened to clearing a landslide off a road, rather than rolling a rock" [13]. That may be an overgeneralization—the size of craters on Mathilde and others suggest that they are single rocks—but we won't know the composition of the Goldilocks asteroid till the spacecraft get there. A major seismic survey will be needed, using penetrators and impactors to map the interior, before we can be certain of the best direction in which to apply thrust: it would be worthwhile sending a specialized probe ahead of the Solaris fleet, to have that survey done before they arrive. The answer will determine whether they have to fire from up-Sun or down-Sun of the asteroid, so the flexible turrets are highly necessary, to make sure the denser part of the plume doesn't hit the primary mirrors (Figs. 5.22a, b). Even at that they may need armored backs, probably thin, flexible Kevlar, with a self-sealing layer below.

In early 2007 Dr. Gregory Matloff, emeritus associate professor of physics at New York City College of Technology, began studies with NASA's Marshall Spaceflight Center on solar sail deflection. Assuming plume velocities on the order of 1 km/s, much lower than those Max later measured, his investigation looked at the dissipation of heat from the 'hot spot' on an asteroid and concluded that it would seriously reduce the deflection effect unless penetration was deep enough to create a trench. His tests with lasers on the Allende carbonaceous chondrite meteorite of 1969 suggested that was true, at least for that type of meteorite, and he recommends a preliminary survey of a target asteroid, using lasers to test for penetrability, before attempting deflection [32].

But if we can launch hundreds of Solaris units to Goldilocks, we have sufficient redundancy to try out different configurations, different constellations and different firing strategies. Major policy controversies may arise, as with Douglas Bader's 'Big Wings' in the Battle of Britain, and there will be time to resolve them. Even if the optimum strategy proves to be sacrificial, and the lifetime of collectors on the firing line is short, there will be reserves to throw into the fray as they become required; robot repairers may be able to refurbish them and restore them to action, as John Braithwaite envisaged; and if success needs more mirrors or bigger ones, there is another launch window 6 years after discovery in which what's needed can be supplied.

At the very least we can say that with so many options, and time to try them out, there is a very good chance that the solar deflection scenario will succeed. But if it isn't succeeding—if the asteroid is absorbing the thrust, say, or if the plumes are destroying the mirrors—then our question 'what would we do?' becomes 'what *else* would we do?' And that, too, has to be asked at the outset, so that in 6 years after discovery we are ready to do it.

6. The Second Scenario: Manned Mission

I am the risk and the purchase of the world,
carry me with you.

—Robin Williamson, *The Song of Mabon* [1]

If the light-sail constellation is accomplishing its task, or better yet has already done so, then there might be no need for further launches to the asteroid in the second window, 6.6 years after its discovery. Looking at the pebbles jinking wildly in reaction to the jets stimulated by Max Vasile's laser beams, it's easy to imagine that the whole problem is solved. To every action there is an equal and opposite reaction, and the energy imparted to the asteroid by the lasers must have some effect.

However, maybe it isn't that easy, for instance if the target is covered in a layer of pebbles forming a regolith. If Gordon Ross's mental picture is correct, the vaporized material will leave behind a glass-lined trench while scattering solid particles into space, all creating a powerful thrust acting inwards. But if there's more regolith below, will it pass that thrust into the solid body of the asteroid or will it simply compress, as it does on the Moon under an astronaut's footprint? If the thrust does reach bedrock, will it propagate through the body of the asteroid to propel it into a new orbit, or will it simply be liberated as heat in the upper layers of it? The Deep Impact collision was too slight, in relation to the mass of the comet, to answer those questions.

In either of those situations (or worse still a combination of both) it seems all too likely that a nearby nuclear explosion would repeat the failure on a larger scale. It still might be tried, but because we don't know what a single massive blow would really achieve, at this stage it should still be considered a last resort (see Chap. 7).

D. Lunan, *Incoming Asteroid!: What Could We Do About It?*,
Astronomers' Universe, DOI 10.1007/978-1-4614-8749-4_6,
© Springer Science+Business Media, LLC 2014

A more gradual, controlled method is still to be preferred, and it seems obvious that it should be directed in real-time by on-the-spot human supervisors, if it can be. However, that judgment isn't universal. Even after multiple satellite rescues, five successful maintenance flights to the Hubble Space Telescope and the construction of the International Space Station, there is still a perception in some quarters that all work done by humans in space is clumsy and ineffective.

In the time available, there is only one available deflection method that meets the criteria above, and that is the use of mass drivers. But if the crushability of the outer layers has been the problem, the mass drivers will need to be anchored to bedrock, or to a grid structure that spreads the forces throughout the body as a whole. It would need to enclose the equator of the asteroid (if it has one axis of rotation), or the whole surface if it has biaxial rotation, as many asteroids do, or even triaxial, or if it rotates chaotically, like Saturn's 'sponge moon' Hyperion. A grid structure would make it much easier for the astronauts to move around (see below), and as noted in Chap. 5, the acceleration applied to the asteroid will not overcome the gravity holding it together. But if the solar collectors have been partly successful, they may by now have encased Goldilocks in a glassy shell that will be solid enough for the mass driver purpose— leaving aside problems that might arise in drilling through it.

MADMEN in Space

Mad, is he? Then I wish he would bite some of my other generals.

—George II of General Wolfe [2]

In Arthur C. Clarke's *The Hammer of God*, the so-called 'mass driver' is actually a giant rocket with 200,000 t of liquid hydrogen as reaction mass, mined from Jupiter's icy moon Europa. It's intended to change the orbit of Kali, a dumbbell asteroid 1,295 m by 656 m, with "a velocity change of a few centimeters per second," enough to miss Earth [3]. The term mass driver is more often used for an electromagnetic launcher advocated by the late Prof. Gerard K. O'Neill, of Princeton University, in his 'High Frontier' scenario for orbiting space settlements [4]. His mass drivers were electromagnetic launchers, to deliver lunar soil to construction sites in space by packing it into aluminum slugs and firing them like shells or bullets.

But for moving consignments around in space, or for moving the construction sites themselves, he proposed using mass drivers as rocket motors with lunar material for reaction mass. Instead of being launched into space, the conducting aluminum shell could be shaped like a bucket, braked to rest at the top of the 'gun-barrel' and returned to the 'breech' for re-use.

Prototype mass drivers at the Space Studies Institute in Princeton reached very high accelerations, and operational ones in space could achieve 150–1,000 g, so there's no danger of the expelled rock returning to machine-gun hapless ships or installations (as happens with slower-moving detritus in the late James White's story "Deadly Litter" [5]). Mass drivers and other electromagnetic launchers featured strongly in *Man and the Planets*, especially in Chap. 4, "All Done by Electricity," in which the discussion group set itself the exercise of doing away with rockets for all applications within the Solar System [6]. In 1979–1982 parts of that discussion fed directly into a linked discussion project, Project Starseed, on nuclear waste disposal and space solar energy (see Chap. 8), and correspondence with Prof. O'Neill led to the author's presenting a poster paper on it at the Space Studies Institute's Space Manufacturing Conference in Princeton in 1985.

In 2007, well after we had decided that a manned mission with mass drivers would be the second attempt to divert Goldilocks in Bill Ramsay's three-hit scenario, Max Vasile drew our attention to a significant paper by SpaceWorks Engineering, Inc., of Atlanta, Georgia [7]. Their study of small automated mass drivers began in 2004, with NASA sponsorship, and was worked out in detail for two cases: D'Artagnan, a fictitious 130-m stony asteroid in a near-Earth orbit with 5 years' warning time to impact, and Apophis, whose diameter was taken to be 250 m but whose orbit was quite well known (see Chap. 4). Each MADMEN unit (Modular Asteroid Deflection Mission Ejector Node) would be 15 m in length, anchored by mechanical barbs to the asteroid surface (assumed to be strong enough for the purpose), and powered by a small nuclear reactor (Fig. 6.1). It would drill rock from the landing site (assumed to be accessible at depth up to 6 m), and would fire masses below 1 kg at 570 m/s.

To allow the drill to operate continuously, a regime of three shots per minute was assigned, powered by capacitors recharged by

Fig. 6.1 MADMEN on the surface of an asteroid (© SpaceWorks Engineering, Inc.)

the reactor between shots. In the initial specification, "Mass driver operations at the asteroid should be limited to 60 days or less in order to satisfy public anxiety and yield a short action time after which the outcome of the deflection mission would be known," but that was later relaxed to a year. On that basis, D'Artagnan would require five MADMEN units to deflect it, but the much smaller deflection required for Apophis to miss its 600-m 'keyhole' in 2029 meant that only two MADMEN units would be needed. The D'Artagnan mission required an Ares V launch, but the Apophis one needed only a Falcon 9.

In relating the D'Artagnan mission to our Goldilocks scenario, one major element not stressed in the MADMEN paper has to be noted. In Clarke's *The Hammer of God* the mass driver is attached nose-first to the asteroid on the equator, and it can be used for only 10% of the time as 'Kali' rotates on its axis in 3 h, 25 min, "thirty minutes of Kali's brief, four-hour day." (If it was docked to the pole, firing would be continuous.) There are to be five MAD-MEN units on D'Artagnan, and each is capable of being rotated up to 30° away from the vertical in any direction—surprisingly far,

FIG. 6.2 Deploying MADMEN units (SpaceWorks Engineering, Inc.)

though seemingly the Apollo Lunar Lander could take off at up to 45° from the vertical [8]. If they were equally spaced along the equator, they could cover 300°, five-sixths of the potential launch time. We decided to be more conservative with the allowed deflection, giving each mass driver a 30° arc of coverage, giving us 12 units in all.

But the MADMEN paper states, "The mass driver operates according to a schedule or 'duty cycle' that is determined by the frequency that the mass driver viewing angle can be brought into alignment with the firing vector—generally opposite the asteroid's heliocentric velocity. For an irregular and rotating asteroid, a duty cycle of only 15% is assumed for any individual lander." So the units are neither clustered at one of the poles nor spaced along the equator. If the asteroid has a single axis of rotation and the desired direction of thrust is more than 30° from it, the landers must be clustered in one hemisphere; but if they are equally spaced over the surface (Fig. 6.2), the assumption is that D'Artagnan has more than one axis of rotation.

If we assume the best case, that Goldilocks has a single axis of rotation and the firing vector lies on or near it, then 12 MADMEN

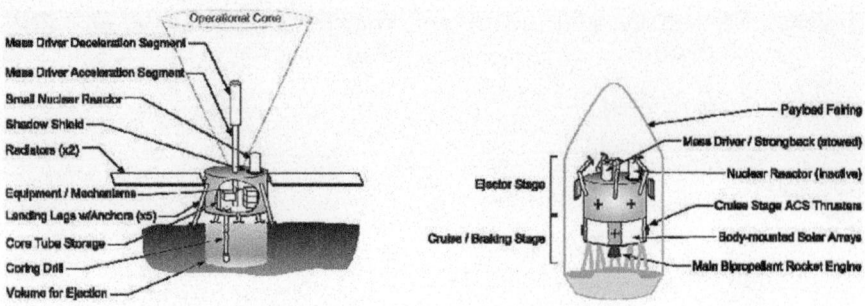

FIG. 6.3 MADMEN diagram (© SpaceWorks Engineering, Inc., 2007)

units operating for 3 years will be more than adequate for our task. But if one of the other cases applies, relating the MADMEN numbers to the Goldilocks dimensions indicates that each of those 12 units needs 64 times the thrust of a MADMEN one. This could be achieved by firing 64 kg payloads instead of 1 kg, or firing 64 times more often, but either requires a higher rate of rock extraction, with a larger drill as well as going deeper—a lot deeper. It could also be attained by stepping up the reactor output, to give a higher acceleration, but calculation reveals that the MADMEN unit already fires at approximately 1,700 g. If all other values were unchanged, 64 times the launch velocity would require increasing the length of the barrel eightfold, and the structure would have to be strengthened to take the greater loads and stresses…

Careful study of the MADMEN unit is needed at this point (Fig. 6.3). The much larger mass driver of Fig. 7.1 is basically a tube with superconducting magnetic coils wound along its length, with the direction reversed at the top end to decelerate the 'bucket' while the projectile mass flies on. The MADMEN unit main section contains the intriguingly labeled 'equipment/mechanisms,' while above it extends the 15-m tube, 65% of which is labeled acceleration segment and the thicker upper part deceleration segment. Its length relative to diameter makes it look like a guide rod in models and artwork, but it's actually a tube, whose telescopic segments presumably fit together like a tent pole rather than an actual telescope or an automobile radio aerial, and whose comparatively small diameter lends it strength. Its inner diameter has to be slightly greater than the bucket's, which is in turn slightly greater than the drill core. Changing any of its dimensions, rate of fire or capacity requires big changes to all the rest. "The complex interactions among the independent variables and the sub-systems create a difficult non-linear and non-smooth optimization problem" involving over 200 equations [7].

Deflecting an asteroid the size of Goldilocks might require a complete redesign, or else flying 320 MADMEN units. Fortunately, they each have a mass of only 1,650 kg, well below a ton, and we can allow much larger mass drivers, or many more small ones, in the much larger mission outlined below.

Wiser Councils Prevail

What, quite unmann'd in folly?

—*Macbeth*, Act III, Scene 4

If the MADMEN scenario has a weakness, it's that it places total reliance on technical operations that have never been tried off-planet, much less robotically. The only one that has been tried, for the first time in March 2013, is rock drilling; and although Curiosity performed it successfully, that's for a small scientific sample, not for 1-kg masses, and into solid rock, while both the vehicle and the rock were firmly placed on the surface of Mars, not in high vacuum and near-zero gravity. Drilling into regolith proved more difficult on the Moon, at least for Dave Scott on Apollo 15, [9] though the Apollo 16 and 17 crews were more successful.

The mass driver launches may be straightforward, though even that is problematic when the bucket has to be stopped and returned to the breech of the launcher each time. But every stage of the operation has moving parts, which are the ongoing bugbear of operations in space, sticking or even welding together at the slightest opportunity. The timeline of space exploration is strewn with stuck fairings and with solar panels and antennae that either failed to open or failed to lock. Even with two advanced robotic arms on the International Space Station, deploying solar panels, removing failed pumps and moving experimental packages often require extra-vehicular activity and human intervention. On board spacecraft, the tape recorders and the gyroscopes have normally been the first components to fail. The overworked camera platform on Voyager 2 regained its mobility during the Saturn flyby, as lubricant worked back into the bearings that had spun dry, but the high-gain antenna on the Galileo Jupiter orbiter remained half-open and useless for the duration of the mission.

Ingenuity and dedication on the part of the controllers have often been able to work around such problems to save the missions—SOHO in Chap. 1 is a spectacular example. But there's no robotic counterpart yet to the Apollo 17 Rover, on which the astronauts repaired a broken fender with maps and sticky tape.

To move the greater mass of Goldilocks, the mass drivers will have to be bigger and run for longer than the comparatively light-weight ones of the MADMEN reference missions. They will have been built under pressure of time, for departure from Earth within 6.66 years of Goldilocks's discovery, and as Prof. Terence Non-weiler pointed out, lecturing in 1970 on the lessons of Apollo 13, [10] time pressure can lead to bad design decisions, like putting all the vital tanks of the Apollo Service Module in a single bay of the structure. Had the explosion occurred on Apollo 8, or on any of the missions coming back from the Moon after the Lunar Module was discarded, the astronauts could not possibly have survived.

Probably modifications will have been introduced during manufacture, and no two mass drivers will be identical. Still more changes will have been made as a result of what's learned about the asteroid from the light-sail mission, and that too introduces the potential for unexpected problems. The solid-fuel Payload Assist Modules for taking satellites to higher orbits from the space shut-tle worked perfectly until STS-41b, when two new nozzles failed and left the Palapa and Westar satellites stranded [11]. As astro-naut Donald Slayton remarked, "It's a perfect example of how we improve ourselves into trouble" [12]. And if any part of the MAD-MEN processing train stops working, it all becomes unusable.

At the Glasgow seminar, Jay Tate expressed surprise that we should think the mass drivers on Goldilocks needed human super-vision. Surely automation, multiple units, redundancy would be adequate and easier to achieve? Bill Ramsay was concerned that developing manned deep-space capability might slow down the development of the mass drivers themselves, and both foresaw at least a year of argument on the manned vs. unmanned issue.

To this author, it seems more likely that the machinery will need constant real-time monitoring, tweaking and cosseting.

An instructive case for comparison is the space shuttle's main engine (SSME), the largest reusable rocket engine in the world (still). In principle a rocket motor is a very simple device, especially one running on liquid hydrogen and liquid oxygen, which ignite spontaneously on contact. But that simplicity disappears when you consider the rates at which super-cooled fluids with awkward physical and chemical properties have to be pumped continuously into the high pressure, high temperature environment of the combustion chamber without disrupting the flow processes going on inside. Some of that fluid has to be pumped around the jacket of the nozzle and combustion chamber as coolant to keep the engine from vaporizing; that fuel can be reused, and was in the F-1 engines of the Saturn V, but new versions of that design have sacrificed that gain in thrust for the advantages of greater simplicity [13].

In the complex mechanism of the SSME, too, that fuel is pumped into the combustion chamber for use. Upstream of all that, the turbines that power the fuel pumps are running on the same LH_2/LO_2 mix. That outflow can be reused, but until recently this was a trick only the Russians had mastered, for the N-1 second stage engines of their Lenin booster for manned lunar missions, hidden away for decades and now being used to power the Atlas V (In an extraordinary turn of history, Soviet engines are powering the latest version of a booster built originally to destroy the USSR).

On top of all that, but unlike all other large rocket motors to date, the SSMEs had to be reusable for multiple missions, while retaining the extraordinarily high reliability needed to qualify as 'man-rated.' Originally each shuttle orbiter was to have had a dedicated set of engines and spares, which would become tuned to the idiosyncrasies of the airframe—and no two of those were alike, as they were hand-built with ongoing modifications. In 1979 the author had the privilege of walking through the *Discovery* while it was laid out for assembly on the shop floor at Rockwell International in Palmdale, CA. It was like seeing a huge model aircraft kit but also, because of the precision manufacture, like the parts of a gigantic Swiss watch. We were also briefed on how different the process was from production-line assembly, which would have generated 80% cost savings had the *Endeavour* been ordered up front instead of as a later replacement for the *Challenger*.

Due to budgetary constraints, there weren't enough SSMEs for the whole shuttle fleet, and they had to be swapped between orbiters and between flights. The manufacturers, Rocketdyne, took a page from the Rockwell book and assigned a dedicated team to each engine, following it alone, refurbishing it after flight and retuning it each time it was moved to another orbiter. When Richard Feynman served on the Rogers Commission, he decided to run an independent check on the management of the SSMEs in comparison to the solid boosters, and he found the same pressures to meet launch dates were leading to similar compromises on safety [14]. In the summer of 1986, 6 months after the loss of the *Challenger*, the author met a main engine team head who had been under so much stress that he suffered a nervous breakdown on hearing of the accident, even though he learned that 'his' engine wasn't the cause of the disaster.

Relating these lessons to the Goldilocks priorities, our conclusion was that each of at least 12 mass driver/drill units should be accompanied to the asteroid by a dedicated team of three engineers with overlapping skills, preferably in addition to the astronauts responsible for getting them there and back. The MADMEN report admits, "Cooling the drill, removing shavings, applying pressure to the drill bit, and automatically adding core tubes are difficult in a low-gravity, airless environment that may have 15 min of roundtrip communications lag time between Earth and the site of operations." In our scenario time lags could be even longer, but all operations would go much faster with a skilled team on site. If the decision were to go with 320 MADMEN units instead of 12 much larger ones, each team would have to care for 27 mass drivers, and that might make madmen of them all, though the pressure would be somewhat eased by having each cluster in use for only 15 % of the time. The size of the anticipated maintenance commitment will almost certainly govern the decision on whether to go with MADMEN-size or build larger units.

Economically, building 12 ships justifies setting up production lines, and because of the efficiencies that generates, 12 ships can be built for the cost of two made by hand. The extra costs are for launch and for mission control, and we may suppose they will be met when so much is at stake.

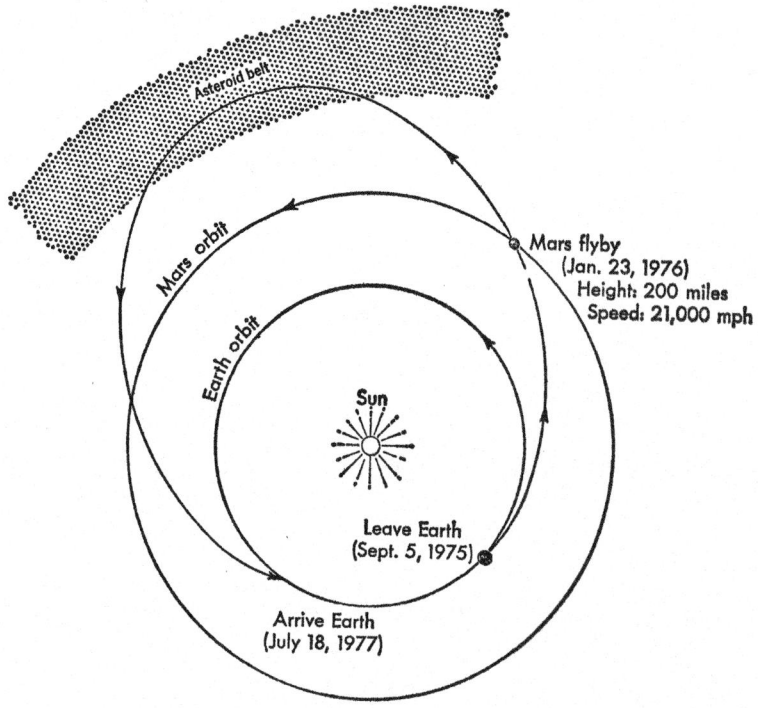

FIG. 6.4 Proposed Mars flyby mission [15]

If we could fly this as a Hohmann mission, on an elliptical minimum-energy transfer orbit cotangential to Earth's orbit and the asteroid's, in energy terms it could be compared to a manned mission to Mars, but it would be more accurate to compare it to a manned Mars flyby. A detailed study of one in 1967 proposed launch from Earth in September 1975, a Mars flyby in January 1976, reaching aphelion at 2 au within the inner part of the Asteroid Belt and return to Earth in July 1977 (Fig. 6.4) [15]. The same mission was later proposed for 1981. Realizing that we would be 36 and 35 then, respectively, John Braithwaite and the author gave serious thought as to how we might qualify to go.

Launch velocity for the Mars flyby would be 27,000–28,000 mph, higher than for a minimum-energy Hohmann transfer to Mars, but with a much lower characteristic velocity for the total mission than the 35,000 mph required for a Moon landing, let alone for landing on Mars. The first great advantage is that there's no significant gravity well to fight at the destination, so rendezvous with the target is a great deal less complicated. It may be more costly in fuel, because there's no air braking in the Martian atmosphere, but that's more than offset by not having to lift off back to Mars orbit, and no return trajectory injection. Since the asteroid is on a collision course with Earth, after rendezvous with it the expedition would already be on a 'free return' trajectory, which would bring it back to Earth without any more powered maneuvers.

Wernher von Braun's first study of a manned expedition to Mars was *Das Mars-Projekt*, early in the 1950s, imagining a fleet of ten chemically fueled Mars ships, carrying 70 people with winged landers, land transport, etc [16]. It inspired the second and third of the BBC's *Journey into Space* serials by Charles Chilton, *The Red Planet* and *The World in Peril*. *The Exploration of Mars*, by von Braun and Willy Ley, illustrated by Chesley Bonestell (Sidgwick & Jackson, 1956), outlined a smaller expedition of twelve people with two ships, one the lander and the other the return vehicle [17]. A more sophisticated version featured in the George Pal/Bonestell film *The Conquest of Space* (wrongly said in the credits to be based on the Ley/Bonestell book of that name). Using chemical propellants, the main ships would launch from Earth orbit with a combined mass of 4,000 t, mostly fuel, because it had to cover braking to Mars orbit, landing on Mars and return to orbit, returning one ship to Earth and braking it into high orbit. A one-way mission would require 'only' 1,000 t.

However, as we saw in Chap. 4, our three-hit scenario requires a higher-energy orbit and propulsion system for the manned mission. Soon after *The Conquest of Space* von Braun collaborated with Walt Disney to produce *Mars and Beyond*, the third of their animated films on space exploration, featuring a fleet of ion-drive Mars ships and rockets landing by parachute. Bonestell adopted this method for his artwork in *Mars* by Robert S. Richardson (1965) [18]. However both concepts assumed a much denser atmosphere, with a density equivalent to Earth's at the summit of Mount Everest, rather than equivalent to the top of the stratosphere, as it really is.

By the late 1960s, with the Moon landings imminent, von Braun was working within NASA on a new plan for Mars exploration. As originally planned Project Apollo would have performed two series of Moon landings and flown a series of space stations culminating in the two Skylabs (the second of which is in the Smithsonian Air and Space Museum). In the 1960s the United States had a nuclear rocket program, Project Rover, which was intended to produce a nuclear upper stage for the Saturn V (Saturn V-N), as well as a nuclear shuttle for operations between Earth and the Moon and manned missions to Mars using an advanced engine called NERVA (Nuclear Engine for Rocket Vehicle Application, Figs. 5.12 and 5.14c). In the post-Apollo program, the space shuttle would be used to build a 12-person space station in the second half of the 1970s, and the Saturn V would then launch a 100-person space base with a nuclear reactor and artificial gravity, using the S-2 second stage of the booster as a counterweight. The nuclear shuttle, also launched by Saturn V, would then move the Space Station into circumlunar orbit as a control center for surface activity in the early 1980s.

The NERVA mission to Mars would have used two modified space station modules, powered by the solid-core nuclear engine of the lunar shuttle (Figs. 6.5a, b). Two additional nuclear shuttles would have been used as boosters to launch each of the ships from the Earth-Moon system. Some versions of the concept involved spinning the ships around their common center of mass while linked nose-to-nose by a tether. Although only lunar gravity at most would be achieved, it could make a big difference to the crew's state of health on reaching Mars. Another option was to return by way of a Venus flyby, which would have required more fuel but paradoxically brought the crew back to Earth faster. Crucially, with the higher thrust of the nuclear engine, each ship would have launched with a mass of 2,000 t, rather than the 4,000 t of the earlier proposals.

The two ships would each have had a crew of six. The mission was intended to fly in 1986, by which time there would have been 200 Americans in space—100 on the space base in low Earth orbit, 50 in geosynchronous orbit, and the rest on the Moon or en route to Mars. Another NASA design of the time included a much larger Mars ion-drive mission with three crew modules, two of them to be spun for artificial gravity in transit (Fig. 6.6) [19]. But the whole program was dependent on the Saturn V booster developed for the

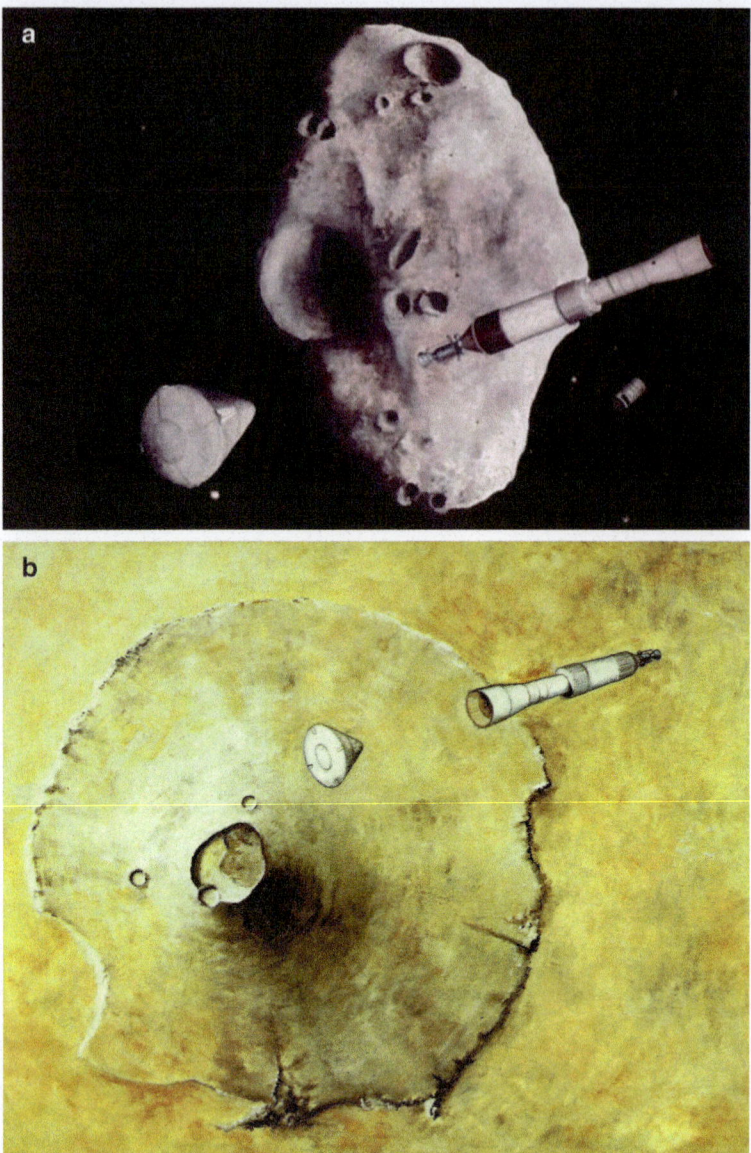

FIG. 6.5 (a, b) The two NERVA mission ships over Phobos and Olympus Mons (© Gavin Roberts, 1975, for *New Worlds for Old*)

Apollo program, and was canceled with it at the end of 1973. One sad thing is that the design of Mars landers hasn't changed since.

The year 1986 provided the closest opposition of Mars to Earth until 2003, and that year the author visited Jim Oberg at the

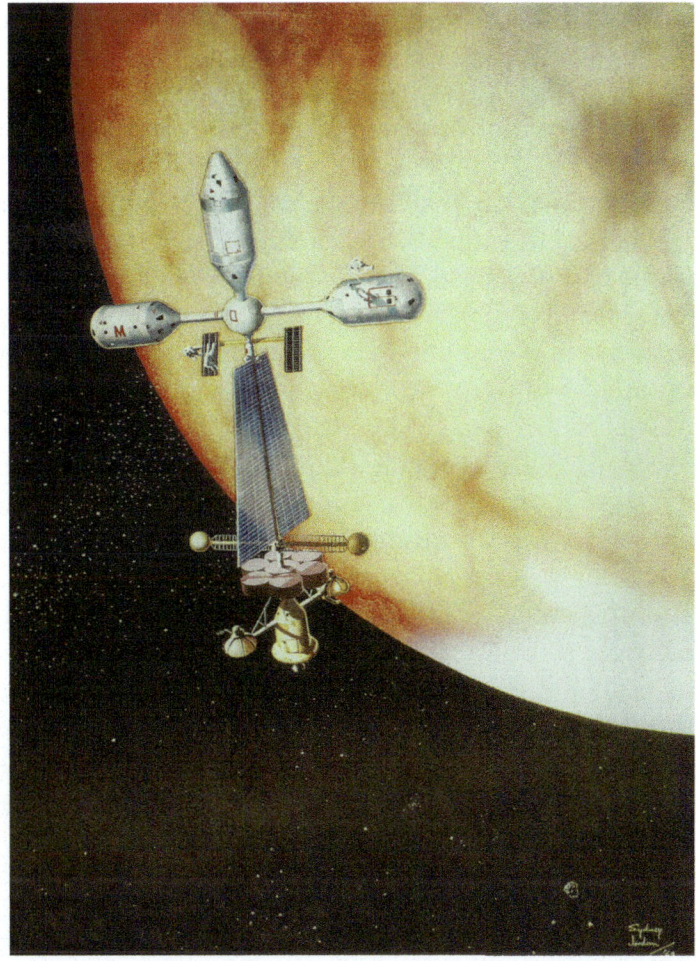

FIG. 6.6 Advanced propulsion system for Mars missions (© Sydney Jordan, 1969)

Johnson Space Center in Houston. We went to the vehicle display park under cover of darkness, on a crystal clear starry night, and as we walked along the side of one of the last two Saturn V's, it was pointing straight at Mars, where it should have been going, shining over its nose. When we came level with the J-2 engines of the second stage, Oberg picked up a stone and threw it. There was a great chorus of hooting, and a huge owl came out and flew over our heads across the Milky Way. "They're nesting in the engines," said Oberg. "They know how to make use of it, which is more than we did." However, the J2 engine is being reborn as J2-X, part of NASA's architecture for a space launch system, to return to the Moon and go on to Mars.

The Soviet Union continued its activities in Earth orbit with the Salyut space stations in the 1970s, while Europe was developing its own station, Columbus, eventually modified to become part of the International Space Station and finally launched in 2007. Japan's Kibo space station was also intended to be independent, but later became part of the ISS. In the early 1980s Bob Parkinson of the British Interplanetary Society designed a multinational mission to Mars, with modules supplied by the various spacefaring nations and with liquid hydrogen, solar-sail or ion-drive propulsion [20].

By 1986 NASA had a new 'architecture' for the Moon and Mars, to set up a large operation to supply liquid oxygen from the Moon to support chemically fueled exploration of both, with nuclear-powered ion-drive freighters to support the creation of the Martian base. It was described in detail to the author on his visit to Houston mentioned above, and was partly incorporated into the report *America at the Threshold* for George Bush Senior's administration, though it recommended a return to nuclear thermal propulsion for the Mars missions [21]. The cost was so extreme that there was no prospect that either of the major parties would support it, and the proposal became known to its critics sarcastically as Battlestar Galactica.

However, one element in it called for detailed study of Phobos and Deimos, which appeared to be captured asteroids. Prof. Fred Singer had proposed a manned 'Ph-D' mission [22], and the author attended a seminar on it at Boulder, Colorado, in 1984 [23]. Flybys by the Viking orbiter spacecraft had shown that both moons had a very low density and could be carbonaceous chondrite asteroids, high in water content and carbon compounds from which to make rocket propellants (not the case, as we now know, though there may be hydrated rock among the material expelled from the Martian surface).

Scientists in the Soviet Union had the same idea. As far back as 1979, the author had pointed out that the long stays of cosmonauts on the Salyut stations, and unmanned test flights of vehicles such as Cosmos 929, seemed to be working towards the capability for a manned Mars mission [24]. In 1989 the Soviets amazed the world with the first launches of their Energia booster, fueled with liquid hydrogen and potentially more powerful than Saturn V. With four clip-on boosters and an upper stage, it could put 250 t into low-Earth orbit. In 1990 Dr. Anders Hansson of Commercial Space Technologies revealed that the plan was to assemble a manned Mars mission with four Energia launches. With chemical propellants, that 1,000-t mass is only adequate for a one-way trip, so the plan was to refuel on Phobos, assuming it to be a carbonaceous chondrite from which water could be extracted to make fuel [24]. In 1989, the USSR had launched two probes to Mars, generally called Phobus 1 and 2 to distinguish them from the moon itself. Sadly both failed, Phobus I due to operator error in flight and Phobus 2 on approach to the satellite. The collapse of the USSR put an end to both programs, and in a recent attempt to restart them with the Phobos-GRUNT mission, the spacecraft was stranded in Earth orbit when its empty fuel tanks failed to separate, falling back to Earth in January 2012.

Europe's Ariane V, Japan's H-II or Russia's Proton booster could in theory be used as the baseline for Earth orbit assembly of a manned mission to Mars, but the 50 Proton launches or 100 of the others would be impractical. Both H-II and Ariane V use cryogenic propellants (liquid hydrogen and liquid oxygen), like Energia, but neither vehicle was designed for on-orbit storage, and they don't have the strong thermal shielding of the space shuttle's External Tank. A 1992 study supposed the use of a heavy-lift booster with 150–200 t payload [25], but even then there are big technical problems, on which ASTRA was briefed by Max Vasile in April 2006. Without better cryogenic storage on-orbit than was currently available, by the time the seventh rocket reached orbit, the tanks of the first one would be empty. Hypergolic propellants that ignite on contact can be stored at 'room' temperature, but they provide less energy per unit of mass, and a prohibitive number of launches would be needed to build a 2,000-t mission.

After a study conducted in Munich, on-site refueling on Phobos was dropped because the energy requirements were too high, and to finish the exercise the participants had to assume that the Russians contributed a nuclear-thermal rocket equivalent to NERVA and that the Energia was reborn to launch it. The Russians did have a nuclear rocket program, now said to have been restarted [26], and without time to develop more exotic alternatives, most probably that and/or NERVA will be the propulsion system for the Goldilocks mission.

The ESA study considered ion drive or plasma jets as backups to nuclear-thermal propulsion. Both systems have been used on communications satellites for attitude control and station-keeping in geosynchronous orbit, and ion-drive missions have now been flown successfully to the Moon and to comets, but it's questionable whether they could be scaled up for full manned interplanetary missions in so short a time. Although they generate very low accelerations, they produce a great deal of thrust per unit mass of propellant expended, and because that thrust is continuous they can reach very high speeds on interplanetary transfers.

An ion-drive mission could reach Mars in 3–6 months, once it reached interplanetary space. The drawback is the long time it would take to climb out of Earth's gravity well. Unless launched by some other means, the climb would add at least a month to the journey, probably more, and bring the transit times more into line with those of nuclear rocket missions. One option would be to put the crew aboard after the long spiral out from Earth, using high-powered shuttle vehicles, once escape velocity was reached. The shuttle would have to achieve a transfer velocity that would take it to Mars on its own, but wouldn't have to carry consumables for the whole trip because they're already aboard, nor would the deep space ships have to carry them for the first month. But if for some reason the rendezvous failed, the crew would be beyond hope and beyond rescue.

Europe's SMART-1 probe reached lunar orbit 14 months after launch and 2 months ahead of schedule because the ion drive had provided higher thrust than anticipated. It would be good if that happened on the way to Mars, but a lot less desirable if the thrust proved to be too low. On a continuous low-thrust acceleration transfer the shortfall might not become apparent until it was too late for a simple return to Earth, and the best option might be to move into an orbit that would re-encounter Earth on the second swing around the Sun. All this supports using NERVA for the main propulsion system, but there might also be a case for using ion drive or plasma jets as auxiliary propulsion (see below).

In first draft, this chapter went on to consider the full range of exotic alternatives, including fluorine and monatomic hydrogen as alternative chemical propellants, gas-core nuclear reactors (like in *2001, a Space Odyssey* [27]), nuclear fusion (continuous or pulsed, as in Project Orion [28], Project Daedalus [29] and Project

Vista [30]), and solar sails, possibly magnetized or electrostatically charged [31]. Some have been demonstrated on a small scale, all have major possibilities for future applications, and the references have been left in for interest. But to restate the urgency, we have 6.6 years, comparable to the time from Kennedy's speech to the launch of Apollo 8, in which to mount an expedition whose purpose is to save the world. It adds force to a recent comment by General Tom Stafford, chairman of the *America at the Threshold* report above: "We know NERVA works." It went past the prototype stage, working examples were built and test-fired, and under pressure its development can be completed in time for our second hit.

Working in 2002, the ESA study group above considered a Mars mission in 2030, but preferred 2043, when the two planets would be closer. The payload mass would be more than 120 metric tons, 70 metric tons for the habitation module carrying five people. If each mass driver is carried by a ship with a crew dedicated to it, as above, and we need 12 ships to be sure we get the job done, the scale of our mission is comparable to a one-way Mars-Projekt without the gliders, but with NERVA propulsion, so 250 t of fuel for each should suffice. With those ships, their tanks and the extra modules that will be needed when they get there—see below—it would take at least 25 launches of the SLS heavy-lift booster to assemble the fleet in orbit in under 6 months, even if the problems of storing liquid hydrogen on-orbit have been overcome.

In addition, each ship requires two nuclear boosters to send it on its way. Since the NERVA propulsion module is standard, and they'll have to cluster together at their destination (see below), they could boost one another if the insertions to transfer orbit can be staggered sufficiently to let them return and refuel. But the last ship needs two more boosters, so probably at least six would actually be needed, bringing the launches from Earth to at least 31.

Four rockets could be assembled at once in the Vehicle Assembly Building at Kennedy Space Center (KSC) in Florida, but the demands on the two existing launch pads 39A and B might be excessive. Originally there were to have been five pads, and the present 39A is on the position first designated 39C. With it likely to be reassigned to commercial launches, a new Pad 39C would be necessary, and D and E as well in case there's a serious accident at one of the others.

The load on KSC would be greatly eased by opening up another launch site. Vandenburg Air Force Base in California might be considered, but launching eastward would take the rockets with their nuclear payloads over populated areas—perhaps not the best option, even if the reactor cores have not been activated and aren't dangerously radioactive [16]. The best option would be Kourou in French Guyana, and since a Soyuz launch facility has already been established there, the Energia program might be restarted for launches from there—again something Bill Ramsay and the author suggested in 1991 [32]. Almost on the equator, launching eastward from there has a 20% advantage over KSC, and would allow 25–30% higher payloads than launches from Baikonur. It would raise very interesting possibilities for operations later, if the asteroid deflection is successful (see Chap. 8).

Although the ESA Mars study had a crew of five people, three, six or twelve were considered as alternatives, possibly with an inflatable structure to increase the living space in transit. Taking 50 days for the transit each way the mission could be limited to 114–130 days, but as the 14–30 days at Mars would include time spent orbiting the planet, landing and preparing for return, a 900-day mission spending 600 days on Mars seemed more worthwhile. However the study group expressed strong reservations about maintaining the crew's health and safety during a mission of such length.

Biological Considerations

> The terms 'dose' and 'relative biologic efficiency' cannot really be used to describe the biological effects of the HZE particles encountered in space. More appropriately, the terms 'ion kill' and 'microbeam' are used to describe their radiobiological characteristics...
>
> —Apollo 17 Preliminary Science Report [33]

Max Vasile's conclusion, from the ESA study mentioned above, was that there were too many unknowns for manned missions to Mars to be considered at the present time. The biological issues were his biggest concern—the uncertainties of maintaining life support for up to 3 years on spacecraft tens of millions of miles from Earth.

Traditionally, the biggest danger facing interplanetary voyagers was meteor impacts. While not to be ignored, the danger has proved easy to counter: a double skin for a spacecraft. The so-called meteor bumper first proposed by Prof. F.L. Whipple has proved adequate to protect both manned and unmanned vehicles from the dust-sized particles that are thinly scattered through the inner Solar System, emanating from the comets and from the Asteroid Belt.

Radiation hazards are more of an issue. The short-term problem is the risk of solar flares and coronal mass ejections, which generate deadly streams of protons, electrons and heavier nuclei particles beaming out from the Sun like a jet from a hose, contained by the magnetic field that they themselves generate. A foot of lead, or an equivalent thickness of water or other absorbent materials, would be needed to protect astronauts in space, or on the surfaces of the Moon or Mars. Metal walls are to be avoided because when the particles hit metal atoms they generate X-rays and showers of secondary particles, which are carcinogenic. All-around shielding is needed, because the deadly subatomic particles spiral around the magnetic field lines.

A storm cellar—a safe zone for your food and water storage—would suffice to counter the danger of solar flares, which last up to 2 weeks, but the cumulative damage caused by galactic cosmic rays, which are much more massive, much more penetrating and carry much higher energies, are enough to kill any living cell they strike. It used to be thought that central nervous tissue does not regenerate; it does, but only very slowly, and cosmic rays would produce noticeable brain damage after 3 years' exposure, or twice that in unshielded surface domes on the surfaces of the Moon or Mars.

The danger was first recognized when the Apollo astronauts reported the flashes caused by cosmic ray particles passing through their eyeballs. On space stations, in low Earth orbit, Earth's magnetic field protects against all but the most intense cosmic rays (known technically as HZE particles), but protons trapped in the Van Allen radiation belts are accelerated to high energies, and the stations pass through them in a region called the South Atlantic

Anomaly, where the belts dip closer to Earth. In his stay on the Mir space station, U. S. astronaut Jerry Linenger found that,

> Some nights on Mir I would be awakened by bright flashes in my eyes, caused by heavy particles penetrating my closed lids and then striking and exciting the nerve endings on my retina. Turning my head ninety degrees, the particles would move right-to-left and leave behind temporary, ghostly contrails. Although I would try to reposition myself behind lead-filled batteries for protection, more often than not the light show would continue unabated, and I would be irradiated. Feeling helpless from the onslaught, I would return to my sleeping wall and try to fall asleep despite the disturbing flashes. After ten minutes or so, the space station would zoom away from the defect in the Van Allen magnetic belts...We would once again be shielded by their deflective force field, and my closed eyelids would once again provide darkness [34].

U. S./British astronaut Michael Foale found that during solar flare events, when the Van Allen Belts are saturated and radiation leaking into Earth's atmosphere causes auroral displays, "He was aware of it even when his eyes were closed in sleep. The flashes would reach his brain regardless, whether through his eyelids or directly to his cerebral cortex he wasn't sure, interrupting and sometimes preventing sleep" [35].

On the ISS, the Russian Matroyoshka experiment with mannequins simulating human bodies have found that protection against 'soft' cosmic rays such as protons is comparatively easy. The bunkrooms of the Russian section offer significant protection, and even damp napkins packed in polythene bags provide still more: "[A]t a depth of 5 cm into the human body, the dose of radiation reaches almost background levels, and then becomes even less noticeable." Results from NASA's Lunar Reconnaissance Orbiter indicate that plastic shielding is decidedly better than metal [36].

Long-term accumulation of damage by primary cosmic rays, the heavy nuclei, would produce noticeable effects in about 3 years due to destruction of central nervous system tissue. The effects of impaired speech, vision and coordination, loss of memory, etc., would be very much like being punch-drunk. A Moonbase would require up to 10 ft of rock or soil shielding fully to counter the danger.

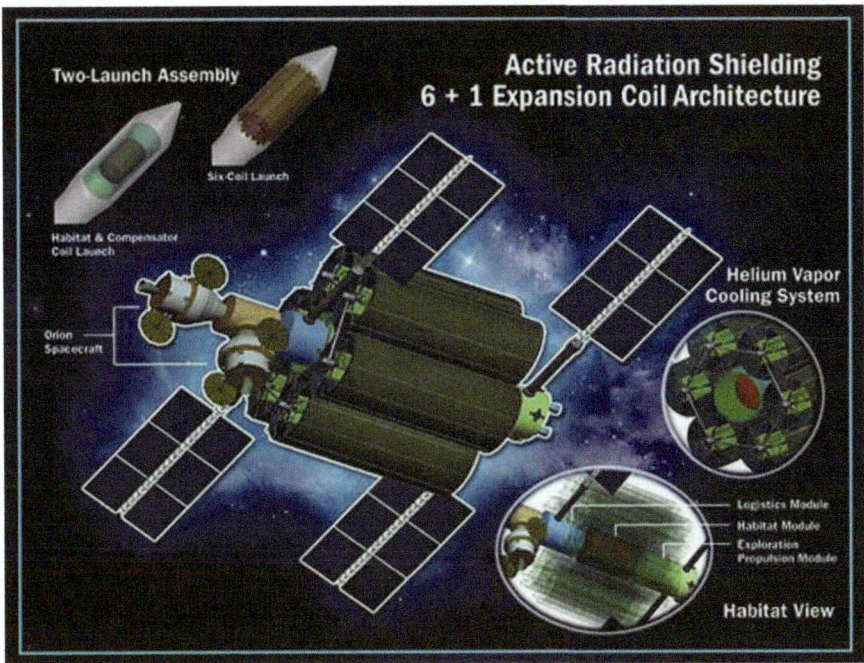

FIG. 6.7 Electromagnetic radiation shielding (NASA)

Electromagnetic shielding is possible in theory, but you need to encase the outside of your spacecraft or habitat with superconducting magnet coils, and with present technology, they need to be immersed in liquid helium. Back in the 1970s, it was reckoned to be so bulky that it would make quite effective shielding even when switched off. (A bit like the electrical shark-repelling system NASA developed for the Mercury capsules in the early 1960s. According to astronaut Scott Carpenter, the navy tested it and reckoned it was "mildly repellent to sharks, with the switch in either the on or off position") [37].

There have been big developments since in superconductors, but seemingly the ones developed so far that work at higher temperatures aren't ductile, and so can't be drawn into cables, and are brittle, which doesn't lend itself to launching on rockets. In 1996 scientists at Glasgow University seemed to be on the track of an organic room temperature superconductor [38], and according to some sources work is continuing in related areas. In current designs, the coils would still be as heavy as the rest of the spacecraft and require a separate SLS booster to launch (Fig. 6.7).

The other idea was that you could do it electrostatically; this was for space habitats, and there was some clever thinking about how spaceships could dock with them, because the charges required are immense [39]. So there seemed to be some pretty big risks if anything went wrong while doing that. However, it was noted that you could get some extra thrust from a solar sail if it was charged up enough to repel the solar wind, and a lot more if you magnetized it, maybe so much that you could take away the sail you first thought of, maybe replace it with a plasma cloud, all working around to the idea that maybe you could have a magnetic propulsion system within the Solar System that would generate a cosmic ray shield as a by-product [40]. You're still talking high energies, so you'd need big ships, at least until we get those super-conductor breakthroughs.

Late in preparation of this book, a news story broke that might have major implications for this discussion. One of the instruments carried by the Mars Science Laboratory Curiosity is the Radiation Assessment Detector, which was intended primarily to measure particle radiation on the Martian surface. It was realized that in flight the device would be shielded to levels approximating those inside a manned spacecraft, so on the way to Mars it measured the dosages of both solar protons and galactic cosmic rays. The total exposure was then converted to an equivalent dosage in milliSieverts and suggested that acceptable risk levels for cancer would be matched or exceeded on a 360-day round trip to Mars [41], not counting time spent on the planet itself. The highest counts came in five peaks during solar storms, but these were not particularly severe.

Relative to the average dosage in a human lifetime, a 360-day trip would increase the risk of cancer from 21% to 24%. Robert Zubrin of the Mars Society was quoted as saying, "What it shows is that the cosmic ray dose on a Mars mission is not a show-stopper. This is a modest proportion of overall risk. Therefore, what it means is that we don't need to delay a humans-to-Mars program until we have a miraculous advanced propulsion system that can get us there faster." Taber MacCallum, the chief technology officer for the Inspiration Mars Foundation, which rather ambitiously plans to launch a married couple on a 501-day flyby of Mars in January 2018, came out with somewhat startling comment: "Those numbers are less than the risk a lot of people take in sports. They don't seem that unreasonable to me" [41].

That seems casual even for a mission extended to 501 days, let alone the 4 years in space we envisage for the asteroid deflection. Remembering the extraordinary sacrifices made in the immediate aftermath of the Chernobyl explosion, volunteers could no doubt be found to do it, but with time to plan we would hope to do better by them. Fortunately there are expedients that could reduce the dosage to below the 501-day equivalent (see below), though not down to the 360-day one. But our ships will be carrying more consumables than a Mars mission, so better shielding can be provided.

Also, converting the *total* energy measured by the RAD instrument into the equivalent in milliSieverts seems to obscure the 'ion kill' factor above. As noted above, galactic cosmic rays are not carcinogenic, because they kill cells rather than causing damage that generates mutations, and 95 % of the incoming energy measured aboard Curiosity was in that form. The dosage rates for those were measured away from Earth on the later Apollo missions and presumably are unchanged, so the precautions outlined below should still prove effective.

The other major area to address is the effect of microgravity. As soon as spaceflights became extended into period of days and weeks, it became apparent that the human body undergoes quite rapid changes in response to the near zero-g environment. James Lovell gave a fairly harrowing account of his experiences on the 14-day Gemini 7 mission [42], but many of the problems encountered stemmed from the cramped conditions on the spacecraft. With more space in the Apollo capsule the lunar astronauts fared much better. While 12 of them walked on the Moon, the Command Module pilot had space in which to move around, and despite their privations even the crew of Apollo 13 were able to walk from the recovery helicopter and to stand for the service giving thanks for their return.

At the same conference Pete Conrad described his experiences in Gemini, on Apollo 12 and in the first crew on Skylab. They underwent a 14 % heart muscle loss, 18 % loss of red blood cells, and corresponding losses in leg muscles and blood volume, from which it took 30 days to recover. More exercise in orbit was indicated, and learning from that, the 56-day crew returned in better condition than the 28-day one [43].

The ongoing experiences of cosmonauts and astronauts since the 1970s have confirmed those findings. Valery Ryumin spent 175 days on Salyut 6 in 1978, returned to Earth and went back for another 185 days the following year, simulating a mission to Mars without lasting problems. On the Mir space station the fitness equipment included a treadmill and an exercise bicycle, which was facing a window so that the cosmonauts could watch Earth as they cycled around the world. At the end of her stay on Mir, which was unexpectedly lengthened by 6 weeks to 188 days, Shannon Lucid was determined that when she was retrieved by the space shuttle she would join the crew on their honorary walk around the vehicle after landing. Relentless exercise enabled her to do it, though she swore afterwards that she would never go jogging again. But even after 6 months on the ISS, again unexpectedly extended due to the loss of the *Columbia*, when the Expedition 6 team's Soyuz landed short of the recovery force in 2003 they managed to fend for themselves, to exit from the spacecraft, stand and walk, at least for short distances [44].

The previous year, Paul Lavin of the Mars Society assured us that the problem of microgravity adjustment had been solved [45]. The answer, he told us, was a system called 'Magic Fingers,' which Friendship Inns were advertising as an attraction 30 years earlier. More sophisticated versions are actually available as orthopedic beds today. The system vibrates the bed to ease back pain, and it seemed that the microgravity accelerations it provides could alleviate the bone loss problem, at least. Evidently it hasn't been put into practice so far, or didn't work if it was tried; for example Chris Hadfield, one of the most active astronauts on the ISS, was quoted on his return in 2013 saying he needed to learn to walk again.

Nevertheless it was noticeable that in interview a week later, already he seemed much his former self; and he stressed that his crew had been able to perform a demanding EVA and emergency coolant pump repair immediately before returning to Earth after months in space [46]. Research continues on exercise techniques, dietary supplements and special clothing [47], including replacements for the Russian 'Penguin Suit' and the Lower Body Negative Pressure garments of the past, to help astronauts adjust to conditions in space and minimize the harmful effects on return

Nevertheless our crews are going to be in space for up to 4 years, much longer than any long-stay mission to date and without even a break in Martian gravity. We have to assume that they will need some artificial gravity to stay fit. Like the 1980s NERVA Mars ships, ours can be linked in pairs by tether and rotated around the center of mass for spin-generated 'gravity.' How much g that provides will depend on the exact lengths of the return capsule garage, the auxiliary propulsion module and the 120-m mass driver when folded, but it should be lunar gravity at least. For comparison, the much larger ship in the BBC *Space Odyssey* had arms 75–100 m in length and at 3 rpm, they provided Mars-equivalent gravity, one-third g [48]. We are not going to be able to provide that, even on the smaller ion-drive ships, with artificial-gravity booms folded for launch, which Ed Buckley envisaged launching on super-boosters larger than the SLS or Energia (Fig. 6.8a, b), or as Chesley Bonestell depicted on a somewhat similar ship in Willy Ley's *Beyond the Solar System* [49].

On arrival, thorough seismic mapping of the Goldilocks interior will be needed. It may have been done remotely by penetrators and impactors in the solar collector phase, but those sensors may have been destroyed by the laser beams, and the fleet must at least be prepared to complete the exercise. Thereafter the ships will surround the asteroid and deploy the mass drivers in whatever configuration seems best—clustered around one of the poles or at a preferred direction of thrust near to one, spread around the equator if the preferred direction lies in that plane, or spread over the whole surface of the asteroid, as its rotation and internal composition dictate.

If the mass drivers are found to be successful, but not sufficiently so, one option would be detach half of the barrels to double the length of the remainder, thus increasing the momentum of each outgoing shot fourfold. That would also increase the forces on the base structure fourfold, so at least half of them would have to be strengthened still further in anticipation of the possibility. The power requirements would also be scaled up, so the modification would be a lot easier if the mass drivers were running on a grid instead of individual reactors (see below).

Once the mass drivers are extended and operational, ideally the ships should move into whatever configuration is best to minimize the problems caused by weightlessness and galactic cosmic rays, for the next 3 years-plus in which the asteroid swings out from the Sun to the Belt from which it came, then

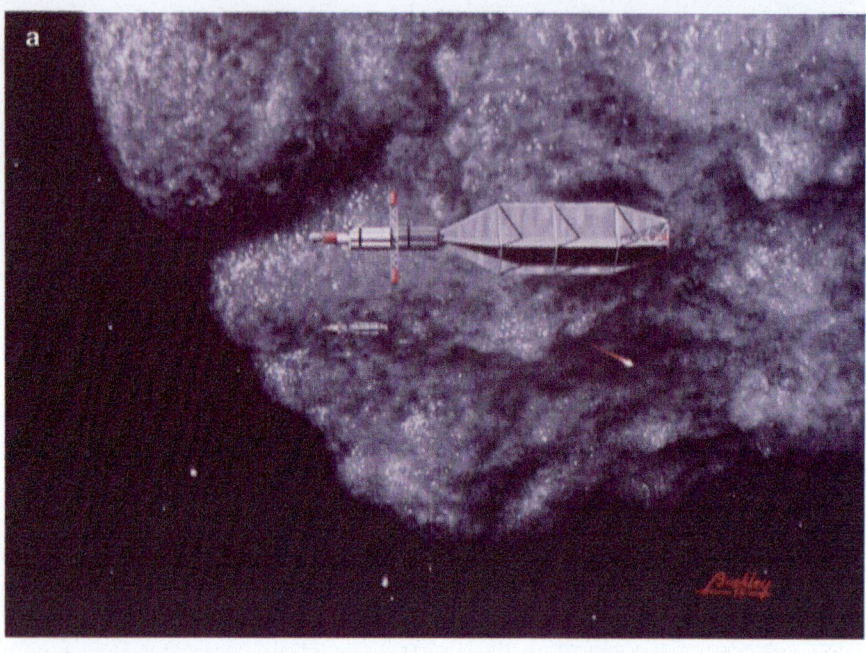

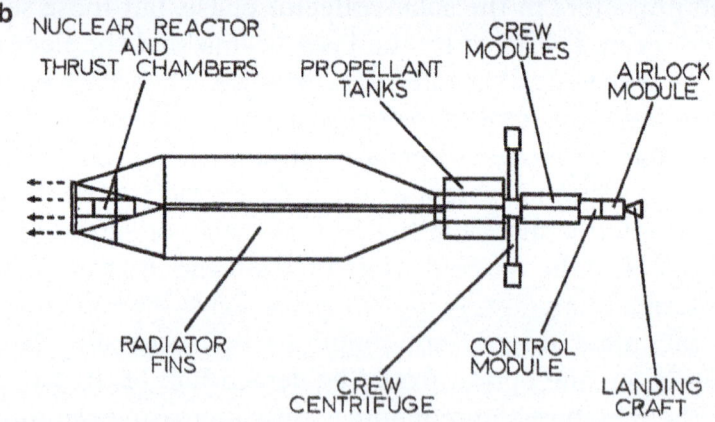

FIGS. 6.8 (a, b) Ion-drive spacecraft for manned missions to the asteroids (Designed by Ed Buckley for *New Worlds for Old* and *Man and the Planets.* © Ed Buckley, 1975)

back to the meeting with Earth, which we hope to change from an impact into a flyby.

The first big factor is the shielding from cosmic rays, which is provided by the asteroid itself. The closer we can park to the asteroid, the better. If we can actually land on it, or better yet shelter in a crater or between ridges, it could reduce the total dosage from up to 4 years, unacceptably long, to the equivalent of a manageable 2 years or less. Unfortunately there is a very good reason not to do that—see below—and although we have an answer for it, first we have to look at the workload the astronauts have to face.

Working on an Asteroid

> The ATLAS mass driver has now been detached from *Goliath*...and we're now engaged in the delicate job of installing it on Kali. Fortunately the asteroid's gravity is so feeble—about one ten-thousandth of Earth's—that ATLAS weighs only a few tons. Don't let that fool you, though. It still all has its mass, *and its momentum*. So it has to be moved very, very slowly and carefully....Believe it or not, the main tools for the job are old-fashioned winches and pulleys, anchored on Kali.
>
> —Arthur C. Clarke, *The Hammer of God* [3]

A major illustration of getting the dynamics wrong was provided in the second episode of the TV series *Defying Gravity*, where an astronaut was stranded outside the ship on the end of a tether. As her air ran out, her colleague in the airlock hauled desperately on the line to bring her in; strangely, instead of reeling in it piled up behind him, but the 'reel' problem was his continuing effort. Though weightless, she and her suit still had mass and inertia to overcome—but once in motion, Newton's Second Law would be in force, and she would retain her momentum. If he continued to pull as shown, rather than less dramatically coiling the line (even if that was needed), she would continue to gain speed until she arrived with enough speed to injure both of them and probably damage the airlock, which would be no joke. On one EVA from Mir the cosmonauts had serious problems because the hinges on the Kvant-2 airlock had been damaged when it opened prematurely under residual pressure. Fortunately another airlock was available [50].

To heighten the TV drama, another astronaut watched in despair from the control cabin while the crisis played out. It looked at one point as if his character would come to his senses, but the moment passed, while knowledgeable viewers continued to shout at the screen, "Just move the so-and-so *ship!*"

Improving the design of spacesuits would make a big difference to all EVA activity. Designers normally begin with the gloves, which are so complicated that they have been described as spaceships in miniature. In the 1950s, some suit designs actually were miniature spaceships, called coke bottles, with thrusters below the feet and mechanical arms worked from inside. In his autobiography *Always Another Dawn*, rocket pilot Scott Crossfield describes the breakthroughs that led to the suits worn on the X-15, Project Mercury and later Project Gemini [51]. The suits worn on the Moon had to be more robust and cumbersome. There were fears, eventually resolved, that if an astronaut fell on the Moon he would be unable to get up again. For the suits worn on the space shuttle the designers removed the flexibility at knees and ankles, simplifying the design at cost of reducing the feet and legs to almost useless appendages.

Working in a spacesuit, the biggest problem to deal with is internal pressure. Soviet spacecraft were pressurized with oxygen and nitrogen at 'normal' pressure, and after the first spacewalk, Alexei Leonov had to depressurize his suit, at risk of oxygen starvation and the bends, before he could get back into Voshkod 2 [9]. The Gemini and Apollo astronauts were breathing pure oxygen at lower pressure; even so, Gordon and Cernan suffered severe exhaustion on their Gemini EVAs for lack of handholds and footholds, and on Apollo 15 Dave Scott found his hands turning black from the effort of working in the gloves [9]. For space shuttle EVAs and more recently on the ISS, astronauts have had to purge their blood of nitrogen and revert to pure oxygen to keep the suit pressure low enough for work.

When speaking about spacesuits, the author often begins with a photo of himself on 'EVA' in the cargo bay of the space shuttle trainer at Houston, wearing a suit—a denim suit (Fig. 6.9a). His apparently small size, compared to the astronauts (Fig. 6.9b) emphasizes both the size of the cargo bay and the bulk of the suits. What makes them so bulky is that water has to be circulated inside, to draw off the body heat. (think of Shirley Eaton's death scene at the beginning of *Goldfinger*.) As Wally Schirra said, "The only time all the Mercury astronauts wore spacesuits at once was for a publicity shot. In *The Right Stuff* the actors playing us wear them like tuxedos. I don't know how they do it because those things are airtight. They'd better be airtight — once you're in one,

FIG. 6.9 (a) The author in the cargo bay of the Space Shuttle trainer, Johnson Space Center (© Jim Oberg, 1986). (b) Story Musgrave and Donald Peterson, STS-6 (NASA)

the worst sound you can hear is sssss....That can spoil your whole day" [52].

Yet human skin is nearly adequate for protection in a vacuum, as Arthur C. Clarke pointed out repeatedly since the early 1950s [53]. Our temperature control system via perspiration is better than most mammals [54]. Even a light pressure bracing could provide humans with a flexible second skin, and the body could achieve its own temperature control as long as it could sweat through the fabric. A 'space leotard' for use on Mars was studied by Paul Webb in the 1960s, and the concept was revived as the Biosuit by Dava Newman and Jeff Hoffman at MIT in 2007 [55] (Fig. 6.10). Although the flexible suit shows great promise, one wonders how well it would cope with handling tools and rocks that could be very hot in sunlight and cold in shadow. Again, NASA's asteroid parked at L1 could be a useful testing site.

For our mission, potentially a still more promising idea is for a suit that would function in some ways like a powered exoskeleton, allowing astronauts to move larger masses, but which would impose 'natural' loads on bone and muscle to keep up muscle tone and alleviate the other effects of microgravity, using flywheels to adjust the loads on the suit limbs. The proposal comes from the Draper Laboratory's Human Centered Engineering Group, at Cambridge, Mass., in partnership with NASA's Johnson Space

FIG. 6.10 Dava Newman (now Professor Newman) in Biosuit, design by Guillermo Trotti, A.I.A., Trotti and Associates, Inc. (Cambridge, MA), fabrication by Dainese (Vincenca, Italy) (MIT photo, Douglas Sonders, 2007)

Center and MIT, and it was thought that an operational suit could be produced within a decade, in time for asteroid missions [56].

EVA on Goldilocks is going to pose extra demands, above what's been encountered so far on the Moon and in space. Since it has roughly one-thousandth the volume of Eros, which has roughly ten times its average diameter and one-thousandth the surface gravity of Earth, the surface gravity on Goldilocks will be about one ten-thousandths of Earth's, less at the equator if it's rotating rapidly. In *The Hammer of God*, Clarke's character takes nearly 2 min to fall a meter to the surface in similar gravity, but although it's very low the pull will still affect operations. Habits develop in work environments, and sometimes have unexpected results elsewhere: in World War 2, for example, technicians used to tether tools to their belts to work on Royal Air Force flying boats on the water. As a result they developed the habit of simply letting go after use—and anyone who dropped a tool in a hangar on land was liable to be called 'a seaplane mechanic' (expletive deleted) because they kept doing it when they came ashore.

After returning from orbit, astronauts are notorious for breakages because they too let go of things in mid-air, expecting them to remain parked in microgravity. Within spacecraft, the ones that are forgotten have to be retrieved from the guards over the air filters. On EVA, it's easy to knock an object out of reach and lose it—in 2008–2009, one could go to www.heavens-above.com and look up when a tool bag lost from the ISS would be passing over. On Goldilocks, such stray objects will come back again, but it will take a long while and undoubtedly be where retrieval is least convenient—especially if the asteroid has glassy linear scars or even a crust from the previous attempt to deflect it with solar collectors. It seems quite possible that there might be sharp edges that would be dangerous to spacesuits, and the material couldn't be trusted as a foothold or an anchor for a safety line.

On the asteroid, although 0.0001 g doesn't sound like much, unattached objects will move inexorably towards the center of mass, possibly creeping downhill on the surface like the boulders on Eros. But tests in ESA's A300 parabolic flight zero-g simulator (the equivalent of NASA's 'Vomit Comet') have discovered a potentially more serious problem. Regolith in vacuum has a 'fairy castle' structure, with the particles of broken rock in contact at only a few points. Lunar dust can levitate if electrostatically charged [57], and it's been suggested that Transient Lunar Phenomena ("Moon-glows") might be due to electrical discharges during landslides [58]. The new studies suggest that small disturbances in the thick regolith on asteroids could ripple around them in slow-moving avalanches, possibly with quite large effects at the antipodes [59]. Even if the mass drivers are rooted to bedrock, the shock of each one's firing could be enough to disturb or damage the rest. Fortunately, we have a backup plan if that problem makes them unusable—see below.

Manned Maneuvering Unit backpacks (MMUs—Fig. 6.11a) haven't been used since the satellite rescues of the 1980s, with the astronauts using handholds or being positioned by manipulator booms for external work on the space shuttle and the ISS. Since the Multi-Mission Space Exploration Vehicle (see Chap. 4) already has manipulator arms (Fig. 6.11b), probably they will be lengthened to provide secure footholds.

Underwater training for asteroid missions has already begun with astronauts from the United States and Japan (Fig. 6.12), and

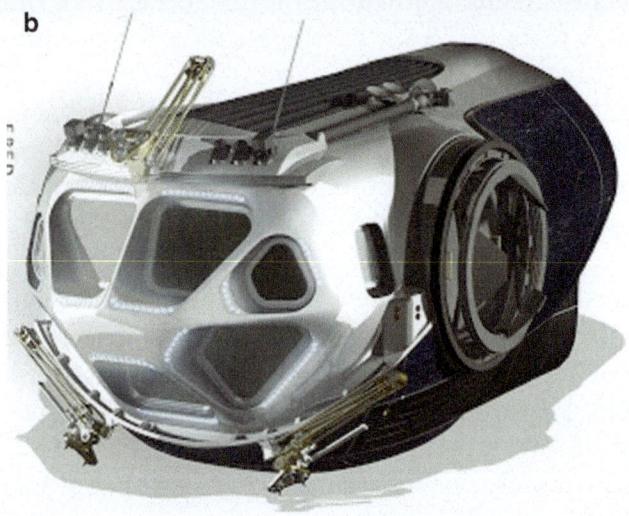

FIG. 6.11 (a) Astronaut with MMU on asteroid EVA from MMSEV (NASA).
(b) Multi-Mission Space Exploration Vehicle (NASA)

from ESA, including Britain's first official astronaut, Major Tim
Peake (Fig. 6.14). In October 2011 NASA mounted the NASA
Extreme Environment Mission Operations (NEEMO) using the
underwater Aquarius habitat off Key Largo, Florida, and deep water
submersibles to simulate the SEV. The mission had to be terminated
early due to the threat at the surface from Hurricane Rita [60], but a

Fɪɢ. **6.12** Takuya Onishi (JAXA) on NEEMO mission, October 2011 (NASA)

second one in June 2012 ran its full 12-day course [61] and included Peake, who went on to train for a long-stay mission on the ISS.

In simulations of NASA's L1 mission, the astronauts are shown taking samples from the asteroid through a small aperture in the capture bag (Fig. 4.29b). That seems very limiting, and a more promising idea is that they could make their way over an asteroid's surface by net (Fig. 6.13).

That raises another interesting possibility, not considered so far. The MADMEN units were equipped with individual nuclear reactors because the drilling environment was likely to be too dusty for solar cells to be used, but they were expected to be on full load for only 15 % of the time, and on a body only 1 km in diameter they won't be far apart. With the operation we're planning, a free-flying solar power satellite could supply them directly with power, either in turn as required or by switching power through

FIG. 6.13 Asteroid EVA with net (NASA)

the grid of cables. If it proves necessary to double up the lengths of the barrels, as above, it will be much easier to switch more power to the longer ones than it will be to give at least half the units twice as much reactor power.

Astronauts moving about on the surface would have to be careful not to link their tethers to the 'live rail,' but that's minor compared to the care they had to exercise on the May 2013 ISS repair. Even as they were going out through the airlock, NASA TV watchers noticed Mission Control adding more and more to the list of things they mustn't touch, clip onto or kick—all of which they successfully avoided (Fig. 6.14).

The power could be supplied by surviving Solaris collectors, or by new ones taken with the expedition. Just two of Max Vasile's 20-m mirrorbees could gather 630 KW, more than even 320 MADMEN units would require. It might be hazardous for the astronauts if high-powered lasers were beaming to the units while they were working on them, but again that could be avoided by beaming to central collectors and distributing the power through a grid. It would definitely be preferable to working around live

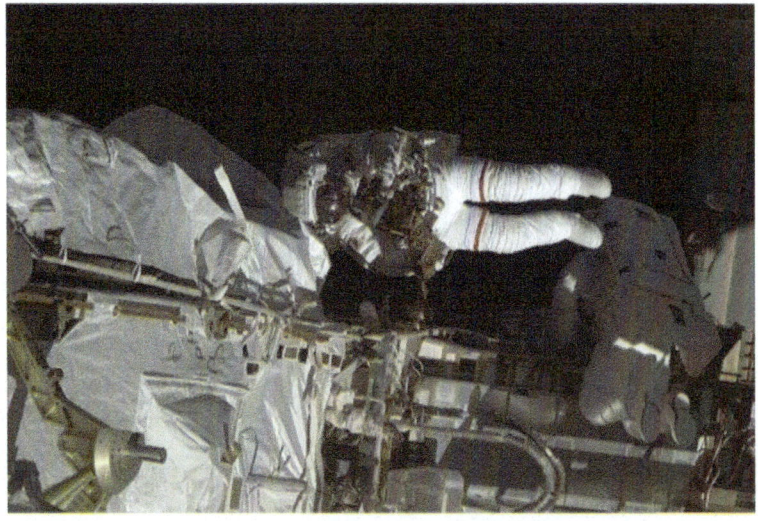

Fɪɢ. **6.14** Ammonia pump repair, ISS, May 2013 (NASA)

nuclear reactors, which would then need shielding that wasn't necessary in the original, automated proposal.

Gravity Tractor Reconsidered

And it's row! It's row, bullies, row!
Them Liverpool Girls, they have got us in tow.

—Traditional

(Some modern singers substitute 'Liverpool Judies' in the two songs with this chorus, but the 'Liverpool Girls' were currents in the River Mersey—still strong today—which impeded the ships' boats when pulling them out into the main stream.)

It's normally said that the gravity tractor method of deflection will require 20 years to achieve its purpose, but as we've seen in Chap. 4, that involves a single unmanned spacecraft with a mass on the order of 20 t. In this case we have a fleet of 12 ships and a lot more mass. For his calculations in *Expedition Mars*, Martin Turner gave his Earth-Mars transfer vehicle a dry mass of 70 t [16]. Our NERVA ships will arrive with little fuel left, since they're then on a free-return trajectory to Earth, but they will be carrying more consumables for a longer stay in space.

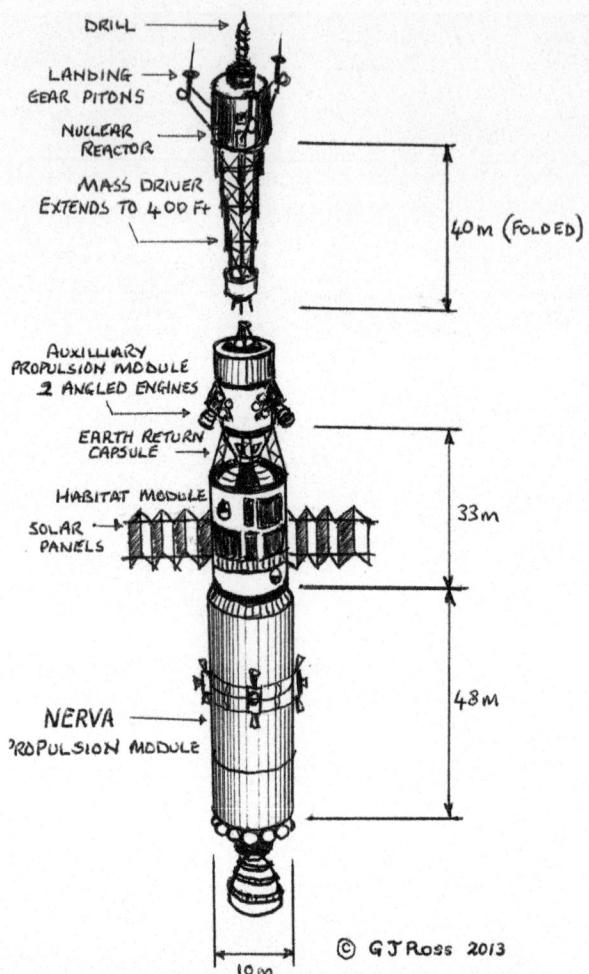

DRILL

LANDING
GEAR PITONS

NUCLEAR
REACTOR

MASS DRIVER
EXTENDS TO 400 Ft

40M (FOLDED)

AUXILLIARY
PROPULSION MODULE
2 ANGLED ENGINES

EARTH RETURN
CAPSULE

HABITAT MODULE

SOLAR
PANELS

33m

NERVA
PROPULSION MODULE

48M

© GJRoss 2013

10m

FIG. 6.15 NERVA mission to Goldilocks (one ship of twelve) (Design by Duncan Lunan, drawing Copyright © Gordon Ross, 2013)

Using the equation from the original paper on gravity tractors by Lu and Love, it appears that we could deflect Goldilocks in 3 years with a tractor mass of 1,500–1,600 t. If our 12 ships reach Goldilocks with most of their propellant used up, but each of them has the mass of the space shuttle, 100 t, and can maintain low-thrust propulsion once they get there, then their mass could generate 75% of the pull needed for deflection. Figure 6.15 shows the configuration on arrival; the main vehicle is a NERVA nuclear thermal rocket, as in the proposed 1980s

Mars missions (Fig. 6.5a, b). Its dimensions are taken from the Mars mission architecture that Wernher von Braun presented to the Space Task Group in 1969; afterwards he gave a copy to the BBC's space correspondent, Reg Turnill, who printed it as an appendix to *The Moonlandings* [62].

Our return-to-Earth capsule would be smaller, since it doesn't have to lift off from Mars, but it does need propulsion capability. If the asteroid deflection is successful the capsules have to shift back onto a return trajectory; if it isn't successful, they have to leave earlier and outpace the asteroid to get back before the impact.

The major addition to the design is the auxiliary propulsion module, whose plasma thrusters have to be angled at 65° (Fig. 4.20b), so that the beams miss the asteroid; otherwise they'll negate the gravity tractor effect. It has to be auxiliary propulsion because it would require a big scaling up from propelling comparatively small space probes at present, to pushing 100- to 150-t ships on the main mission.

There might not be time to develop large exotic engines, but the tractors have a less demanding requirement. Xenon plasma motors generating milliNewtons of thrust are already available. We need 2,000 N to deflect the asteroid directly in 3.3 years, but with the tractor we only have to move the 12 ships and let the mass of the asteroid do the rest of the work, so the thrust required is well within the state of the art. Indeed it appears that existing plasma thrusters could achieve it at 10% thrust or less, which seems surprising until it's recalled that the velocity of the ships has to be changed by only a few centimeters per second, not by kilometers, as in the main propulsion for an interplanetary mission.

If the strengthened mass drivers massed 33.3 t each, the ships and their payloads could in theory move the asteroid by gravity tractor effect alone, without having to deploy the mass drivers to the surface. That generates some interesting options. It's not likely that the mass drivers *would* be left attached and unused. Although there doesn't seem to be any reason why the gravity tractor wouldn't work, relying as it does purely on the laws of physics, the weak point is that the thrusters have to keep firing for at least 3 years. Ion drives and plasma drives run at high electrical potentials, as does the arc jet by definition, and erosion of beaming grids and nozzles can shut down the propulsive part of the system even when, frustratingly, everything else is still functional. Presumably the mass drivers would be deployed, and retrieved to add to the tractor masses of the ships if they didn't succeed—in opposite pairs, for preference, to keep the cluster balanced.

The mass drivers might be scaled back down to the design in the MADMEN paper, but more probably they'd be built to be capable of doing the job on their own if need be. If the auxiliary drives fail, or even underperform, then it would be nice to know that the mass drivers were on site and doing the job anyway, rather than have all civilization's eggs in one basket. If going to the expense and effort of creating this mission, the more options the better. Another option would be to launch three to five more ships, if we can do it, forgetting the mass drivers and using tractor only. Then maybe only three of the ships need be manned and the others can be slaved to them.

Here, however, we'll assume that all 12 ships are manned, and that mass drivers and gravity tractor will work in tandem. As the asteroid rotates, the mass drivers will fire in sequence in the required direction of thrust. The ships will be clustered on the other side, in the direction of deflection, roughly over the terminator, at right angles to the Sun. They'll be 500 m off the surface, all pointing the same way, and with auxiliary propulsion on.

Spinning paired, tethered ships end-over-end to provide artificial gravity for the crews will no longer be an option. We now have three apparently incompatible requirements: the gravity tractor, the need to use the asteroid as a radiation shield, and the need for artificial gravity. Fortunately there is a potential answer within the mission architecture, and the new element it requires is the use of inflatable modules.

In one of Sydney Jordan's Jeff Hawke stories, set on an asteroid, it was reached by a ship with a big wheel on one side, containing the control room, obviously lightweight because it was balanced by a fairly small communications array on the other side, but not explained. The author's guess was that it was inflated and rotating [63]; Sydney has so far maintained a dignified silence on the matter. But Robert Zubrin is a strong advocate of inflatable modules for living space on missions to Mars, including his own Mars Direct scenario; and Bigelow Aerospace has already placed two Genesis inflatable capsules in orbit, with a range of module sizes in development (Figs. 6.16 and 6.17) and the intention to deploy one called BEAM to the ISS in 2015 [64].

FIG. 6.16 Genesis 2 module in orbit (© Bigelow Aerospace Inc., 2007)

So the answer to the microgravity problem could be, say, one or more pairs of Bigelow inflatable modules, joined by tether, rotating about the axis of one or more of the ships. But for the gravity tractor, the optimum towing distance is equal to the radius of the asteroid (see Fig. 4.20b). If the asteroid is a kilometer across, the ships are 500 m out, and if the tethered spin radius is 50 m or more, the habitats will lose a lot of the shielding from cosmic rays that the asteroid would otherwise provide. When the crews are to be in space for 4 years, that's a serious issue.

In addition, tethered rotating capsules would cause big problems when the ships have to be tightly clustered for gravity tractor effect. Even getting the thruster beams not to strike the other ships would be difficult. The answer is to cluster the ships around an inflatable core module (Fig. 6.18), which would have to be 29 m in diameter but would provide an effective and spacious 'storm cellar' against coronal mass ejections, and perhaps a workshop for shirtsleeve maintenance of the MMSEVs and other hardware. In that configuration the plasma thrusters would have to run at higher thrust but still within the present state of the art, and an auxiliary module can be turned around to use the second thruster

FIG. 6.17 Bigelow Aerospace modules (NASA)

if the first one fails. Then the inflatable cabins can be towed on a tether 350–400 m in length, keeping them 100 m clear of the surface to avoid ridges and peaks as it rotates under them, and still get most of the radiation shielding benefit (Fig. 6.19).

To get in or out of them will require an EVA. or docking with the MMSEV. The spin would have to be stopped for ingress or egress, but the easiest way to do that is with a counter-rotating flywheel, which should allow some ingenious commonality with the tether capstan.

Recreation

I liked John, and besides I would have flown by myself or with a kangaroo — I just wanted to fly. All that stuff about crew compatibility is crap. Almost anyone can put up with almost anyone else for a clearly defined period of time in pursuit of a mutual objective important to each.

—Michael Collins, *Carrying the Fire* [65]

Nevertheless, if the crews are to spend a substantial part of 4 years in the rotating capsules, they will need ways to pass their

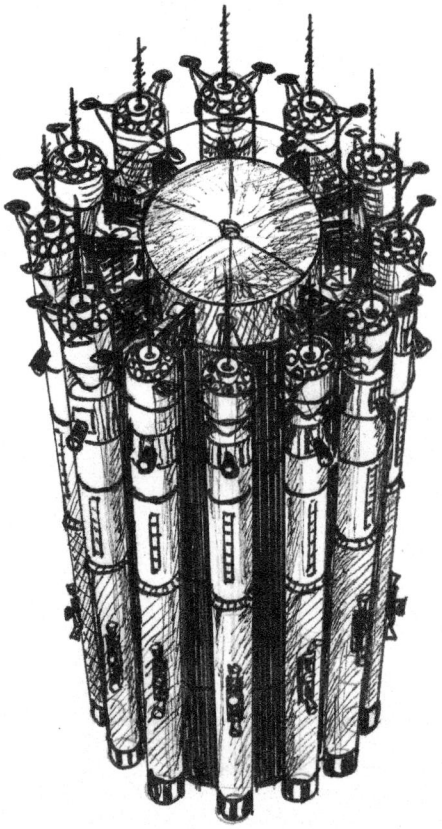

12 ship cluster
around
inflated core.

Fig. 6.18 The 12-ship fleet clustered around an inflatable core (Design by Duncan Lunan, drawing © Gordon Ross, 2013)

time, though much of it will be spent in exercise or sleep. The favorite recreation of crews in orbit—gazing at Earth—will not be available, and one suspects looking at the asteroid will lose its appeal, even though the view will change constantly as it rotates and the Sun angle alters from week to week.

Reading, listening to music and watching movies will continue to be featured, and over that period of time tastes will broaden for the sake of variety. On the relatively short Apollo 9 mission Dave Scott hid Rusty Schweickart's Vaughan Williams tape [66], but the

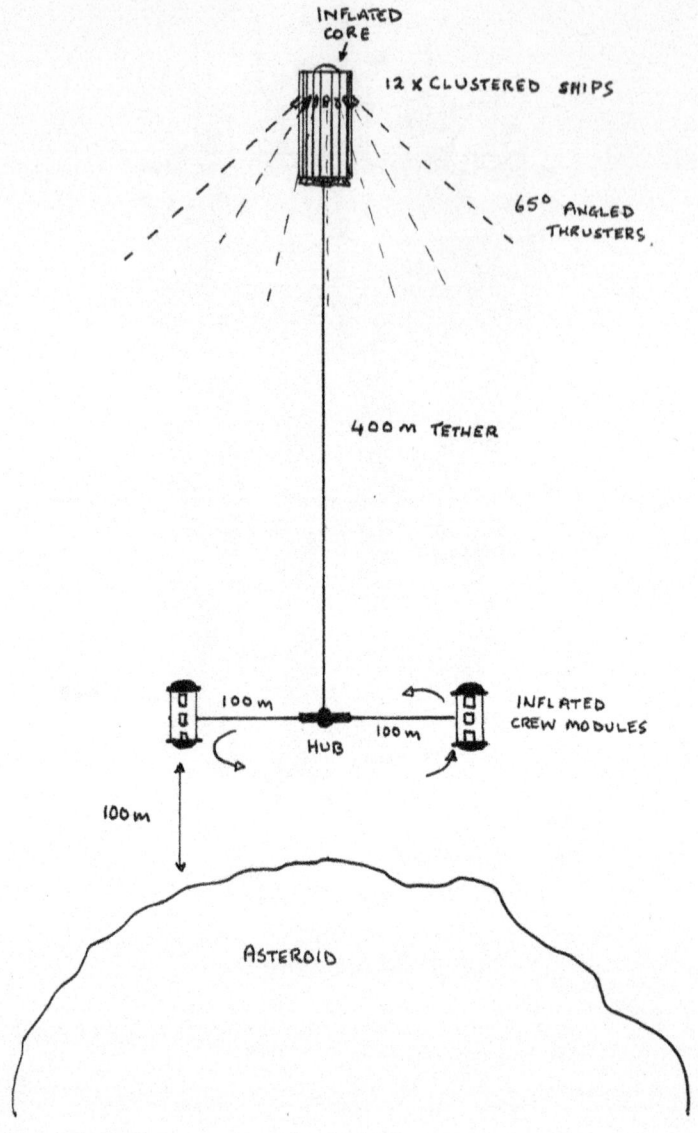

FIG. 6.19 Space fleet in gravity gradient configuration (Design by Duncan Lunan, drawing © Gordon Ross, 2013)

late David Proffitt, RN, found that when his duties took him to sea, his collection of traditional music records would become popular after a few weeks—not because his shipmates particularly liked it, but because their own pop and Country and Western choices began to pall, and classical music was 'officer country.' Seemingly most of the astronauts took C & W to the Moon, though on Apollo 11 it was Neil Armstrong, not the thoughtful Michael Collins, who chose the *New World* symphony and Samuel Hoffman's *Music Out of the Moon* [66].

Food will be even more important. Experience in long-duration submarine missions, polar bases, etc., reveals that a sustained unfamiliar or monotonous diet is a big cause of psychological problems [67]. Meals in simulated gravity may assume particular importance if it provides relief from the blandness that food takes on in microgravity, due to congestion of the nasal passages. Some astronauts have deliberately taken food items that were acquired tastes, so they wouldn't have to share, but many have found the menus of different cultures a welcome source of variety as long-stay missions go on.

In a podcast compiled with the proprietors of Tested.com, Chris Hadfield revealed that computer games and magnetized Scrabble™ were among the most popular recreations on the ISS, but he had found a physical game that could be played safely within the crowded ISS [68]. In his first novel *The Sands of Mars* (1951), Arthur C. Clarke wrote:

> There are not many games of skill that can be played in space; for a long time cards and chess had been the classical stand-bys, until some ingenious Englishman had decided that a flight of darts would perform very well in the absence of gravity. The distance between thrower and board had been increased to ten meters, but otherwise the game still obeyed the rules that had been formulated over the centuries amid an atmosphere of beer and tobacco smoke in English pubs [69].

The darts that Chris Hadfield improvised were giant-sized, with heads like cubesats and Velcro tips. They did indeed have a flat trajectory, but flew slowly and with noticeable wavering, probably due to air currents (there are powerful fans throughout the ISS to prevent pockets of carbon dioxide from accumulating). The board for an actual game would have to be large, especially if it were to accommodate trick shots like the regional variant 'three in a bed equals game' [70]. Perhaps the large internal volume of the inflated cylinder in Fig. 6.18 would allow more energetic games such as zero-g squash, and the lunar or Martian gravities of the rotating cabins would permit low-g badminton, table tennis, quoits, skittles, or billiards. The rotation would introduce some intriguing effects, and no doubt our 'seaplane mechanic' counterparts would adapt to them. In a break between experiments on ESA's Airbus A-300, equivalent of NASA's Vomit Comet, scientists found that American-style football could be played in microgravity, given enough space [71].

In the 1960s Andy Nimmo devoted considerable thought to games for long space missions, and he and Edmund Lavallet came up with 'Nimla Chess'—not the 3-D variant famously played by Leonard Nimoy in an early episode of *Star Trek* [72], but a multiplanar version for up to six players. With just two players it was reduced to classic black v. white, but when colors were added things became more complicated. The author is one of the few to have played Nimla Chess, in its simplest three-player form and can testify that it is both absorbing and time-consuming.* With three volumes to date of more advanced games, one might be concerned lest the crews couldn't tear themselves away to work... but as there would have to be a shift system, probably the games would be played slowly, over months or even years, evolving like giant crossword puzzles (another possibility for off-duty hours).

[*Nimla Chess is being played again by dedicated volunteers, and will be relaunched in public and online at the Satellite 4 SF Convention in Glasgow, April 18–24, 2014.]

> And pursuant to this idea of a holiday, he insisted upon playing cards....Strange mind of man! That, with our species upon the edge of extermination or appalling degradation, with no clear prospect before us but the chance of a horrible death, we would sit following the chance of this painted pasteboard and playing the 'joker' with vivid delight. Afterwards he taught me poker, and I beat him at three tough chess games. When dark came we were so interested that we decided to take the risk and light the lamp....My folly came to me with a glaring exaggeration. I seemed a traitor to my wife and to my kind; I was filled with remorse...
>
> —H. G. Wells, *The War of the Worlds*

And yet, how utterly human that reaction is. With years in which to contemplate the consequences of failure, the astronauts will need all the distraction they can find in their time off. If their mission is succeeding, the atmosphere will grow steadily lighter; but if not, there are big dangers ahead in the last year before the impact.

7. Final Options

If they're shooting at me, it's a high-intensity conflict.

—Former Commandant, U. S. Marine Corps [1]

The major reason why the solar deflectors and mass drivers might fail to achieve the deflection required is that the asteroid has absorbed the kinetic impulses applied to it instead of changing course. The solar deflectors may have proved too vulnerable to the debris plumes, or the mass drivers may have succumbed to mechanical failure, avalanches, or the sheer difficulty of the task. If the gravity tractor has failed, it seems it could only be because the auxiliary drives couldn't sustain the thrust. But if we are down to a year before the impact, more drastic options have to be considered—especially if the kinetic methods have proved partially successful, and there's reason to think more powerful versions might succeed.

If Colin McInnes's kinetic deflector has been working its way around into retrograde orbit, it will now be out there in front of the asteroid and coming for a head-on collision at 60 km/s. With the heavy-lift boosters now in mass production, however, Andy Nimmo suggested that once the expedition was on its way, the next 2 years could well be spent building a large mass driver in low Earth orbit, to launch the kinetic deflector directly into retrograde solar orbit, once other methods were known to have failed, or not to have worked sufficiently.

The mass driver in Fig. 7.1 was intended to launch 1.4 metric ton payloads out of the Solar System, requiring a launch velocity of 8.7 km/s on top of Earth's orbital velocity and the mass driver's orbital velocity around Earth when it was on the night side of the planet [2]. On the other, sunward side of Earth it could launch to retrograde solar orbit at 21 km/s approx., 10 km/s on leaving the Earth-Moon system, so the projectile would hit with half the velocity and roughly one-eighth the kinetic energy of McInnes's one

D. Lunan, *Incoming Asteroid!: What Could We Do About It?*,
Astronomers' Universe, DOI 10.1007/978-1-4614-8749-4_7,
© Springer Science+Business Media, LLC 2014

FIG. 7.1 Mass driver in low Earth orbit. Note trefoil symbols on cargo canisters (see Chap. 8) (© Sydney Jordan and Theyan Rich, 1998)

(one-quarter, if the mass could be doubled). So ten of them could provide the same effect, but more gently and with less chance of fragmenting the asteroid. Psychologically, doing it that way and under control of the crews in the ships is probably preferable.

Obviously now the nuclear option has to be reconsidered. Nigel Holloway said that a year would be adequate for that [3], and by this time the internal composition of the asteroid will be known in detail, so there will be less uncertainty about whether it would work. Warheads would have to be launched from Earth well in advance, and would have to be steerable, while we must assume that any equipment on the asteroid surface will be destroyed, so the ships would have to stand off to provide final guidance. It could mean that for a substantial part of their time at the asteroid, the crews would be working with the knowledge that the impactor was already in flight and that nuclear weapons might be coming, or were on their way.

Jay Tate considered that to be no psychological problem. All military personnel are trained for it and experience it in live-fire practices. But if the contractors have provided civilian teams for the mass drivers, as in Chap. 6, they may view the matter with less equanimity—especially if individual nations take unilateral action outside the international consortium, or lose their nerve within it. A year ahead of impact, there may be known or suspected rogue warheads coming in; if it became clear that the solar deflectors weren't going to work on the asteroid, some of them might be kept in reserve in hopes to provide protection for the task force.

At the end of all that, we came around to thinking that the manned ships should carry a range of nuclear devices, in order to place them exactly where needed, if (literally) push comes to shove. If the U. N. treaties have to be broken, better to break them in controlled circumstances rather than by unilateral action. Keith Llewellyn of ASTRA made the interesting suggestion that the expedition should also carry a shield, steerable and with hardened electronics, like the pusher plate of the 1960s Orion nuclear pulse concept but parabolic, mounted on and protecting an unmanned ship, on the same side of the asteroid as the detonation, to increase its effectiveness by focusing back the blast and radiation that would otherwise be wasted.

Obviously the method would be to start with the smallest available yield, and work up as it became clear whether the asteroid could take the blows. As the asteroid moved along the redline, John Braithwaite's point—that if the asteroid does shatter at least some of the fragments would miss the planet—might become more significant. As the predicted impact point neared the rim of Earth, more and more of these fragments would indeed pass by.

In *Deep Impact*, the first attempt to shatter the incoming comet only breaks it into two. (We won't pause to consider why, since they didn't know the internal composition of the object, they wouldn't have used all the nukes they had on that first attempt.) We need to consider briefly what might be achieved by deliberately breaking Goldilocks up. In *The Hammer of God*, Kali is dumbbell-shaped, and when all else fails, Clarke's characters use a kinetic impact (a 1,000-megaton warhead that fails to explode) to break it into two pieces, one of which misses Earth altogether [4]. Many of the Earth-grazing asteroids are paired or 'contact binaries,' and if Goldilocks were one of those, we discussed using techniques to increase its spin rate instead of deflecting it, in order to maximize the separation rate when we severed the asteroid at the neck.

One obvious drawback is that if the asteroid is structurally weak the two masses might separate prematurely, leaving two problems to cope with instead of one. When we're dealing with billions of tons of mass, tethers aren't going to help even if the mutual attraction is only 0.0001 g. In that situation the two pieces definitely will meet again, one orbit later, though they may not collide and merge due to perturbations by other Solar System bodies, particularly in the Earth-Moon flyby itself; and there will then be decades or even centuries to find new solutions, or to send an industrial task force as in *Man and the Planets* [2].

While thinking about controlled explosions, however, ASTRA's other ongoing technical project was the research on the Waverider atmospheric entry vehicle (see Chap. 5). In John Baxter's *The Hermes Fall*, the incoming asteroid is described as blade-shaped [5], and in *The Hammer of God* the fragment that doesn't miss Earth passes through the atmosphere in a highly destructive graze. Could we sculpt the Goldilocks object into a caret or Maryland Waverider that would be deflected by aerodynamic forces into such a path, not striking the surface? Intuitively it feels as if it ought to work—tektites are naturally sculpted into aerodynamic shapes as they re-enter the atmosphere. But asteroids over 100 m in diameter are hardly affected by drag, let alone aerodynamic lift, and it's not at all likely that the final path would pass through the very thin skin of the atmosphere at just the right angle for them to make a difference even if they were effective.

An Earth-orbiting mass driver would have major uses afterwards, assuming that civilization survives. The cargo canister in Fig. 7.1 bears a trefoil symbol because Sydney Jordan and Theyan Rich were illustrating still another ASTRA project, an integrated program for nuclear water disposal and space solar energy [6] (see Chap. 8). If the mass driver could be made operational within a year of the Goldilocks expedition's departure, theoretically Project Starseed might have its first solar-electric power satellite ready for use 2 years later, at the Earth-Moon L2 point before moving it to geosynchronous orbit.

If it's known that the solar deflectors worked, just not sufficiently, the powersat's laser could be used in vacuum as a beam weapon to give the asteroid a harder push. In 2013 professors Philip M. Lubin and Gary B. Hughes proposed DE-STAR, Directed

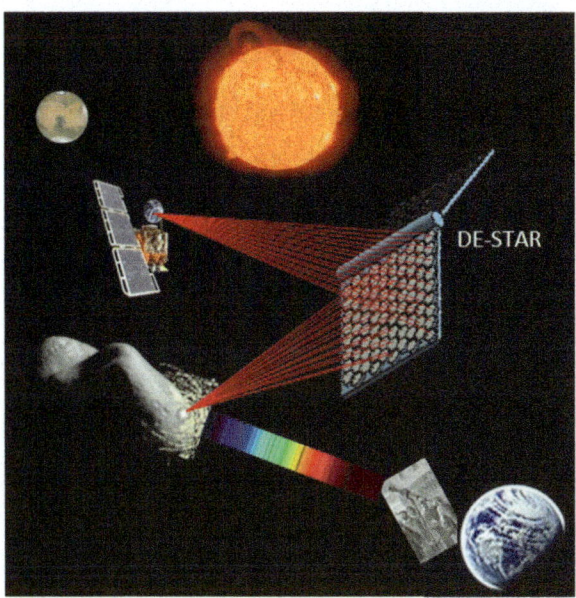

Fig. 7.2 DE-STAR (© UC Santa Barbara, 2013)

Energy Solar Targeting of Asteroids, an exploRation, as a "directed energy orbital defense system" (Fig. 7.2). A 100-m solar array could generate enough power to alter the orbit of an asteroid at a distance of 1 au, whereas an array 10 km across with an output of 10 GW (powersat-sized) could deliver 1.4 megatons of energy per day and vaporize a 500-m asteroid in a year, according to the press release [7], though the accompanying graph appears to indicate 3.1 years (Fig. 7.3). Complete vaporization would remove any worries about the effectiveness of the generated thrust, but it seems it would take 31 years to evaporate a 1-km asteroid even with 35 GW output, so in our scenario we're still looking at deflection, and when it starts it seems like a good time for the ships to separate from the asteroid.

   If the deflection has been successful, by whichever method or combination of methods have done what was required, the auxiliary propulsion modules will detach and the landing capsules will separate from the NERVA ships, which will remain with the asteroid as it passes Earth and goes on. The landing capsules will have to use propellant to get back on a return-to-Earth trajectory, so they might turn the low-thrust modules over and re-dock with

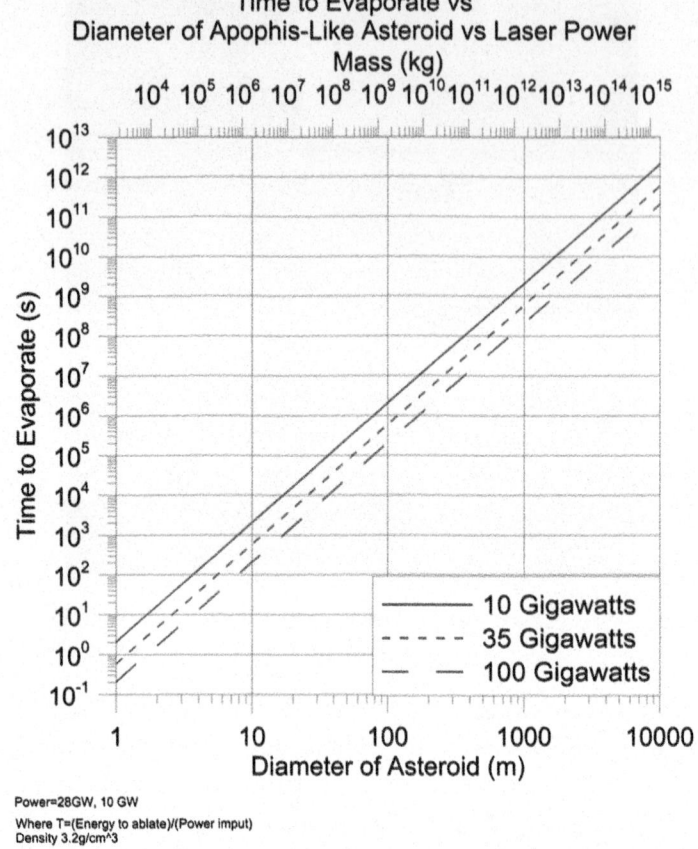

FIG. 7.3 Vaporization times for rock asteroids (© Philip M. Lubin, 2013)

them, as the Apollo spacecraft did with its lunar lander, to use their remaining fuel for the course correction; or, if they separate soon enough, they may be able to achieve that with attitude thrusters alone.

If the asteroid has *not* been deflected significantly, then if the NERVA ships have enough residual propellant, they can be separated from the asteroid to fly past Earth. "It is essential that a used reactor does not re-enter the atmosphere until many years have passed since its last use," [9] though considering what else would be injected into the atmosphere as the asteroid itself struck, the fission products in the fuel rods would not be the greatest problems to cope with.

Impact Unavoidable

Now a' is done that men can do,
And a' is done in vain.

—Robert Burns, *It Was A' For Our Rightfu' King*

The streets and houses were alight in all the cities, the ship-
yards glared, and whatever roads led to high country were lit
and crowded all night long. And in all the seas about the civi-
lized lands, ships with throbbing engines, and ships with bellying sails, crowded with men and living creatures, were standing out to ocean and the north. For already the warning of the
master mathematician had been telegraphed all over the world
and translated into a hundred tongues.

—H. G. Wells, *The Star*

Wherever our astronauts set down, it would be well for them
if they make it to ground the day before the impact, rather than the
day itself or the day after, and especially so if they must land in the
hemisphere of the redline. Even though they've done their best,
the crews' welcome on Earth might be decidedly uncertain. Their
landing zone on Earth could be uncomfortably close to the impact
point, and there wouldn't be much point in stationing recovery
ships there. In *Lucifer's Hammer*, the last words the astronauts
hear from their recovery fleet are "Fireball dead overhead!", and
although they have the option of coming back on land, in a Soyuz
instead of an Apollo, after the impacts they have an interesting discussion about where in the world it would be best to set down [8].

There will be few throbbing engines or bellying sails out on an
ocean that's about to be subjected to 30- to 100-m tsunamis, even
if their effect is somewhat lessened in the open sea; shipping won't
even be available at the coasts, when the owners have had time to
move their vessels into different oceans. Perhaps a page might be
taken from *The Phoenix at Easter* (Chap. 3), for the crews to be
picked up by submarine. Our landers will separate early with the
auxiliary propulsion still attached, so they can speed up and get
back to Earth before the impact, to come down in a different ocean
where ships can still be waiting. Or perhaps Elon Musk's technique
for touching down on land will be perfected. If they must land
on water, an inland sea like the Mediterranean or the Black Sea

would be a good choice and one of the Great Lakes even better—a smaller one better still, if they can achieve that much precision— because after nearly 4 years in space, whatever success there has been in reducing the medical effects of long-term weightlessness, the astronauts will need help once they touch down.

A day after the impact, they and everyone else will need help to survive. Whether the material injected into the upper atmosphere is water vapor or dust, it will spread around the world with terrifying speed. In *The Earth's Beginning*, Sir Robert Ball described the shock to science when it happened with the dust from Krakatoa:

> We previously knew little, or I might say almost nothing, as to the conditions prevailing above the height of ten miles overhead. We were almost altogether ignorant of what the wind might be at an altitude of, let us say, twenty miles.... There was nothing to render the winds perceptible until Krakatoa came to our aid. Krakatoa drove into those winds prodigious quantities of dust. Hundreds of cubic miles of air were thus deprived of that invisibility which they had hitherto maintained. They were thus compelled to disclose those movements about which, neither before nor since, have we had any opportunity of learning.
>
> With eyes full of astonishment men watched those vast volumes of Krakatoa dust start on a tremendous journey. Westward the dust of Krakatoa took its way. Of course, everyone knows the so-called trade-winds on our earth's surface, which blow steadily in fixed directions, and which are of such service to the mariner. But there is yet another constant wind. We cannot call it a trade-wind, for it has never rendered, and never will render, any service to navigation...

The Earth's Beginning is a remarkable book, because its 1901 and 1909 editions bracket two revolutions in science, and Sir Robert recognized both [10]. The foreword leaps from the discovery of radioactivity to the realization that the Sun could be much older than previously thought, and so Earth could have the greater age that geologists and evolutionary biologists were insisting it must have. For good measure, it goes on to accept that the Milky Way is an example of what was then called a spiral nebula, and to a dawning recognition that since there are so many others, the universe must be far more vast than previously imagined.

Sir Robert's imagination may have faltered when it came to using the jet stream for navigation, but if Goldilocks hits a continent, or shallow enough water for it to form a crater in the seafloor, his words might prove prophetic. Jet aircraft can still fly in rain and darkness, but can't fly safely through dust—as illustrated by scares in Indonesian eruptions and precautions during Icelandic ones. The technology for detecting the hazard is improving, but the acronym AVOID (Airborne Volcanic Object Imaging Detector) makes clear the only action that can be taken [11].

Be it due to dust or storm clouds, within a week or two at most, the planet will be in darkness. For many places around the coastal rims it will come even before the waves, and with the waves will come storms like none in the memory of humanity. How many will survive until the Sun breaks through again, and beyond, will depend to a great extent on the thoroughness of the preparation. If Lembit Öpik is right and governments won't do anything to stop it, then it's estimated that 25–50% of humanity will die. But many of his hearers thought that action could and would be taken to minimize that.

Öpik was commenting on an outline of the proposals made at the 2003 seminar by Dr. Arthur Hodkin, by then in retirement after a lifetime of engineering and environmental consultancy. Though he had told us he had a major contribution to make, he had shared little of it with us, and he insisted on opening with the idea that there might be ten mass extinction events in the next 100,000 years, followed by speculation about *Star Trek* technology that Andy Nimmo and the author had urged him to leave out. But then came a major shift in direction. On the wall behind him at the Spaceguard Center there was a display illustrating the environmental effects of a Chicxulub-scale impact, very much like nuclear winter though without the fallout, and ending after 6 months with the words, "90% of the human race is dead." Pointing to it for emphasis, Arthur declared, "I'm going to show you how to reverse that percentage."

One early attempt to assess the longer-term environmental effects of a big impact was by the late Sir Fred Hoyle in his 1981 book *Ice*, though unfortunately he limited his study to impacts on land. His thesis was that ice ages are triggered by rock asteroid impacts and ended by metal ones, and he was considering the material of the impactors themselves rather than the terrestrial material thrown up into the atmosphere [12].

(It may seem extraordinary that Hoyle could have got the dynamics of big terrestrial impacts so wrong, when every text nowadays explains that the material of the impactor is vaporized and lines the inner surface of the initial crater [13]. But in many papers and books contemporary with *Ice* the impact process and effects are mentioned only in passing, if at all [14], and the Tunguska event, where the dust of the impactor *was* blown upwards, is taken to be typical. In the *Scientific American* special issue on the Solar System in 1975, William K. Hartmann noted only that the rate of cratering on Earth and the Moon had remained constant over the last 3.5 billion years; in *New Science in the Solar System* the same year, a similar review by *New Scientist* doesn't mention it at all [15]. In book form, at least, the earliest discussion of the issue in the author's collection doesn't appear until 1981 [16], the same year as *Ice*.)

In Hoyle's argument, the key factor was the very small ice crystals, known to Antarctic explorers as 'diamond dust' at ground level, which form at temperatures below minus 40°C out of pure water, not around condensation nuclei. At higher levels they are responsible for such phenomena as lunar halos, sundogs and moondogs. Whether they form at altitude depends on the amount of heat being transmitted upwards to space. If large volumes of water vapor are evaporating from the sea, the latent heat released as they condense into rain clouds will prevent the formation of ice crystals above. If a large volume of rock dust was injected into the upper atmosphere, said Hoyle, it would take about 10 years to fall back to heights above 12 km; meanwhile it would reflect so much sunlight back into space that ice crystals would form high up, and once formed, they would be persistent.

Reflecting still more heat in turn, they would cool the surface sufficiently for big ice sheets to form below. "We need not search any longer for the conditions that would cause an ice age. Turning only one-tenth of 1% of the amount of water normally present in Earth's atmosphere into fine ice crystals would have a catastrophic effect on the climate. And the resulting disaster would not take long to develop. With most of the 64% of incident solar radiation that now penetrates to the lower atmosphere being reflected back by the high atmosphere into space, the temperature of the land would collapse within weeks and the temperature of the oceans within a few years."

In Arthur C. Clarke's short story *The Forgotten Enemy*, in 1953, he had the glaciers return to London after 30 years of cooling, and in the preface to the anthology containing it he apologized for the liberty taken with geological timescales—"but what is a factor of 10^3 among friends" [17]? Today, the thinking is that while there wouldn't be the overnight freeze of *The Day After Tomorrow*, 10 years could be enough.

Hoyle didn't extend his argument to consider water injected into the upper atmosphere, though the implication seems to be that as that condensed the latent heat released would keep ice crystals from forming higher up—at least initially, though they would then form above the descending dust. But it seems that Hoyle's 10-year timescale for the dust's descent is much too long. Luis Alvarez initially made the same mistake in his impact model for the dinosaur extinction [18], and apparently it stems from an overestimate in the Royal Society's classic report on Krakatoa [19]. One to two years of darkness is often seen as the aftermath of a Chicxulub event [20], but the time may have been a great deal shorter though [18], [20] dust from the Chelyabinsk event was still at altitude after 3 months.

With regard to the water vaporized in the impact, information is hard to come by. Most accounts of the Chicxulub event focus on the dust. Shoemaker wrote that "In an oceanic impact… huge quantities of water must have been driven into the upper atmosphere," [16] and as Enever pointed out, just the dissolved solids released from the vaporized seawater could be enough to create temporary darkness [22]. Allaby and Lovelock go so far as to calculate how much water was involved—6,000–60,000 billion tons—but they go on to say that a great deal of it would escape into space, and the rest would quickly condense and fall back as rain [21], so they don't discuss it further, and most of the other sources cited don't discuss it at all. By contrast Shoemaker thought that enough would remain in suspension for the mass extinction to be caused, not by cold but by a temporary pulse of heat due to the greenhouse effect of water vapor [16]. Could enough of it have survived at altitude instead, as 'diamond dust,' to prolong the cooling caused by the silicate dust? There wasn't major global cooling in the aftermath of Chicxulub, nor in the Ordovician bombardment

FIG. **7.4** Noctilucent cloud over the central belt of Scotland, summer 1978
(© Ian F. Downie, 1979)

described by Ted Nield (Chap. 4), but it is being suggested that the
Eltanin impact triggered a new ice age cycle (Chap. 2).

There could be a connection with the mystery of noctilucent
clouds, which goes back to the International Polar Year in 1882.
In April that year the Astronomer Royal for Scotland, Charles
Piazzi Smyth, reported strange 'night-glowing' clouds in the northern sky. They reappeared in 1884 and since then have been seen
yearly in June and July, when astronomical twilight lasts all night
in Scotland. They're about 60 miles up and shine by reflected sunlight from below the horizon, so their spectrum shows only the
elements that are present in the Sun, and we don't know what the
cloud particles are made of. Also it's not clear why the ones seen
in 1882 should have been much higher up (Fig. 7.4).

The author first heard of them after an incident in the early
1960s when a flight of Lightning fighters, taking part in a night
exercise out of the air base at Leuchars in Fife, were ordered to
climb above the clouds (assumed to be normal) and couldn't do it!
The British Astronomical Association has been gathering observations and photos from amateur observers at least since then, and

the clouds seem to have become more frequent over the last 100 years, leading some to suggest a correlation with global warming. But apparently there are very few observations from the southern hemisphere, where they should appear in the south in December and January, so the BAA is keen to have more—another role for the Falkland Islands observatory discussed in Chap. 4, should it ever come to pass.

An enhanced water content has remained in the stratosphere of Jupiter, above the Shoemaker-Levy impact sites, for 20 years to date [23]. Apparently it is the form of individual molecules, not 'diamond dust' or noctilucent clouds, presumably because conditions aren't suitable in the 2-millibar pressure range. Noctilenticular clouds do form at corresponding heights above Earth, coarser in structure than true NLC, so what does that tell us?

If the Goldilocks impact wasn't in deep water, there will have been wildfires caused by falling rock, or just by the glare from the sky as ejecta re-entered the atmosphere, but those effects might not be global for a 1-km rock impact. Major disaster relief would be needed on the periphery. But if the strike was in deep ocean, and the surrounding coasts had been evacuated before the tsunamis, the first problem afterwards might be, as suggested by Arthur Hodkin, that rain was too saline to drink. People would be dependent on deep wells, springs and stored water, but the salt should wash out of the atmosphere in a year or so.

For his presentation in Powys, "Stockpiles for Survival," Hodkin assumed that survival had to be achieved for a period of up to 2 years of darkness, with high-altitude dust a problem for up to 10 years. With a 1-km impact the period of global darkness might be only 6 months to a year, and as seen above, some estimates put it still lower. But as Lovelock and Allaby pointed out, the definition of darkness is subjective. Many animals would be able to see quite well when we were helpless, but for plants darkness is still total if their leaves are covered by ash, or snow. We can expect to lose at least one growing season in at least one hemisphere whatever happens, and possibly a good deal longer, with or without diamond dust overhead. Farmers can't instantly produce crops if the land is covered with snow and ice, and scoured by floods as that melts.

In Jay Tate's view things will return to normal in a decade, and he may be right, but the first 2 years at least will be rough.

There will be a transition year, possibly the second year after the impact (depending on the seasons), in which potatoes and oats may grow but other cereal crops will not. Rice economies will be badly affected, although insect predation, especially by grasshoppers, will be much lower. In the subtropics things may not be much worse than in a year of drought and famine, and product dumping from the developed world will no longer be a problem for local farmers. The worst will be over in 3 months, and 3 months' supplies can be built up with 2–3 years' notice.

In temperate latitudes, most people could survive, especially in shelters, if enough oil and gas is preserved. In Norway, drought and early winter had once reduced hydroelectric power, forcing the country to buy electricity from Germany at 20 times the price, and in 2003 the French nuclear capacity had been reduced by drought, for lack of cooling water. Coal and oil-fired power plants would be essential for survival in the 'asteroid winter,' with nuclear power as a backup. With no need for air conditioning, the output of the power stations, including their waste heat and carbon dioxide, could be concentrated on the shelters and the artificially lit greenhouses that will help to sustain them with some fresh food. In the 10 years before the impact, Hodkin foresaw major investment in retrieving coal and gas from deep reserves, with technology such as liquefaction and fracking, as well as in geothermal, wind and tidal power, all of which will still be available, even if hydroelectric power will be lost until after the thaw.

Before the impact electricity generation and grid capacity should be built up to 50% above present levels of peak demand, half of the new capacity from nuclear power and renewables, the rest from oil and gas. Scotland and Wales would be robust, and the United States and Canada have huge potential for survival, with 25% of the world's energy reserves on tap. The buildup elsewhere could be financed by U. N. energy bonds, with the assumption that 90% of the investment is for oneself and 10% for the less fortunate. Everywhere, at least a year's supply of oil and gas needs to be stockpiled.

To prepare for the worst there should be plant conservation in botanical gardens and seed banks, similarly for animals with sperm and embryo banks. Livestock forms cannot be allowed to die out. An extra year's stock of the main cereals and sugar should be laid down, and a 6-month supply of deep-frozen milk and dairy products, likewise meat; fruit, vegetables and potatoes, dried or canned; mineral and vitamin supplements. No one needs to starve, with reasonable reserves. Unless output is increased, to stockpile a year's supply of food and fuel over a decade will require 10% of production to be diverted, 20% if 2 years' supplies have to be put by.

During the inevitable periods of rationing, the authorities must keep a sense of proportion, maintain calm, and try to prevent rioting and hoarding by building up individual domestic stores and storage centers, particularly for drinking water and portable fuel such as propane. Building up reserves costs money but doesn't waste resources, investing in alternative sources stretches global reserves of fossil fuel and creating extra generating capacity just brings investment forward. In the actual crisis, where possible people should survive in their own homes, for instance using individual generators; though if there hasn't been a devastating ground shock, gas and oil pipelines will be intact and aboveground electric cables can restrung once the storms are over. Meanwhile basic food packs should be available with 3 months' worth of supplies, to be returned at 6-month intervals and then distributed to hospitals, schools, etc., as new ones are reissued.

In summary, without preparation, up to 25% of the human race would die in the aftermath of a 1-km impact, as predicted; with concerted effort to ensure survival, the losses could be less than 5%. Yet as quoted in Chap. 4, Lembit Öpik's view was that governments wouldn't do it because the measures would be too unpopular; individuals would have their own stockpiles. That view aroused a wide range of reactions. Bob Graham said that spacefaring powers might put all their trust in technology, as Öpik predicted, but the others would stockpile. John Braithwaite said governments' hands would be forced. Economic disruption would set in long before the impact, as exports crashed and imports dried up. Once people became afraid, economic measures would have to be taken to prevent hyperinflation and cornering resources. Craig Binns said there was a moral duty to introduce rationing, as the British government did in 1939. There were starvation deaths in the UK in World War I because it wasn't done.

Once established and continually updated, the stockpiles would be invaluable in other disasters. Hurricanes, earthquakes, etc., are not going to stop just because there's an asteroid coming. Jay Tate argued that comparable provision had not been made for nuclear war; Craig Binns replied that neutrals had done it, in non-aligned Yugoslavia, Sweden and Switzerland. "Only they could do it because under Mutually Assured Destruction it was considered provocative." Hodkin particularly recommended study of the Swiss program of nuclear shelter provision, on which he had based the 'contingency stores' list in Appendix 2.

In the UK there is even some evidence that during the Cold War the government clandestinely stockpiled up to 3 months' worth of food in what were considered vital areas, even if it was cynically calculated that a 40% reduction in overall population would help to stimulate the economy. Post-war government control would be achieved through rationing of luxuries, not necessities, because the risk of civic disorder would otherwise be too great. But if 40% of the populace was dead, consumption was reduced from 80% to 60% of output and the other 40% was reinvested, in theory the economy would take 12 years to recover. "In practice, of course, recovery will be likely to be slower than that because of damage not only to the parts of the economic machine, but also to the way they fit together" [24].

To help to preserve energy supplies and distribution networks, Graham Dale suggested that everything should be switched off before the impact. Similar precautions are now in place, or at least contemplated, if satellites warn us of a coronal mass ejection from the Sun of the scale of the one on September 1, 1859, the largest ever recorded, which would be catastrophic now if it happened without warning [25]. Gregory Beekman suggested that perfluorocarbons, currently suggested for use in terraforming Mars, should be stockpiled in exhausted oil fields to raise the ground temperature after the impact. But what if the darkness lasts only 6 months and there's a quick recovery—what if the greenhouse gases get loose unnecessarily?

Many of the UK's assets, built for conventional or nuclear war, would be valuable in the process of recovery—assuming the impact isn't in the Atlantic. Peter Laurie's book *Beneath the City Streets* goes to some lengths identifying underground car parks, etc., and aboveground structures that have been designed to survive airbursts and be quickly restored to use afterwards [24].

Even at Hiroshima, it was noticed that concrete buildings survived as shells when all else had gone. In the west of Scotland, constructions like Prestwick Airport and the Arches in Glasgow preserve some of the preparations Churchill and Eisenhower made for the campaign to follow the Nazi invasion of southern England. But very little of it would survive the ground shock of a land impact, or inundation by a 100-m tsunami.

To hold the world together in the aftermath, transport and communications will be essential. Structures like the Post Office Tower in London, and less famous counterparts like the one in Glasgow, can quickly be put back into commission. The microwave dishes on the outside are easily replaced. If there hasn't been a devastating ground shock, cables including fiber optics will still be there to reconnect. Satellite links will be out due to the devastation of the ionosphere, and the initial chaos in the Van Allen Belts will take out many of those nearer to Earth, but the Galileo, GPS and Iridium constellations would probably survive, and comsats in geosynchronous orbit almost certainly would, even if they couldn't be accessed for a time.

As already noted, if there was a land impact piston-engine aircraft will come back into their own, but if it was in deep water jets would still be usable. Most of the world's shipping would have been moved to different oceans and would still be on call. Heavy rain and snowfall might make roads impassable, but rail track would be easier to keep in use, as John Baxter pointed out in *The Hermes Fall*, or to restore even if the railbed is broken up by earthquake. Prudent governments will have stockpiled track. It can be replaced at rates of 20–30 miles per day, and the preparations made for maintaining and restoring links after nuclear war would be effective here too [24].

Poland, Japan and Russia were all much harder hit than the UK in the course of the Second World War, yet all three recovered within 10 years. Peter Laurie ended his relevant chapter with the words, "One cannot categorically say therefore that nuclear war would be the end of everything." And the biggest difference after the asteroid event will be the lack of nuclear fallout.

The social effect may be more lasting. "Government will cease to exist," said Jay Tate, and he and Lembit Öpik both foresaw martial law being universal in the years before the impact.

How long would it continue? In his unpublished novel *Martians*, the late Chris Boyce saw it continuing 80 years after the invasion in Wells' *War of the Worlds*; and there is the apocryphal story of the Chinese historian, asked late in the twentieth century about the most important consequence of the French Revolution, who replied, "It's rather too early to tell." The restoration of democracy may take so long that new forms of government evolve in its stead; but such questions lie beyond the scope of this book.

Part IV
The Aftermath and the Present

8. The Starseeds Grow

Sic itur ad astra – This is the way to the stars.

—Virgil, *The Aeneid* Book IX, Line 641

If we have saved the world, the time will come to take stock. If the programs discussed in the last chapter have succeeded, we will now have at least one mobile factory in space, stationed at the Earth-Moon L2 point—at least for now; an industrial lunar base; at least one solar-electric power satellite, either at L2 or already in geo-synchronous orbit; and six or more nuclear shuttles in Earth orbit, capable of taking large payloads to the Moon; and a functional inter-planetary spaceship, already in production, capable of reaching Mars with the shuttles' help, or more productively, reaching near-Earth asteroids with useful resources. And perhaps most importantly, we will have the production lines, the launch facilities and the rest of the capability needed to take industrial civilization into space.

The first steps in that direction may have been taken even sooner. On April 25, 2013, the author had two calls from BBC Scotland, to appear on radio discussing the new proposal to mine the asteroids. As mentioned previously, a new consortium of entrepreneurs, including the film producer and adventurer James Cameron (lately returned from the Marianas Trench), intends to launch exploratory probes to Earth-grazing asteroids, with a view to moving at least one into orbit around the Moon and extracting its resources.

Three important points arise (two of which we got to cover in the program). The first concerned an assessment of the possibilities. The essence was that even if the asteroid was 10% platinum, like something out of Jules Verne, it wouldn't be economical to reach it, extract it and return it to Earth. The most useful product at this stage of development would be water, which is difficult and expensive to launch from Earth. You can't compress it, and

D. Lunan, *Incoming Asteroid!: What Could We Do About It?*,
Astronomers' Universe, DOI 10.1007/978-1-4614-8749-4_8,
© Springer Science+Business Media, LLC 2014

it's heavy as well as bulky. But up there, it can provide rocket fuel and radiation shielding as well as life support, and has many other useful applications. Fueling and provisioning a space program from the top, instead of from ground level on Earth, would make an enormous difference to the economics of exploration and development.

The second major point, which the author has yet to see mentioned, is whether the consortium intends to honor the existing U. N. treaties concerning extraterrestrial resources. Under the 1967 Treaty on the Peaceful Uses of Outer Space, all such resources are stated to be 'the common heritage of mankind', and if they are exploited it should be for the benefit of all nations. Most U. N. members signed it, including Britain and the United States. Neither signed the much stronger 1980 Moon Treaty, drafted under heavy Soviet influence, but France and Canada signed it and so did most of the developing world. For years now, some groups in the United States have been calling for the nation to withdraw unilaterally from the 1967 treaty, supposedly because it's a brake on private enterprise.

Their claim is that nobody can make money out of land unless they own it. Historically this is untrue. Technically, nobody in Scotland owned any land until the Scottish government recently abolished the historical relic of 'feu duty.' Some years ago the author organized a meeting between Scottish opponents of the treaty, and Professor Angus McAllister of Paisley University, an expert on the law of leases, who politely but firmly put them straight. Extracting minerals and using other resources under license is very common practice and mining leases, especially international mining leases, is a specialized area of law in itself. In *The Moon: Resources, Future Development and Colonization*, Bonnie Cooper and her co-authors make out a detailed case for a lunar administration analogous to port authorities on Earth [2]. If Planetary Resources intend to honor the treaty, they will attract hostile criticism from many in the United States; if they don't, they can expect massive hostility from much of the rest of the world—and under the treaty, the U. S. government will be held responsible for the consortium's actions.

The third point is that if the technology to extract materials from near-Earth asteroids is developed, we then have the means to protect Earth from incoming comets and asteroids. As noted in Chap. 4, before the possible means of deflecting hazards had been thought of, our discussion group thought that industrializing the asteroids was perhaps the only answer. Just send a task force to the threatening asteroid, use up its valuable resources and release the rest as dust! Even if new methods of deflection proved effective, they're only short-term answers. It can be shown mathematically that any object whose orbit crosses Earth's must hit Earth, the Moon, Mars or Venus within a maximum of ten million years.

The mineral resources of the asteroids will be of great importance in the future. The total mass of the asteroids amounts to 10% of Earth's mass at most, but all of it is accessible, in microgravity and vacuum, allowing large-scale extraction. When we do come to industrial extraction from the asteroids, the conventional SF image is of loners, outliers of a 'Belter' civilization centered on Ceres, who go to great lengths to shun the gravitational fields of planets. But in a classic article [3], Jerry Pournelle showed that orbital dynamics require the Belters' capital to be Earth, and if they have enough power and reaction mass available to fly between asteroids, then transfers to and from the planets are easy by comparison.

The mineral wealth of the asteroids could meet the entire needs of an industrial civilization far more advanced than our present one. But there is a catch about mining them for precious metals such as gold, silver or platinum. In his contribution to the ASTRA *Man and the Planets* discussions, the late A. T. Lawton remarked that the iridium and platinum content of a typical asteroid could be worth as much as $50 billion. "What would that do to the economy if you brought it to Earth?" he asked. Chris Boyce replied, "The value would plummet," and Robert Shaw added, "… especially at the point of impact!" We hear a lot about the amount of gold, platinum, etc., in the asteroids, but many of those metals can be found on the Moon, and the first need from the asteroids will be for compounds of carbon, nitrogen and hydrogen which the Moon lacks but are essential for life support.

As we now know, the Moon formed from the superheated crust of a protoplanetary body after a grazing collision with the proto-Earth, an event that would have been visible at a distance of 400 light-years [4]. The cores of the two bodies merged to form the large nickel-iron core Earth has today, and the material that fell into orbit around Earth formed the satellite we have today.

Although Moon rock is chemically similar to Earth's crust, there are significant differences, and at first the Apollo samples indicated a complete absence of volatile compounds. The temperature the lunar material reached in the collision could be calculated from the vaporizing temperature of the minerals that remain; but all the compounds of hydrogen, carbon and nitrogen that are essential to life as we know it appeared to be missing. We now know that there are deposits of water ice at the lunar poles, from temporary atmospheres formed by impacting comets, and they will make settling the Moon a lot easier; there are also exciting signs that there may be layers of hydrated rock deep in the Moon's crust [5]. But inevitably, most of the life-support material will have to be imported—perhaps from Mars, but more probably from the carbonaceous asteroids, initially from Earth-grazers and later from the main belt. Just after that was written, in June 2013 Earth was passed at six million km by 1998 QE2, a binary asteroid of a previously unknown type—markedly red in color, like many asteroids in the outer belt, and presumably rich in organic compounds [6].

Shepherding Asteroids

Personally I would not support the idea of moving any asteroid into orbit around Earth.

—Jay Tate, Glasgow, 2012

In the nineteenth century there was considerable interest in the idea of small, asteroid-sized moons orbiting Earth. A detailed account of an imaginary voyage to one was published in 1811 [7], and Jules Verne's characters encounter one, specifically described as a captured asteroid, in the second part of *From Earth to the Moon* [8]. In the 1950s Clyde Tombaugh, the discoverer of Pluto, made a detailed search for such objects without success [9].

As an aside, in some circles it continues to be alleged that Tombaugh did find something, and the story has become confused with Prof. Ron Bracewell's suggestion that a probe from another civilization may have tried to contact us by radio in the 1920s [10]. James Strong of the British Interplanetary Society suggested that the probe might be located at either the L4 or L5 point in the orbit of the Moon [11] (not in orbit around it, as widely reported later in the press), and subsequently a strong correlation was found between 'long-delayed radio echoes' (LDE) and the movements of the Lagrange points [12].

In 1972 the author suggested a translation of the 1920s echo patterns [13], but key points in it proved to be invalid, and most of the suggestions had to be withdrawn [14], although some intriguing possibilities remain [15]. A. T. Lawton published a suggested natural explanation for LDE [16] that has been widely accepted, but leaves many issues unresolved [14]. Since 1975 the LDE effect has apparently been much less common, and two subsequent optical searches failed to find anything in the vicinity of the Lagrange points, down to the size of the Pioneer 10 spaceprobe [17].

The issue is compounded by stories about a mystery satellite found in polar orbit in 1961 and designated Black Knight, presumably by confusion with the British rocket being used at Woomera at the time [18]. Between 1959 and 1972 the United States orbited a series of Discoverer satellites, later renamed Corona, which were spy satellites returning film to Earth by parachute. The search for one such payload off Spitsbergen inspired Alistair MacLean's *Ice Station Zebra*. In December 2012, a release of previously classified documents revealed that another was lost in the Pacific near Hawaii in 1971, and recovered in 1972 using the submersible *Trieste 2*. The capsule had separated from the parachute during descent and broken up, so the recovered film was unusable [19]. In 1959 and 1961, the Discoverer 5 and Discoverer 23 capsules ended up in higher orbit, instead of returning to Earth, because the retro-rockets fired in the wrong position. It's now claimed on numerous websites that the author translated signals from Black Knight and that it was the Bracewell probe, although the decade is wrong, and the 'lost' Discoverer capsules and Black Knight were in near-polar orbit between 600 and 1,000 miles up [20], nowhere near the distance or the orbital plane of the Lagrange points.

Before Tombaugh's search, there had already been discussion of how useful such a minor moon would be, and in the late 1950s there were suggestions that small asteroids might be brought into Earth orbit for the purpose [21], although rocket motors seemed the only way to do it and would require vast amounts of propellant

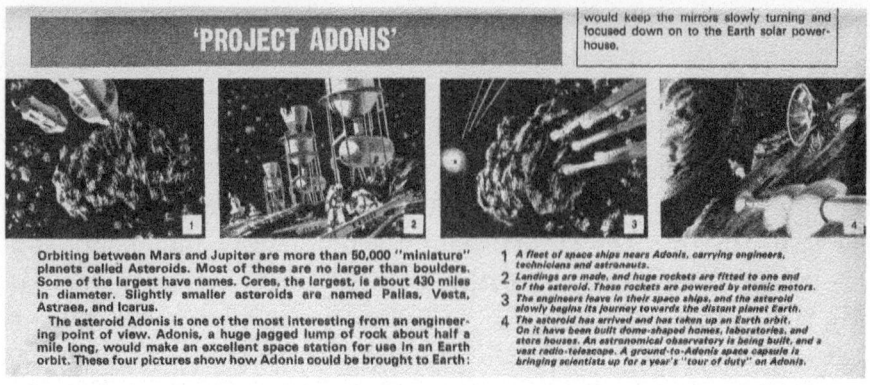

FIG. 8.1 Proposal to bring the asteroid Adonis into Earth orbit [21]

(Fig. 8.1). In 1951 L. R. Shepherd suggested that using an ion drive (again the only possibility at the time), asteroids might be hollowed out and used as multi-generation 'interstellar arks' [22] (Fig. 8.2a, b). The idea caught the imagination of BBC producer Charles Chilton, who used it in the third of his *Journey into Space* radio serials, *The World in Peril* [23]. So it struck some unexpected chords when Colin McInnes declared to us in 2011, "The future lies in shepherding, not asteroid deflection" [24].

The three-body problem applying to the Sun, Earth and an asteroid was probably exquisitely sensitive, McInnes argued, and capturing one into a high elliptical orbit should be easily possible. Von Braun had proposed exactly that maneuver for his returning ship in *The Exploration of Mars*. His crew were to be recovered using the Moonship he had designed for his second movie collaboration with Walt Disney [25]. If life was found on Mars, similar options were considered to quarantine the returning NERVA missions; and something very similar happened in 2002, when the S-IVB stage of Apollo 12 came back from orbiting the Sun to a temporary, unstable orbit around Earth [26]. An object with a mass of 10^7 g could be captured with a small change in velocity, given time. Smaller ones could be slowed into Earth orbit by aerocapture, perhaps breaking up within the Roche Limit to make exploitation easier. Sunlight pressure would sort the fragments by size. An M-type asteroid would be best for metals, including rare earths, but just water-bearing rock would be very valuable, even in low Earth orbit.

FIG. 8.2 (a, b) Converting an asteroid for interplanetary or interstellar flight (*Rocket*, 1956) [22]

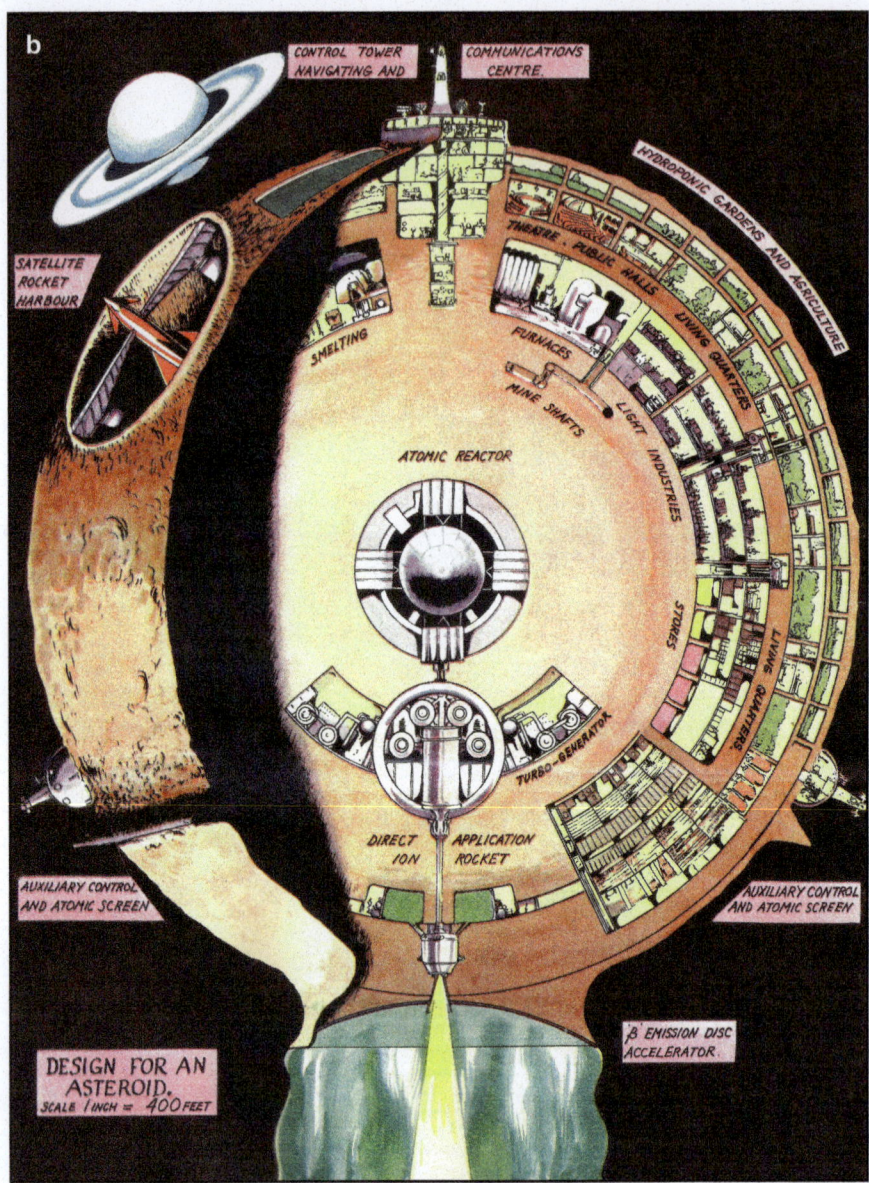

F**IG.** **8.2** (continued)

Reactions to the proposal were very mixed. Jay Tate's is quoted above. Bill Ramsay thought it was to be welcomed, as he had been saying all along that asteroid mitigation should be driven by economic motives, redirecting the military-industrial complex.

The author wondered if the protests against flybys by spacecraft carrying radioisotope generators would find a new target. Andy Nimmo suggested bringing asteroids into orbit around the Moon, to allay anxieties, but a satellite of a satellite is in an unstable configuration and would need constant station-keeping. It's thought that the Moon had several sub-satellites in its early history, whose impacts formed some of the larger features on the surface [27], and the ones placed in orbit by the later Apollo missions were very short-lived.

On further consideration, McInnes added that potential impactors were probably not the best candidates for capture, which require an orbit with a very low energy relative to Earth's, and while such an impactor might turn up, there's no guarantee that it will, while many would have higher relative energies [28].

By 2013, further study had led to still more sophisticated options. The ongoing study of near-Earth objects had divided them into the four distinct classes of Apollos, Atens, Amors and Atiras (see Chap. 2) plus several subclasses, including those with exotic orbits such as horseshoe orbits, Earth's Trojans (Chap. 2) or objects that for a short period of time naturally become weakly captured by Earth, referred to as natural earth satellites, or temporarily captured orbiters [29]. But most of those sub-categories had few known members.

From among them all, NASA had begun publishing and continuously updating the Near-Earth Object Human Space Flight Accessible Target Study, identifying potential candidates for crewed missions to asteroids, and thereby providing "an objective, quantifiable and ordered classification of objects in NEO space that allow feasible return missions." That allowed McInnes and his colleagues to propose a new subcategory of EROs—Easily Retrievable Objects—defined as "objects that can be gravitationally captured in bound periodic orbits around the collinear libration points L1 and L2 of the Sun-Earth system under a certain Δv threshold, arbitrarily selected for this work at 500 m/s." Δv, velocity change, is the key to understanding orbital dynamics in astronautics. In this case, the researchers were narrowing the asteroid study to a subclass that could be brought into the Earth-Moon system within the capabilities of existing rocket vehicles. Examples were found in all four of the main families above.

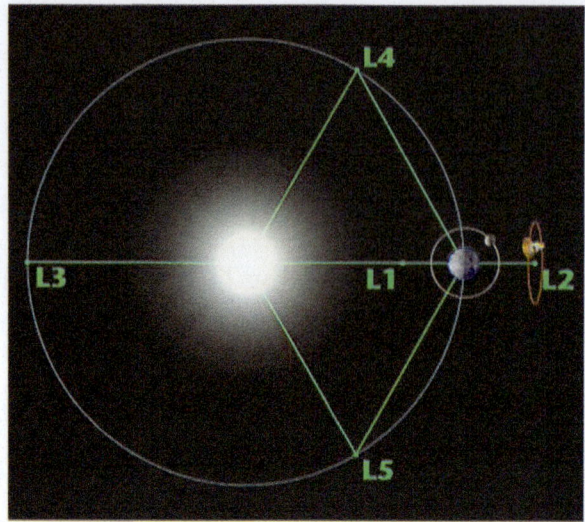

FIG. 8.3 Sun-Earth Lagrange points, with L2 halo orbit at right (NASA)

As previously explained, the five Lagrange points are special solutions of the 3-body problem in Newtonian physics. In theory, objects placed in them will remain there forever, but not in practice, because Earth and the Moon are not perfect spheres, their orbits are not perfect circles and the pulls of the other planets act on all of them. Objects like Trojan asteroids will drift around L4 and L5 points, and objects at L1 or L2 points will be perturbed away. Robert Farquhar, the architect of the ICE mission (Chap. 1), identified a class of 'halo orbits' in which a satellite could be made to move around an L2 point, with minimal station-keeping, as if it were responding to a gravitational field centered there, although the path that it followed would be the result of the interacting fields of Earth and the Moon, or Earth and Sun (Fig. 8.3).

Now it turns out that halo orbits are themselves a special case, "3-dimensional periodic orbits that emerge from the first bifurcation of the planar Lyapunov family" [29]. EROs can be nudged into them with low expenditure of energy, turning the Sun-Earth L1 and L2 points into "gate keepers for potential ballistic capture of asteroids in Earth's vicinity" from there, transfers into the Earth-Moon system via the Earth-Moon L2 point can be achieved, again with comparative ease—more easily than from the Moon, in fact, let alone from Earth's surface [29]. Consider Fig. 2.21a and imagine reversing the velocity vectors, as Robert Farquhar did when he redirected the ISEE-3 mission to become International Cometary Explorer (Chap. 1).

Twelve candidate asteroids were found, although because of their small size and the difficulty of detection there could be many more. It's estimated that only one in a million of objects in the 5- to 10-m size range have been detected to date. Multiple lunar or Earth slingshots could make more candidates accessible. 2006 RH120, 2.3–7.4 m in diameter, came top of the list for accessibility, from a close pass in 2028; twelfth in the ranking was the intriguing 1991 VG, which reflects sunlight as if it has solar panels [30]. This can't be identified with any known human spacecraft, and might—just *might*—be the hypothetical Bracewell probe, if that left Lagrange orbit in 1975 for some reason. Alternatively it might have be an object ejected from the Moon in the apparent impacts of 1178 [31], which would be equally interesting to study.

In addition to the scientific interest, however, any of them could make up to 10^{14} kg of material available for exploitation— volatiles (including propellants), metals, and bulk radiation shielding, not at all to be sneezed at, as we've seen in the previous chapter. Ten of the 12 candidates were in the 4–12 m range and could be retrieved by spacecraft comparable to the Cassini mission to Saturn (and therefore to the canceled CRAF asteroid mission).

McInnes and his colleagues went on:

Regarding the safety of such a project, there could be a justified concern regarding the possibility of an uncontrolled re-entry of a temporary captured asteroid into Earth atmosphere. A migration through the unstable invariant manifold leading towards the inner region around Earth could result in homocline or heteroclinic transits between L1 and L2, some of which intersect the planet. An active control would be required....It is however a less serious concern due to the small size of the considered EROs. Objects smaller than 5 m have a low impact energy (especially if we consider the lower velocity impacts that would result from a transit orbit when compared to a hyperbolic trajectory)....Statistically, one object of a similar size impacts Earth every 1–3 years with limited consequences [29].

That evaluation might have to be reconsidered in the light of the Chelyabinsk event. Their paper adds, "If larger objects were considered, additional mitigation methods would be required." But once the principle is established, and if it shows significant economic return, that seems almost bound to happen. The paper discusses techniques for masses of 400 and 1,500 t, and the eleventh of the 12 objects listed is 2000 SG344, much larger at 20.7–65.5 m, well up in the nation-buster range, and of such a size that it couldn't readily be deflected if passage through the 'gateway' went wrong. Aiming for lunar orbit is mentioned here, too, as a precautionary measure, but if lunar settlements have been established in the meantime, they could be at risk from moonquakes. The Moon's rigid crust rang like a bell after the comparatively minor impacts of Lunar Module Ascent stages and S-IVB stages during the Apollo program.

In his article "Just How Dangerous Is the Galaxy" [32]? David Brin noted that if there is a 26-million-year periodicity to mass extinctions, it falls within the estimated 20–30 million years, which an industrial civilization might take to fill the Milky Way (see below). Perhaps Earth has been occupied several times by settlers who took advantage of the Solar System's asteroid riches, only to wipe themselves out with misdirected asteroids due to greed, overconfidence, incompetence, or Sagan's insane tyrants. They could be here for millennia before the accident and still not leave a visible blip in the geological record. In *Man and the Planets* we suggested that there should be absolute computer overrides, under government seal, to prevent any large body being put even temporarily on a course threatening any large body or satellite [33]. To gain access to asteroid resources, we should go to them; and with the scenario we now have, after Goldilocks has been diverted, it will not be too difficult to do so.

On July 10, 2013, the issue gained immediacy with an announcement that NASA's asteroid missions using the Orion spacecraft had been brought forward. Instead of planned lunar flyby missions, the first all-up test of the Orion with its European service module would be an unmanned launch to the Earth-Moon L2 point, and the second mission in 2021 would be crewed and would rendezvous with the asteroid brought to the Earth-Moon system (see Chap. 4), raising the 10-day mission originally planned to 25 days. The rendezvous would be at L2, not at L1 as previously specified [34].

By September 2013, there was to be an unmanned mission to L2 before the Moon-asteroid one; the list of target EROs had been narrowed to three [33]; and the rendezvous might be in high retrograde orbit around the Moon instead of at L2, removing the risk that the asteroid might collide with Earth later. But flight to and from it remains at least as dangerous as Apollo 8, even if rendezvous and docking go smoothly. In the 1990s, Colin McInnes urged ASTRA's solar sail group (Chapter 5) not to incorporate more than one untried technology at a time.

However, if NASA is going to do this at all, and the target is the Earth-Moon L2 point, presumably the asteroid will come in through the Sun-Earth L2 gateway. The risk of the unmanned spacecraft with the bagged asteroid falling into an Earth-impacting trajectory is very small, and if it did happen, the incoming asteroid would come in much more gently than the Chelyabinsk one—but it's a dangerous precedent to set.

Project Starseed Reconsidered

Every time the Shuttle takes off you discover a new asteroid.... [I]t's the only type of asteroid which is known to have liquid hydrogen and liquid oxygen.

—Andrew Cutler, California Space Institute [35]

Project Starseed evolved within ASTRA, in Interplanetary Project discussions between 1978 and 1983, leading to this author's *Man and the Planets*. Its details were still being worked out when the book went to press, and a more complete, formal version was published as two papers in the *Journal of the British Interplanetary Society* [36], and a popular version in *Analog* [37], with updated versions in various magazines since [38]. The proposal took some of the favorite ideas of space advocates, such as space habitats and solar power satellites, first advocated by Gerard O'Neill [39], and rearranged them into a format that looked much more practical. Prof. O'Neill asked the author to present Starseed at his Space Studies Institute's Space Manufacturing Conference at Princeton University in 1985, but unfortunately he was so tied up with the launch of his company developing a precursor to the Global Positioning System that the feedback hoped for never took

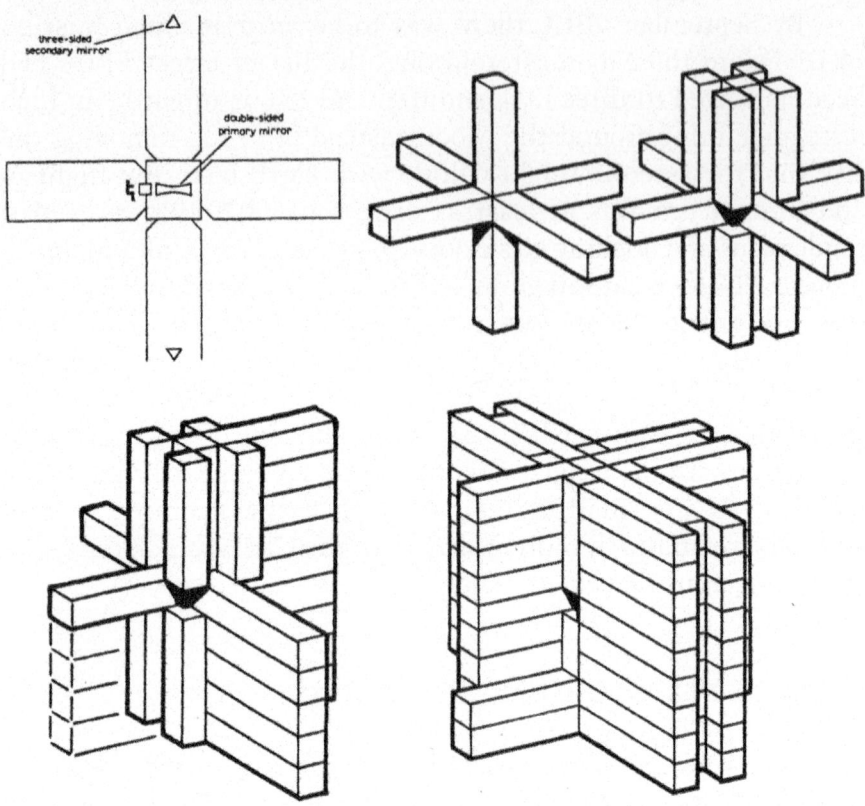

FIG. 8.4 Starseed 1 and Outgrowths to the midpoint of Outgrowth 3 (Design by John Braithwaite and Duncan Lunan, drawings © Ed Buckley, 1981)

place. (The first satellite was built, but launch delays allowed GPS to gain dominance of the market, and in the end O'Neill's satellite never flew.)

Starseed began with a proposal in 1979 by John Braithwaite and the author for a manned orbiting observatory (Fig. 8.4, top left), to be constructed from four Space Transportation System (space shuttle) External Tanks and an Aft Cargo Carrier, to go on the end of one of them, which had already been designed by Martin Marietta but was never used. At the same time ASTRA was engaged in a study of nuclear waste disposal in space at the request of the Third Eye Centre in Glasgow, and a major seminar was held on it at the Glasgow Film Theatre during the ASTRA/Third Eye exhibition called The High Frontier, A Decade of Space Research in October 1979.

The scenario that emerged was for vitrified nuclear waste to be carried to a mass driver in low Earth orbit, using a heavy-lift booster with two STS External Tanks, two main engine clusters and four solid boosters, proposed by Bob Parkinson of the BIS (Fig. 7.1) and probably flown from an equatorial launch site. We were advised on safety and other aspects by Capt. Chester Lee of the STS Office at NASA headquarters [39], and the astrodynamical calculations were performed by Archie Roy, a top expert in the field. The numbers proved remarkably good, and it seemed that the world's anticipated output of nuclear waste in the 1990s could be launched into interstellar space with one heavy lifter launch every 3 days. But even with economies of scale from mass production, the most optimistic estimates came out at 25 times the cost of burial, and could only be justified if space disposal would lead to a long-term solution of the world's energy problems.

O'Neill's scenario for that was the Solar Electric Power Satellite, or powersat, the brainchild of Dr. Peter Glaser. The standard version of it was an array of solar cells ten miles long, built of lunar materials shipped out by mass driver, stationed in geosynchronous orbit, and broadcasting 5–10 gigawatts of energy to Earth by laser or microwave beams. It would take up to 500 of them to meet Earth's anticipated energy needs in the twenty-first century, although for environmental reasons much of that usage might itself be off-planet in orbiting factories. O'Neill saw powersats as the major export of space habitats at the L4 and L5 points in the Earth-Moon system. In the 1975 Summer Study organized by NASA's Ames Research Center and Stanford University, it was concluded that powersats could be produced by year 13 of the program and the energy needs of the United States could be met by year 25 [40].

In the Starseed paper, the author compared the Stanford torus habitat to the Sea City designed by the Pilkington Group in the 1960s as a base for exploiting the resources of the North Sea [41]. When the time came to do that, what was built was not a city but a flock of bigger, stronger oil rigs, and when powersat building comes, he said, there will be no delay for building space habitats. There were 'space-rigs' (generally called construction shacks) in the space habitat scenarios, but usually they received little attention, though they would be far more ambitious than any space station then planned.

However, with the steady addition of External Tanks, in pairs to keep the structure symmetrical and avoiding the end-on couplings whose flexings are the bane of large structures in space,

FIG. 8.5 Starseed at the midpoint of Outgrowth 2 (© Tom Campbell, 1985)

the Starseed could develop through a succession of 'Outgrowths' (Fig. 8.4). In its final configuration it would have 198 tanks, 180 of them with artificial gravity, which at the maximum allowable spin rate would be one-sixth g, equivalent to gravity at the lunar surface. O'Neill had calculated that a workforce of 2,000 would be needed to build powersats, and if 100 tanks were converted to living quarters, with ten decks 4 m high, then people would be living just two to a deck, in much more comfort than on the average oil rig.

Based on figures from the Glasgow Parks Department's Methane Digester Project in the late 1970s, it seems that a single tank would be adequate for reprocessing the biological wastes of 2,000 people and returning the essential elements to the life-support system as fertilizer. Following Soviet experience with 'gardens' on their space stations, it could be expected that a lot of food will be produced by plants that will also help to purify the atmosphere— another respect in which conditions would be better than on the average oil rig.

Tom Campbell's Outgrowth 2 painting (Fig. 8.5) shows two more structural elements. In the distance a solar mirror, bigger than the Starseed itself, is being prepared for mounting on the

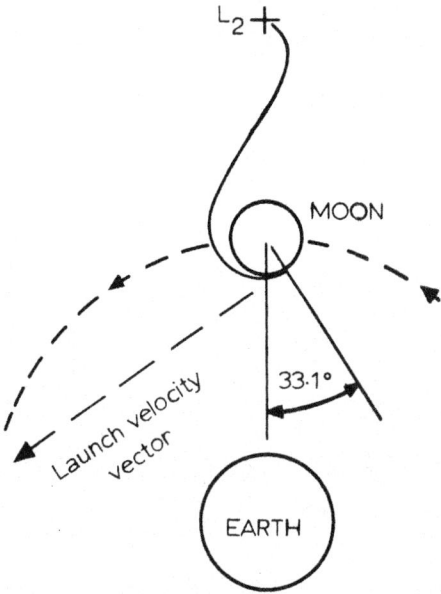

L₂

MOON

33·1°

Launch velocity vector

EARTH

FIG. 8.6 Lunar launch trajectory to L2 (© Ed Buckley, 1981, and NASA)

'top' of the structure. In Braithwaite's view the toroidal version of the Strathclyde flexible mirror would be best for the purpose. Protruding from the bottom of the structure is part of a mass driver engine, which would use ground-up Tanks as reaction mass, and would take the Starseed to the Earth-Moon L2 point.

On the lunar equator at longitude 33.1° east, near the crater Censorinus A, it should be particularly easy to launch payloads of lunar soil by mass driver to L2 (Fig. 8.6). O'Neill didn't place his habitats at L2 because the orbit around it is unstable and needs much more correction than one around L5. But the Starseed is less massive than a habitat—*much* less massive—and could hold the required halo path around the L2 point with much less expenditure of reaction mass, even after it was coated with bagged Moon rock to shield against cosmic rays. Cutting out cargo transfer from L2 to L5, and manufacturing at L2, we could achieve continuous deliveries from the lunar surface and continuous processing on board. We could also launch completed powersats into '2:1 resonance orbit' (Fig. 8.7), for a gentle delivery to geosynchronous orbit.

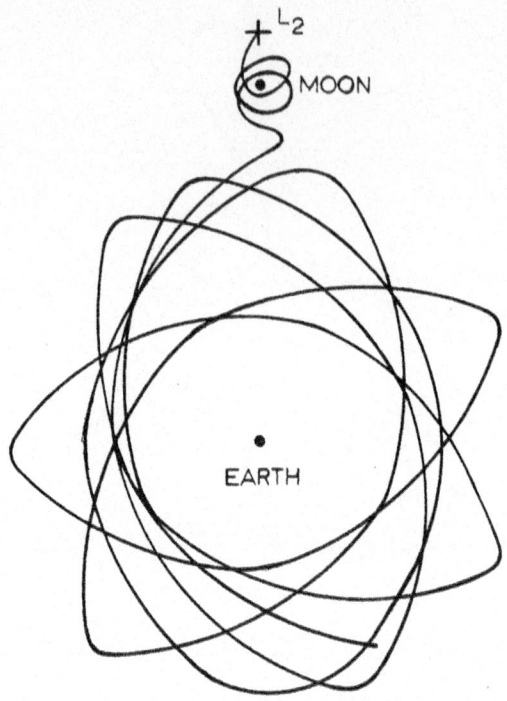

Fig. 8.7 2:1 resonance orbit from L2 to geosynchronous orbit (© Ed Buckley, 1981, and NASA)

By 1985, work at the Space Studies Institute on processing Moon rock had reached the stage where actual samples could be experimented with. A similar dispensation was granted to the Lawrence Livermore Laboratory, for reasons that will soon become obvious. Some processing systems under development, described at the SSI Space Manufacturing Conference in 1985, would melt the rock and use electrostatic fields to separate the oxygen and metals driven off as vapor. Even this, however, would be batch processing, and there's no need for Starseed at L2 to be tied to that. Braithwaite's plan was instead to vaporize incoming rock at the focus of the giant mirror, and separate the products in the hollow core of the Starseed as in a huge mass spectrometer. But if Starseed could reach the output level of a major ground-based steelworks (in this case mainly of vapor-deposited aluminum and titanium), it could build a powersat in a month. If each Starseed took a year to work up to full productivity, and the nuclear waste program was generating another Starseed each year to move out to L2, then after 9 years nearly 500 powersats would have been built—enough for all the energy needs of the world in the twenty-first century, as anticipated in the early 1980s. In that there should be enough commercial return to justify the project. On Earth, environmental and political benefits should be enormous.

In round figures, if 500 powersats were built, each massing 60,000 t (comparable to a large aircraft carrier), then their total mass would be 30 million tons. From the 1975 Summer Study figures, 530 million tons of Moon rock would have to be processed [40]. The 500 million tons of residues would be mainly silicates and iron, which can be processed as pre-stressed concrete—the ideal hull material for space settlements, D. J. Sheppard had argued [42], providing integral radiation shielding rather than the Stanford concept of an aluminum torus rotating within a static rock shield. Those 500 million tons would provide enough hull material to build 50 of O'Neill's Island One space settlements. Recovering the hydrogen, carbon and nitrogen deposits on the surface of the lunar grains from the solar wind would provide 180,000 t of volatiles, enough to make six of those settlements habitable, each capable of supporting 10,000 people.

Reconsidering Starseed in the light of the Goldilocks scenario, many of the elements would already be in place by the time the asteroid passed by in safety. We've assumed in Chap. 6 that the program is based on the Space Launch System currently in development, perhaps with Energia brought back into production, and we have at least 31 launches to make with payloads up to 250 t. Space Transportation System elements could be put back into production to meet the need, and the attraction of doing so is that it would put 60 or more tanks on-orbit—enough to complete Outgrowth 2 and move it to L2. Its 36 tanks with artificial gravity could house up to 1,000 people in comfort, surely enough to build the powersat for the De-Star laser. With the advances in robotics of the last 30 years, probably it could be done with fewer. The Starseed could be simplified still further; perhaps a merger of Fig. 6.18 and Outgrowth 2, with a wider core, opened to vacuum after inflation, and fewer habitation tanks radiating from it. It could then be built out of SLS or Energia upper stages. It will still need the mass driver engine, once it's on station in Halo orbit around L2, but now that it's smaller, the nuclear shuttles (which weren't in the original scenario) can take it there. If none of the upper stages have to be cut up for mass driver reaction mass, then, depending on the redesign and the size of the slimmed-down workforce, the Goldilocks mission launches could provide enough components to build two of them.

If the Goldilocks mission is "like the *Das Mars-Projekt* without the gliders" (Chap. 6), then this slimmed-down Starseed is like *The Exploration of Mars* or its movie version, *The Conquest of Space*. The weakness of the original scenario was its dependency on nuclear waste disposal in space, which was always going to be a hard sell, no matter how effective the safety precautions. The objection that we shouldn't be spreading our pollution up there can be countered [43], but the public fear of radiation and confusion over radioactivity were always going to be show-stoppers. Burial is the obvious strategy. It means burial for millions of years, at least for actinide wastes, but a great deal of thought has been given to that [44], including what would happen if these wastes were excavated by an impact (the answer to which is, if it's big enough to do it, what the impactor digs up is the least of your problems [45]).

Taking nuclear waste out of the equation has allowed the first conceptual redesign of Starseed in over 30 years. The processing capacity has been halved, but the big cut is in the size of the workforce. Scaling down the numbers above, but still with continuous processing, two such mini-Starseeds might build 90 powersats in 8 years, plus hull material for the first nine habitats and the needed volatiles for one of them; though if the Starseeds have proved to be as economically viable and ecologically beneficial as hoped, there may be more of them on station before then and production may have speeded up.

The habitat material would be too bulky to store in Halo orbit, without expending excessive amounts of it on station-keeping with mass driver engines, so it would have to be parked in 2:1 resonance, where the first habitat could be built robotically. Off-duty crews might work remotely on it for recreation, like a gigantic ship in a bottle. Some interesting techniques were suggested in the *Man and the Planets* discussions. One of the most promising followed a Soviet discovery that Moon rock could be spun into a fiber resembling silk [46]. That raises the possibility of spinning a multi-layered habitat like a cocoon.

The much bigger scale of the original scenario assumed a crash program, like the exploitation of the North Sea, where the need for a major new source of energy meant that solar power satellites were wanted immediately. At the moment of writing, something similar seems to be happening with the exploitation of shale oil and gas. The advantage of the crash program was that in theory, all the powersats needed could be created within a decade, and then the Starseeds would be available for other uses, in particular spreading out to Earth-grazing asteroids. The major driver for that was assumed to be providing volatiles to open up the Moon for settlement. That may still be the case if the resources at the lunar poles turn out to be *only* water, without the carbon and nitrogen compounds that are also needed. At the very least, having one or two Starseeds in hand and more in the pipeline means that we can send task forces to the asteroids, rather than bringing them to us, with the misgivings that would entail.

This brief overview and revision of Project Starseed has skipped over a number of issues, particularly the technical and political issues of beaming solar energy to Earth. But now that we are thinking of long-term missions to asteroids, taking years or even decades without the option of systematic crew rotation to Earth, the issues of habitability become crucial. Rock shielding against cosmic radiation has already been mentioned; it might be dispensed with for a short journey to an asteroid, taking months at most, but would have to be replaced as a high priority on reaching the destination. The question of recreational space is much the same as for the Goldilocks mission, but the solutions have to be more permanent.

The habitability of the living quarters needs serious study. Raising that issue at the 1985 Space Development Conference, Brian O'Leary said that a space station was not a military outpost and should not be designed along the lines of a cell, a barracks etc. "Some of my NASA colleagues think I want a Playboy pad up there just because I want a potted palm or something" [47], he said. Since the Starseed does have artificial gravity in the horizontal modules, probably there would be a lot of green plants, to provide recreation, to vary the diet and to help recycle the air and water.

As regards diet, we have to think of growing food on board. The old science fiction staple of a diet based on algae was so thoroughly discredited by the 1980s that dieticians could scarcely believe it when there was a resurgence of interest. Although they can provide useful dietary supplements, integrated systems show much better ratios of output mass to input over time [48]. Animals as well as plants will be needed, with all the complexity that entails. Interestingly one of the best subsistence diets, also helping to meet the need for more highly flavored food in space, is Mexican [49].

The key issue, which is going to have a big impact on the future of the human race outside Earth, is gravity value. The people who crew the Starseeds may now be committed to human expansion into space. If they are, hitherto their efforts have been in the one-sixth g environment of the Starseeds and the lunar surface. But they will almost certainly want more living space, so once they start shipping refined materials back to the Earth-Moon system, like their counterparts at L2 they will probably start building habitats out of the residues.

Following earlier suggestions by Tsiolkovsky and others, Gerard O'Neill's proposals for large free-flying settlements in space were initially for paired, counter-rotating cylindrical 'Island One' habitats, stationed at the L4 and L5 points of the Earth-Moon system, and even the smallest models were to be capable of supporting 10,000 people. Artificial gravity would be provided by the rotation, and the pairing of the cylinders would cancel gyroscopic effects, so the long axes could be kept pointing at the Sun and mirror systems would provide natural daylight. Conditions inside the cylinders were to be as Earth-like as possible, including standard atmospheric pressure and full simulated Earth gravity [38].

The habitats were to be built from lunar materials, without the volatiles needed for life support, and the very large quantity of nitrogen to be imported to provide normal air seemed to be the biggest problem. For his early settlements O'Neill therefore produced a less ambitious 'Bernal sphere' design. Even in that, however, the rotation rates required for 1-g simulation would be high enough to cause vertigo, and the 1975 NASA/Stanford University summer study evolved the Stanford Torus, a ring-shaped habitat a kilometer in radius, built of aluminum or steel and rotating within a stationary radiation shield of rock [40]. D. J. Sheppard later improved the concept, using pre-stressed concrete and glass as the building materials so that the radiation shield was the hull itself [42]. While these designs met all the requirements, the settlement became much more massive, anything up to ten million tons (assumed above in the calculation that 50 Island Ones could be built from 530 million tons of rock and steel).

In the Starseed scenario, the author pointed out that building and fitting out habitats will be much easier if the requirement for full simulation of Earth gravity is relaxed. In the original Starseeds, the maximum rate of rotation without nausea would be enough only to simulate lunar gravity, one sixth of Earth's. But no other solid body in the Solar System has a surface gravity more than one-third of Earth's—with the exception of Venus, whose otherwise hellish conditions rule out settlement in the foreseeable future. If we were to settle on quarter-g as a compromise, giving acclimatized occupants access to every solid world except Earth and Venus, then the slower rotation would allow Island Ones to be built in cylindrical form as O'Neill originally imagined, and would reactivate a proposal to fit those habitats with Daedalus engines and make them mobile.

Project Daedalus was the British Interplanetary Society's design study for a two-stage unmanned probe to Barnard's Star, 6 light-years away, to reach it within an adult human lifetime of 60 years [50]. The engine would run on pulsed nuclear fusion (triggered by converging electron beams) of pellets of frozen deuterium enclosing liquid helium-3. It would take 50,000 tons of propellant, 30,000 tons of it helium-3, to accelerate the second stage to its interstellar velocity of 12% of light speed (Fig. 8.8a), and the Daedalus team proposed to gather these from the atmosphere of Jupiter. The operation was to be run from Callisto, the only one of Jupiter's large moons to lie outside the planet's deadly radiation belts (Fig. 8.8b), on the assumption that human beings could approach no closer to the planet.

Man and the Planets pointed out that just 200 t of propellant would suffice to move a 50,000 tons payload from Earth to Mars in 5 days. It would take a fortnight for a metal Stanford torus at 500,000 tons to move from here to Mars for the same fuel expenditure, and five million tons (the lower mass estimate for the Sheppard design) could make the trip in less than a year.

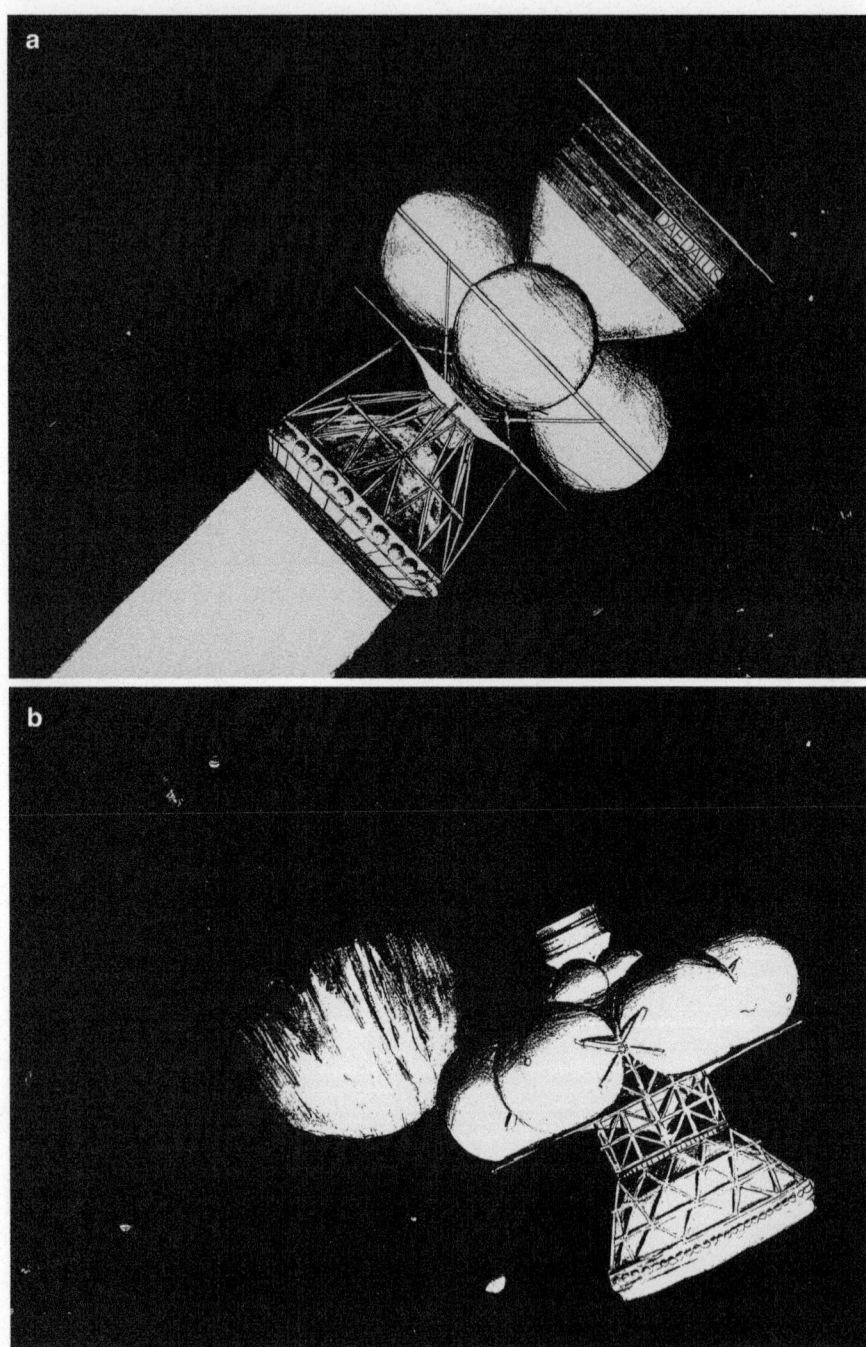

FIG. 8.8 (a) Daedalus second stage under drive (© Gavin Roberts, 1979).
(b) Two-stage Daedalus probe orbiting Callisto, Jupiter in background
(© Gavin Roberts, 1979)

When the author checked these figures with Gerry Webb of the Daedalus team, he replied, "Within the Solar System, Daedalus is pure Flash Gordon in its potential." On his visit to the Jet Propulsion Laboratory in 1986, the author was surprised to have the same figures quoted back and attributed to the Lawrence Livermore Laboratory. The Daedalus propulsion system (or a very similar variant) was then under development there [51], and initial results had been so promising that the laboratory was allowed samples of lunar soil for destructive testing. At first it was reported that the solar wind deposits on the regolith grains contained (relatively) large amounts of helium-3—enough to open up the whole Solar System for development [52]. As deuterium is plentiful in seawater, the program could be financed by importing helium-3 to Earth, if deuterium/helium-3 fusion could be adapted for energy generation.

The author wrote up the story with enthusiasm [53], but that proved premature. Although the Sun is nearly 20% helium, the helium-3 isotope is rarer, and the amount deposited on the lunar surface is a very small percentage of a very small amount. The Stafford report *America at the Threshold* revealed that 10–100 million tons of lunar soil would supply just 30 kg of helium-3 to run a demonstration reactor, if D-He^3 fusion is mastered [54]. To supply 2 t of helium-3 per year, enough to meet 10% of the United States' energy requirements, 200–300 square km of lunar surface would have to be strip mined annually—in 25 years, an area comparable to the crater Tycho and visible to the naked eye from Earth. The huge areas of the maria mined out in the movie *Moon* are not unrealistic in that respect. What's harder to believe is that apparently it's been done with just three robot bulldozers and one human supervisor.

Earth is not sufficiently massive to retain helium, and since it's not chemically reactive, any that was taken up by the planet as it formed was thought to have diffused out and escaped into space before the crust solidified. The few places on Earth where it escapes now are being fed by radioactive decay. But in 2013 a paper in *Nature* revealed a previously unknown process whereby helium can be trapped by a material called amphibole, with still better pathways perhaps provided by other hydrous materials such as serpentine and chlorite [55]. They all require water for their formation, but with evidence for a previous liquid core within Lutetia [56], and for water action on Vesta (Chap. 2), perhaps there may turn out to be helium reservoirs within some asteroids. However, useful as they would be, they can't be expected to provide the quantities needed for Daedalus propulsion.

In 1975, Prof. Gregory Matloff proposed to use O'Neill's original Island Ones as space arks, or world ships as Ed Buckley terms them, for self-supporting communities to cross interstellar space. The 50,000 t of propellant of the Daedalus probe would accelerate an O'Neill habitat to 1% of light speed, giving a 430-year voyage to Alpha Centauri or 600 years to Barnard's Star [57]. Again, the resources of Jupiter would be required; but a mobile habitat, fully shielded as it must be against galactic cosmic rays, could enter Jupiter's radiation belts with impunity to 'mine' Daedalus propellants from the planet while in close orbit—more efficiently than the Callisto based operation the Daedalus team proposed. It suggested collecting deuterium and helium-3 with hot-air balloon stations in Jupiter's atmosphere, but *Man and the Planets* opted for Waverider flying factories [36], studied later in more detail [58].

The baseline of Earth-Jupiter operations is very long, so big reserves will be required to ensure success. In Fig. 8.9 the habitat is a single O'Neill cylinder, with a counter-rotating centrifuge to train pilots for the 2.54 g above Jupiter's visible surface (Fig. 8.10). Jupiter's offset magnetic field generates a particle-free zone just above the atmosphere, rotating like an eccentric cam as the planet rotates and allowing human access to the planet below the radiation belts. Instead of the angled mirrors and rings of industrial units envisioned by O'Neill, the structure has been extended into 'wings' housing the Waverider flying factories, and shuttles—perhaps nuclear variants of Alan Bond's HOTOL or Skylon (Chap. 4)—to bring the cargoes back up.

Getting the habitat to Jupiter is the hard part. Fig. 8.10 shows an unpowered pathfinder for the flying factories, but the mission will require a Starseed pathfinder for the habitat itself. To get the habitat to Jupiter will require at least 75 metric tons of helium-3—plus what's needed for Jupiter capture and transfers between Jupiter's Galilean moons, with energy requirements comparable to those between the inner planets [3]. Gavin Roberts has portrayed it with 'wings,' but one would hesitate to try to have something

FIG. 8.9 O'Neill habitat with Daedalus engine crossing the orbit of Ganymede (*left*), with Europa ahead at right. The artist assumed that the Great Red Spot would continue to fade, as it had done in the mid-twentieth century (© Gavin Roberts, 1979, based on Voyager imagery)

that size captured by aerobraking, as in *2010* or the BBC's *Space Odyssey* [59]. A total velocity budget of 14.5 km/ps, the minimum required [60], would need at least 123 metric tons of helium-3, and to get that from the Moon you'd have to process 4.1 *billion* metric tons of regolith. Yet even if the Jupiter shuttles carry only payload equivalent to the space shuttle's, five trips from a factory to a Starseed will get the job done.

Once the habitat follows the Starseed to Jupiter and regular shipments of deuterium and helium-3 begin, the whole Solar System lies open to development in line with Krafft Ehricke's 'strategic approach,' harnessing its matter and energy resources for the benefit of humankind [61]. Mobile world habitats will soon be spread among the planets and the Asteroid Belt, then the Kuiper Belt (Fig. 8.11). Even if they were restricted to the Daedalus-equivalent 50,000 t of propellant apiece, limiting their cruise velocity to 1% of the speed of light, Matloff calculated that they could be at the furthest reaches of the Oort Cloud in 200 years. Even if none

Fɪɢ. **8.10** Mach 6 'Gothic Arch' Waverider pathfinder in the atmosphere of Jupiter, at a pressure level corresponding to the stratosphere on Earth. The ammonia plume at rear is 50 km in height (© Ed Buckley, 1975, based on Pioneer imagery)

has gone directly to Alpha Centauri from Earth (Figs. 8.12 and 8.13a, b), even if Alpha and Proxima Centauri have no envelopes of comets to provide "stepping-stones to the stars" as Isaac Asimov suggested [62], we already know that Alpha Centauri B has at least one Earth-sized planet, as Archie Roy suggested in 1967 when most astrodynamicists thought double stars couldn't have planets at all [63]. We can be at Barnard's Star, the original Daedalus target, in 600 years.

Epsilon Eridani, a sunlike star at 10.8 light-years, beckons strongly, even though it's too young to be a candidate for supporting

FIG. **8.11** Daedalus habitat in the Kuiper Belt. When this painting was created, it wasn't clear where it was located. The existence of the Kuiper
Belt had been suggested, as had the existence of binary asteroids, but both
remained conjectural. Now, not only are binary asteroids confirmed, but
the Kuiper Belt has many of them (© Gavin Roberts,1979)

intelligent life, as it was believed to be in the late 1950s at the
time of Project Ozma. It has both asteroid belts and gas giant planets (Fig. 8.14a, b), and those are the targets that will most attract
them (Fig. 8.15). Although Figs. 8.12 and 8.13 shows a habitat at an
earthlike planet, its crew will not be adapted to earthlike gravity,
and in the *Man and the Stars* discussions we learned with the help
of John Macvey how difficult it would be to integrate terrestrial
life into the biosphere of a different earthlike world [63]. Making
new ones, either by planetary engineering or creating new mobile
ones, will always be easier.

FIG. 8.12 Interstellar O'Neill habitat with Daedalus propulsion in destination system. Note that the propellant tanks have been discarded (© Gavin Roberts,1979)

Perhaps this is part of the explanation for the Fermi Paradox, in which Enrico Fermi asked, during a lunch break at Los Alamos, "Where is everybody?"—meaning not his colleagues or the cafeteria staff but the extraterrestrials who might have been expected to be present. "Everyone knew what he meant," the participants recall, but they had the advantage of prior discussion [64]. The full question is, given the relative ages of the galaxy, Earth, and the human race, would we not expect that the galactic civilization would have already been established, and have encroached on us (benevolently, for preference) long since?

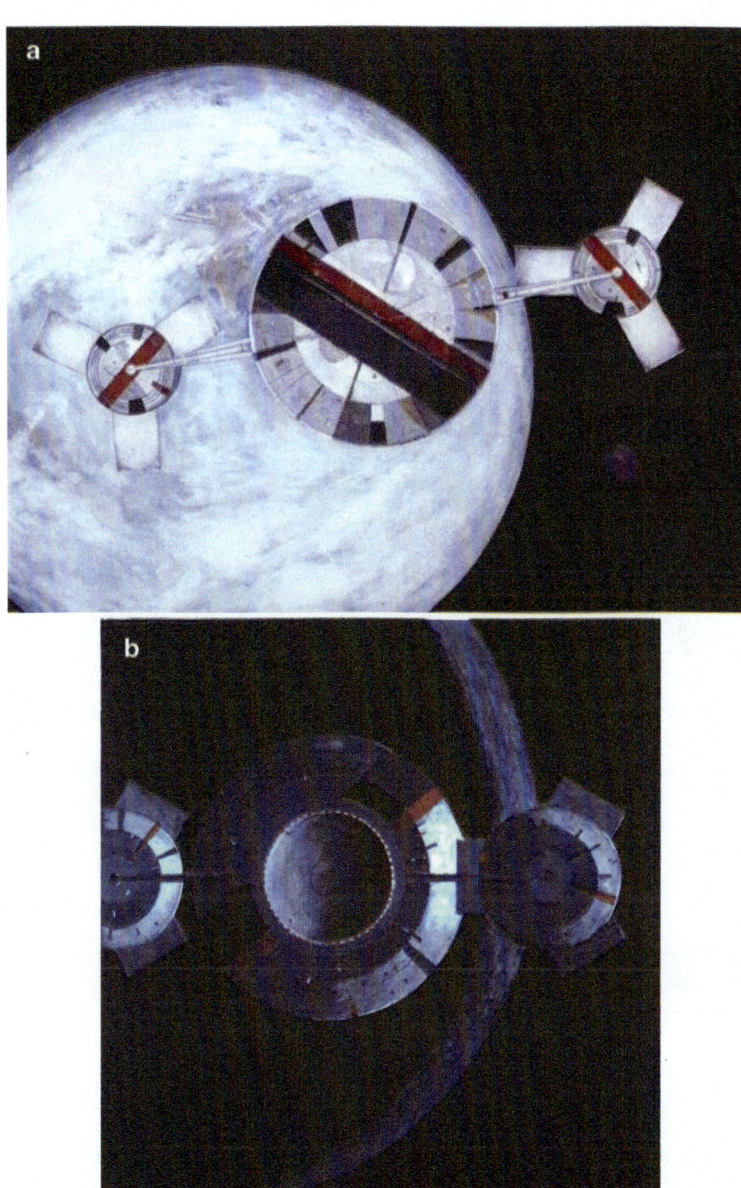

Fig. 8.13 (a) Interstellar habitat, front view, with erosion shield discarded to expose docking ports and mirrors deployed. (b) Interstellar habitat, rear view, showing Daedalus engine (© Gavin Roberts, 1979)

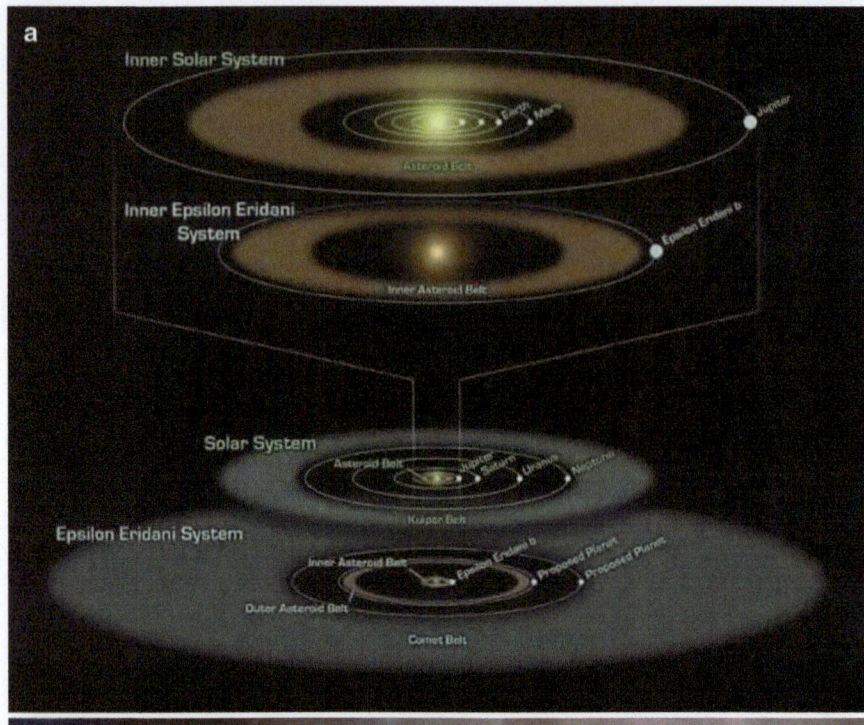

FIG. 8.14 (a) The Epsilon Eridani planetary system compared to our own (NASA). (b) Artist's impression of the Epsilon Eridani system (NASA)

Until recently, the answer you preferred depended on the optimism you felt about the human race itself; about the feasibility of interstellar travel; or about the rarity of life, intelligence and technology. In the last category, the numbers are nowadays

FIG. 8.15 A gas giant planet with airless moons (© Ed Buckley, 1972, for *Man and the Stars*)

outwardly more favorable to Fermi's argument. We now place the age of Earth at 4,600 million years, with primates established on it for at least ten million, while in the disc of the galaxy there are Population I stars like our Sun but up to twice as old (Population I stars formed out of interstellar clouds enriched by supernova explosions, taking up the heavy elements necessary for the formation of Earth-like planets and the evolution of life. Population II stars [discovered later] are older still, but when they formed, to make up the galactic nucleus, halo and global clusters, the only building materials available were hydrogen and the helium formed in the Big Bang).

Those who believe that interstellar travel is possible believe that the galaxy could be thoroughly explored in 10–30 million years. If others attempted it, it could well have been accomplished long since. The author believes in the Zoo Hypothesis, or as Martyn Fogg calls it, The Interdict Hypothesis, that life-bearing worlds are protected and left to go their own way [65]. That status would be much easier to award if the grantors had no wish to occupy life-bearing worlds in any case.

And if we are alone in the galaxy, or at least the first spacefaring species, then in 20–30 million years' time the two waves of human expansion will meet on the far side of the Milky Way from here. They will meet as aliens, because they will have evolved differently on the way around. After we have sorted out the problems *that* causes, we can decide what we want to do next.

Back to the Comets

> Though fast rooted [trees] travel about as fast as we do. They go wandering forth in all directions in every wind, going and coming like ourselves, traveling with us around the Sun two million miles a day, and through space heaven knows how fast and far!
>
> —John Muir [66]

In the excitement of deflecting Goldilocks, and the possibilities that generates for the future, it's easy to forget that there are other kinds of impactors still out there. With continuing improvements in searches, especially in follow-ups of sightings, we may hope that all the current potentially threatening ones will be identified. Problems beyond the scope of this book, such as bigger asteroids, more difficult orbits, rubble piles or metallic asteroids, can be dealt with by the same methods, or by the longer-term ones. Once the catalog of PHAs is complete, we only have to watch for new additions to the list—perturbed into new orbits by Jupiter, or products of new collisions.

Comets are a different matter. They can be very large, they can come from any direction, including retrograde orbit like Halley's and from high inclinations to the ecliptic, and at much higher speed than asteroids. The few that have been visited so far show a wide range of surface conditions, and their interior structure may be equally variable. They can be periodic, belonging to the families of any of the major planets, or very fast visitors from the Kuiper Belt or the Oort Cloud, showing up once in tens or hundreds of thousands of years. They can break up comparatively harmlessly, like most but not all Sungrazers do, or threateningly like Shoemaker-Levy 9 or the 1178 impactors.

Ross's assumption that his original Solaris could deflect comets as they crossed the Moon's orbital distance now looks overly optimistic, but more distant rendezvous may be difficult to achieve. In his strategic approach paper, Ehricke categorized the four possible types of interplanetary transfer as BAT—Brief Acceleration Transfer, as with chemical fuels or NERVA; CLAT—Continuous Low Acceleration Transfer, like ion drive or solar sail; CHAT—Continuous High Acceleration Transfer, like Orion or Daedalus; and PAT, Partial Acceleration Transfer, which is CLAT or CHAT without propulsion continuously used [61]. BAT and CLAT will reach only periodic comets, with years or preferably decades of warning. Only CHAT or PAT with high accelerations can achieve rendezvous with comets in the more difficult cases.

Even then, if the object is tens of miles in diameter, it's not clear what can be accomplished. Even with a smaller one, unless it has a single pole of rotation, creating a jet might be ineffective because once triggered, it will tend to keep firing under the action of sunlight as the nucleus rotates, and the effect may be canceled out, as seems to be the case (at present) with the jets from Swift-Tuttle. Arthur Hodkin's advice, to hope that it won't happen for centuries or millennia and that *Star Trek*-level technology will be available someday, no longer seems so absurd.

Meanwhile, if we can use asteroids for interstellar travel, what about comets? Isaac Asimov suggested that, too, in his article "Steppingstones to the Stars" [62]. The only asteroids that could be used would be chondrites or carbonaceous chondrites, the ones that contain the necessary elements for life support; but comets have them in even greater abundance, so much so that Sir Fred Hoyle and Chandra Wickramasinghe argued that life originates in comets and then seeds planets. A sustained effort to establish a settlement on the nucleus of Halley's Comet features in the novel *Heart of the Comet*, by Gregory Benford and David Brin [67], which should strike some notes of caution. Isaac Asimov pointed out that a body of that size, hollowed out as in Fig. 8.2b, could hold a human population the size of Earth's [68], and John Macvey pointed out the difficulty of integrating terrestrial life into an alien biosphere, as noted above. Benford's and Brin's settlers find the interstices of the comet filled with highly evolved 'Halleyforms,' which react to the heat of their presence as to an unexpected perihelion summer, and because of a distant common origin, find people highly edible. Under that pressure the human settlement splits along religious, racial and political lines, and within 30 years the settlers are at war with one another—a long way from the 'steppingstones to the stars' Asimov envisioned [62], though they get there in the end.

Freeman Dyson (the originator of Project Orion) suggested that genetically modified trees could be grown on comets, to such size that their foliage could absorb enough energy to sustain them even at the distances from the Sun of the Kuiper Belt or the Oort Cloud [69]. Although the incoming energy from starlight would be very low, the collecting area would be as large as a continent, and it could sustain plant life over an interstellar journey. A David Hardy painting of it appears in Ian Ridpath's *Worlds Beyond* [70], and the author drafted a story about it for Sydney Jordan's *Lance McLane* strip, syndicated worldwide as an alternative version of Jeff Hawke (Fig. 8.16a, b) [71].

The trees could be bred to extract metals from the comets and concentrate them so that cuttings could be sold on return to the Sun—at which point the trees themselves, adapted to live in starlight, would presumably need sunglasses. Proper industrial forestry and coppicing of such a tree world would be a job for a full space habitat workforce. In *I Talk to the Trees* the author made his trees sequoias or redwoods, drawing heavily on *Redwoods, the World's Largest Trees* by Jeremy Joan Hewes [66].

Conifers evolved around 200 million years ago and dominated Earth for 30 million years, the Age of Conifers, during the Jurassic (100 million years ago). They therefore lived with and outlasted the dinosaurs. Pterodactyls and Brontosauri probably inhabited the first sequoia forests. The much reduced coverage of redwoods and sequoias in North America now is due to the effects of the recent ice ages (during the last million years). Bristlecone pines can live up to 4,600 years, though they aren't the biggest ones; those are the giant redwoods of the California coast and the giant sequoias

Fig. 8.16 (a, b) Growing trees on comets ('I Talk to the Trees,' Lance McLane strip, *Daily Record*, story by Duncan Lunan. Drawn by and © Sydney Jordan, 1984)

inland, which can live 3,500 years. The largest surviving sequoia was the General Sherman Tree, Sequoia National Park, which was 272 ft high, producing 40 cubic feet of new wood per year, equivalent to growing from a seedling to 50 ft high in a single year. Trees on comets would need to be fast-growing, as they recede from the Sun, to work up enough area of foliage to survive thereafter; and of course, they'd have to be evergreens. In 1981 genetic modification of redwoods was already under way to try to improve commercial forestry.

The heartwood of a giant redwood is inert, providing structural strength to the tree while the surrounding sapwood carries up water and nutrients. If the tree is of sufficient size it can remain standing if the core is destroyed by fires, termites, etc., so the cores are unnecessary in the low gravity of a comet nucleus and can be removed to provide interior living space. Carpenter ants can burrow 20 ft into a tree to make chambers for their young; perhaps they could be modified to burrow outwards to make habitable interiors within the trees.

Plant leaves can seal their pores and stop transpiration at temperatures over 80° F or below freezing, so leaves in a vacuum, far from the Sun, wouldn't transpire but would still collect energy. But there still has to be a process to enable them to manufacture nutrients using the energy collected, probably in symbiosis with other life forms such as vines or ivy inside the trunks—unless the trees can be modified also to produce internal foliage! To maintain that symbiosis there would need to be other creatures inside, as well as humans. Redwoods harbor raccoons, cougar, owls, elk, squirrels, bobcats, coyotes, martens and black bears, so *I Talk to the Trees* turned their whole ecology inside out, putting humans, bats and many other creatures inside the hollow trunks and roots, with bioluminescence provided by fungi.

Squirrels are very important to sequoia propagation by gathering and burying cones, so the humans on the comet would have an analogous role, in a more directed way. The coast redwoods can send up sprouts vertically from their lateral root system, at a suitable distance from the parent trees. This trait will allow the trees to spread in rings over the nucleus of the comet, and give the humans access from one tree to another through the root system.

The interconnectedness of the trees on the comet would inevitably lead to a form of sentience. Trees already demonstrate the trait to some extent, evolving collective defenses and chemical warning systems against predators [72]. For example oak trees produce tannin and phenol to repel attacking caterpillars, and that alerts surrounding trees when the first one is infested [73].

Climbing plants demonstrate an almost paranormal awareness of their surroundings, growing across gaps and kinking at just the right place to stabilize themselves [74]. Prof. Malcolm Wilkins of Glasgow University has demonstrated that plants can also see, hear, speak, tell the time, predict the future and count in their own ways. For instance, the Venus fly trap doesn't close until a *second* touch on its sensors [75]. They also make decisions [76], as in "How does a tree climb a hill? It drops a seed up-slope." David Attenborough demonstrated many equally remarkable feats in *The Secret Life of Plants* [77]. The oldest known tree is an Arizonan bristlecone pine of 4,700 years, and a U. S. huckleberry is 13,000 years old, but the oldest living plant, which reproduces by cloning itself, is a King's Holly in Tasmania that is 43,000 years old [78].

In pioneering experiments in 1919, the U. S. Army's Chief Signal Officer George O. Squier discovered that eucalyptus, poplar, pine and oak all functioned as radio receivers. "From the moment an acorn is planted in fertile soil, it becomes a 'detector' and a 'receiver' of electromagnetic waves, and the marvelous properties of this receiver, through agencies at present entirely hidden from us, are such as to vitalize the acorn and to produce the giant oak." The 'entirely hidden' secret is that salts in the water of the sap make it electrically conducting.

U. S. army engineers used trees as jungle antennae in Vietnam. According to Dr. Mario Grossi of Raytheon, "A tree is just an antenna with an inherently broad bandwidth....You can move the wire up and down the trunk until you find a reasonably good match." A satellite expert in the Indian Space Research Organization, Shiv Prasad Kosta demonstrated 'fair quality' TV reception with coconut, mango, date palm and other trees with large leaf areas [79]. In *Man and the Planets* the author suggested that branches and foliage could be wired up as communication systems of great size, listening for extraterrestrial signals, far from the radio noise of the Sun and the planets, scanning very low frequency waves like the semi-organic system suggested by Arthur C. Clarke in *Imperial Earth* [80], also detecting gravitational waves and any signals in them. Kosta suggested linking up the entire Amazon Valley to search for ET signals [79].

Thinking in terms of really long interstellar voyages, the trees could evolve the equivalent of a central nervous system, and this might be provided by something analogous to the giant fungi that are the largest living organisms on Earth. Even a 'fairy ring' weighs half a ton. The first of the giants was found in Michigan, 1,500 years old, occupying 37 acres and weighing 100 t [81], but it was outdone by one of 1,500 acres in Washington State [82]. Fungi 'talk' to each other by chemical signals through their underground networks of tendrils [83].

As with asteroids, there are comets to make use of in other planetary systems. For example, Tau Ceti, one of the nearest Sun-like stars to us, is ten billion years old, twice the age of our Sun, and Britain's UKIRT infrared telescope on Hawaii has discovered that it has a huge entourage of comets, at least 20 times as many as we have [84]. If Hoyle and Wickramasinghe are right, or if some other civilization has reached Tau Ceti, who knows what might be going on out there?

Even within our own Solar System, Dyson foresaw the comet and asteroid cultures evolving in radically different ways, which he characterized as 'green technology' and 'gray technology,' respectively [69]. Both have their parts to play in the future of humanity, and the asteroid-based Mobile World culture will spread to the stars more rapidly. But on timescales of tens of millions of years there is no need for conflict between the two approaches. The Mobile World dwellers will always know how to settle comets, and there will always be people to whom the alternative ways appeal.

Dyson believed that the future lay ultimately with them, entitling the relevant chapter in *Disturbing the Universe* "The Greening of the Galaxy." As an example of how the two technologies can support each other, in the final chapter of *Man and the Planets* we discussed building solar systems in barren parts of the Milky Way, and described it as the work of galactic gardeners rather than industrialists. Once created, new solar systems would be left to their own devices in hopes of interesting outcomes. The late Iain M. Banks said that at such a level, the test of whether something should be done would not be, "Is it profitable? Is it practical?" or even "Is it dangerous?" but rather, "Is it entertaining?"

In one of his bleaker moments, Dyson wrote, "Intelligence may be a cancer of purposeless technological exploitation, sweeping across a galaxy as irresistibly as it has spread across our own planet" [85]. Nikolai Kardashev characterized three types of civilization: the first would control the matter and energy resources of a planet, the second of a solar system, the third of a galaxy [86]. Dyson's own concept of the Dyson sphere, breaking up the planets of a solar system to form either a solid shell around the star [87] (apparently impossible), or a shell of asteroids [88], is often taken to be the ultimate example of a Kardashev 2 civilization. But our discussions concluded that such a culture was not in control of its resources: Malthusian pressures had forced it to destroy diversity and use everything up. As Bill Ramsay said, "Going to a singularity is always dangerous", and it may be no coincidence but an intuitive judgment that all Dyson Sphere stories known to the author (including his own) [89] have portrayed Dyson civilizations as collapsed.

To attain Kardashev 2 status, in our view, a spacefaring civilization had to adopt a conservationist attitude before even aspiring towards Kardashev 3 level. Looking out at the Milky Way, we see no evidence of the 'purposeless technological cancer,' not even the wreckage of its failures as in Clarke's The City and the Stars [90]. If others are out there ahead of us, making the fullest uses of asteroids and comets, the galaxy may already be greener than we suppose.

9. Keep Watching the Skies

On 18th September 2000 two extraordinary things happened. First the Minister for Science and Technology, Lord Sainsbury, launched the report of the government's Near Earth Object Task Force at a press conference in London. The report verified the impact threat to the UK, and made 14 substantive recommendations for government action. Secondly, nobody laughed at him.

—Jay Tate, lecture abstract, Charterhouse 2001

At the same event Colin Hicks, director of the British National Space Centre (now the UK Space Agency) added that he had written to other European governments requesting cooperation, and after the initial response 'Is this a British joke?', reaction had quickly become positive. "May I congratulate the UK... how different from what we have come to expect..." The *Washington Post* reported, "The Brits picked up the ball NASA has dropped." All 14 recommendations had been adopted, though with reservations regarding the policy on mitigation, i.e., asteroid deflection. ESA's Science Policy Committee had taken small but positive steps in the right direction, and several nations were setting up their own studies. The recommendations were being worked into discussions on missions to the asteroids, BNSC was pursuing the issues in all the forums named in the Response section of the report, and it was getting positive responses throughout [1].

It was not to last. By 2003, Jay Tate reported, only one of the 14 recommendations had been implemented, and that was to set up a UK NEO information center. It wasn't assigned to Spaceguard UK, as one might have expected, but to the new space museum at Leicester, with a budget of £356,000. The remit ended when the British National Space Centre became the UK Space Agency, with nothing about the impact threat in the new agency's charter. But in 2013, the UK Spaceguard Centre finally became the official

D. Lunan, *Incoming Asteroid!: What Could We Do About It?*,
Astronomers' Universe, DOI 10.1007/978-1-4614-8749-4_9,
© Springer Science+Business Media, LLC 2014

UK NEO Centre—without funding, so far, but perhaps with more prestige as a center where actual search work is being done.

Discussing the implications in 2012, Gordon Ross asked if the impact issue should be addressed in terms of disaster relief. It was in fact covered in the keynote speech of an International Disaster and Relief Conference at the Hague, arousing interest, but it seemed the topic was too big for action. The insurance question, which had aroused discussion in the early 2000s, had produced nothing. The insurance industry was no longer interested. "Awareness has moved backwards," said Jay, and again the 9/11 events were largely responsible. Re-insurers were interested in preventative measures, but would need government briefing before anything was done.

Bill Ramsay remarked that the 2010 UK government Defense & Security review had stressed the need to manage risk, including providing resilience for the UK to recover from 'shocks.' Should impacts not be in there as potential threats? Human security is not in the same portfolio as national security, but surely it should be. Maybe, Tate replied, once we are out of Afghanistan and other such burdens, there will be a reappraisal. In the United States the responsibility for dealing with the threat still rests with the Space Command, as discussed in Chap. 4. In 2003 the Space Command was 'desperate' for UK involvement, but the Ministry of Defense did not respond, and a French delegate took the UK's intended place. As of 2012, the MoD still did not recognize any responsibility to protect the UK in this area.

In the same year, however, Tate put us in touch with Prof. Richard Crowther at Oxford, currently chief engineer at the UK Space Agency, former head of the Space Technology Division at the Rutherford Appleton Laboratory, with primary research interests in manmade orbital debris, planetary protection, and near Earth objects; head of the UK delegations to the Inter-Agency Debris Committee and the U. N. Committee on Peaceful Uses of Outer Space, and previous chair of the U. N. Working Group on Near Earth Objects. Crowther also leads the UK delegation to ESA's International Relations Committee and Space Situational Awareness Program Board (see Chap. 4).

Crowther explained that although it wasn't overtly explained in policy documents, the UK strategy is and was to escalate the issue to international level because:

i) It is a global issue. NEOs do not recognize national boundaries, nor are their effects likely to be restricted to one country or region;

ii) Localized solutions are inappropriate, as international response is needed to share resources to identify and counter the threat of NEOs (especially in the current financial environment);

iii) the action of deflection must necessarily involve all nations of the world in deciding how/when/who to respond as the consequences of action or inaction are critical to all.

> So although you will not find a document which identifies UK strategy/plan for responding to NEOs, our objective is to harness/engage the international community and contribute/support where we can most effectively play a role (hence the importance of the EU funded NEOShield program which has significant involvement of UK industry/academia). We also plan (as we drafted the original proposals) to support the frameworks envisaged for IAWN (detection and tracking), MPOG (deflection missions) and MAOG (oversight) envisaged by the ASE DPATM proposals. [See below.] The final report of the UN NEO group in 2013 will trigger that process of states establishing and subscribing to the different frameworks (scientific, technical, political) to address NEOs [2].

Big things were indeed about to happen. At the Third United Nations Conference on the Exploration and Peaceful Uses of Outer Space (UNISPACE III) in 1999, a resolution was passed setting up an Action Team on Near-Earth Objects, under the Scientific and Technical Subcommittee of the Committee on the Peaceful Uses of Outer Space (COPUOS). Its initial remit was to:

(a) Review the content, structure and organization of ongoing efforts in the field of near-Earth objects (NEOs);

(b) Identify any gaps in the ongoing work where additional coordination is required and/or where other countries or organizations could make contributions;

(c) Propose steps for the improvement of international coordination in collaboration with specialized bodies [3].

After submission of an interim report and draft recommendations in December 2011 [4], the following year saw significant progress towards international agreement. The proposals were reviewed in further sessions on February 13–16, 2012, culminating in their adoption by the Subcommittee.

Only days later, on February 25, following a meeting hosted by ESA in October 2010 [5], draft terms of reference were compiled for a Near Earth Object Threat Mitigation, Mission Planning and Operations Group (MPOG), whose brief was to:

1. Recommend and promote key research required for planetary defense. Such investigations can be addressed through NEO observations, computer simulations, laboratory research, and deep space missions;

2. Identify research opportunities for international collaboration on technologies and techniques for NEO deflection. This will help avoid costly duplication of effort, and speed the development of an effective deflection capability;

3. Develop and adopt a set of reference missions addressing a variety of potential NEO impact scenarios and deflection/disruption possibilities. These reference missions will facilitate accurate technical planning and provide a basis for mitigation campaign cost estimates;

4. Develop decision and event timelines for a variety of potential Earth impactors and trajectories identified for mitigation campaign analysis;

5. Evaluate technical maturity and technical consequences of deflection techniques;

6. Recommend to the appropriate authorities, in collaboration with the Information, Analysis and Warning Network (IAWN), criteria and thresholds for action (e.g., notification of a significant impact risk, initiation of observation and/or mitigation campaigns);

7. Recommend a minimum acceptable Earth-miss distance and other criteria for deflection targeting;

8. Recommend operational responsibilities for a mitigation campaign;

9. Coordinate with the relevant actors involved in the implementation of the threat response;

10. Recommend for detailed review any legal issues (e.g., liabilities) that may arise in undertaking NEO mitigation actions or selecting any likely mitigation option;

11. Communicate its activities to the international community;

12. Offer a yearly briefing to the U. N. Committee on the Peaceful Uses of Outer Space (UNCOPUOS) on the status of these activities.

A year later, at the close of the 50th session of the Subcommittee, the report was endorsed [6]. An international asteroid warning network (IAWN) would be established by linking together the institutions that are already performing many of the functions, including discovering, monitoring and physically characterizing the potentially hazardous NEO population, and maintaining an internationally recognized clearing house for the receipt, acknowledgment and processing of all NEO observations. IAWN would also recommend criteria and thresholds for notification of an emerging impact threat, as well as a strategy using well-defined communication plans and procedures to assist governments in their response to predicted impact consequences.

It was further recommended that a Space Mission Planning Advisory Group (SMPAG) should be established by those Member States of the United Nations that have space agencies. This group, of representatives of spacefaring nations and other relevant entities, would set the framework, timeline and options for initiating and executing space mission response activities.

The recommendations were put before the whole Committee on the Peaceful Uses of Outer Space, at its fifty-sixth session from June 12–21, 2013, where the proposals were approved, to be sent to the U. N. General Assembly in October 2013 for review and adoption. The text is sufficiently important to merit printing here in full.

United Nations General Assembly
Committee on the Peaceful Uses of Outer Space
Scientific and Technical Subcommittee
Fiftieth session
Vienna, February 11–22, 2013
Draft Report of the Working Group on Near-Earth Objects

1. Pursuant to paragraph seven of General Assembly resolution 67/113, the Scientific and Technical Subcommittee, at its fiftieth session, reconvened its Working Group on Near-Earth Objects under the Chairmanship of Sergio Camacho (Mexico). The Working Group held seven meetings, from February 15–22, 2013.

2. In accordance with the multi-year work plan under the item on near-Earth objects (NEOs) (A/AC.105/987, annex III), the Working Group reviewed the following items:

   (a) Consideration of the reports submitted in response to the annual request for information on NEO activities and continuation of intersessional work;

   (b) Review of progress on international cooperation and collaboration on NEO observations and on the capability for the exchange, processing, archiving and dissemination of data for the purpose of NEO threat detection;

   (c) Finalization of the agreement on international procedures for handling the NEO threat and engagement with international stakeholders;

   (d) Consideration of the final report of the Action Team on Near-Earth Objects;

   (e) Review of progress made in activating the work of an international asteroid warning network and the mission planning and operations group, and assessment of their performance.

3. The Working Group heard the following scientific and technical presentations:

   (a) "Report of the Action Team on Near-Earth Objects: recommendations for an international response to a NEO threat," by the Chair of the Action Team on Near-Earth Objects;

   (b) "NEO threat detection and warning: plans for an international asteroid warning network," by the representative of the United States;

   (c) "Mitigation of the NEO impact threat (NEOShield)," by the representative of Germany;

   (d) "Recommendations of the Action Team on Near-Earth Objects for an international response to the Near-Earth Object impact threat," by the representative of the United States and the observer for ESA;

   (e) "Fly-by of 2012 DA14: preliminary results," by the representative of the United States;

   (f) "Chelyabinsk event of February 15, 2013: initial preliminary analysis," by the representative of the United States.

4. The Working Group had before it information on research in the field of Near-Earth objects carried out by Member

States, international organizations and other entities (A/AC.105/C.1/106).

5. The Working Group noted that, during the current session of the Subcommittee, technical presentations had been given on close-approaching asteroids, new missions to asteroids to learn about their nature and composition, and the recommendations of the Action Team on Near-Earth Objects for an international response to the threat of a NEO impact on Earth. To elucidate the recommendations, technical presentations were made on plans for an international asteroid warning network, on mitigation capabilities being developed by space agencies and international consortia to respond to an asteroid threat, and on functional aspects of the international coordination needed among space agencies for planning and operating mitigation campaigns in case of a NEO impact threat.

6. The Working Group was informed that in 2012 the intersessional work of the Action Team on Near-Earth Objects had been carried out: (a) on the margins of the 55th session of the Committee on the Peaceful Uses of Outer Space; (b) in a workshop to provide information to the Action Team on the international analysis of the potentially hazardous asteroid known as 2011 AG5; (c) in a teleconference of representatives of entities that could form an international asteroid warning network; and (d) through electronic correspondence. The Working Group noted that a second meeting of representatives of space agencies was held on the margins of the 55th session of the Committee to discuss the terms of reference for the establishment of a space mission planning advisory group. The Working Group would offer recommendations for consideration by member States.

7. The Working Group had before it the final report of the Action Team on Near-Earth Objects (A/AC.105/C.1/L.330), which contained current knowledge on the structure and organization of ongoing efforts in the field of NEOs, including the number and size distribution of NEOs that had been found. The report also identified gaps in ongoing work where additional coordination was required and/or where member States or organizations could make contributions.

8. The Working Group also had before it the recommendations of the Action Team on Near-Earth Objects for an international response

to the near-Earth object impact threat (A/AC.105/C.1/L.329). The Working Group noted that the report contained a summary of the findings on which the Action Team had based its recommendations for a coordinated international response to the NEO impact threat.

9. The Working Group noted that there were three primary components of threat mitigation: (a) discovering hazardous asteroids and comets and identifying those objects requiring action; (b) planning a mitigation campaign that included deflection and/or disruption actions and civil protection activities; and (c) implementing a mitigation campaign, if the threat warranted it. The Working Group emphasized the value of finding hazardous NEOs as soon as possible in order to better characterize their orbits. This would help to avoid unnecessary NEO threat mitigation missions or facilitate the effective planning of missions, should they be deemed necessary.

10. The recommendations that follow are meant to ensure: (a) awareness among all nations of potential threats; (b) the coordination of civil protection activities by nations that could be affected by an impact, directly or indirectly; and (c) the design and coordination of mitigation activities by those which might play an active role in any eventual deflection or disruption campaign.

11. Upon consideration of the two reports referred to above, which were presented by the Action Team, the Working Group recommended that the following actions should be taken:

(a) An International Asteroid Warning Network (IAWN), open to contributions by a wide spectrum of organizations, should be established by linking together the institutions that were already performing, to the extent possible, the proposed functions, including discovering, monitoring and physically characterizing the potentially hazardous NEO population and maintaining an internationally recognized clearing house for the receipt, acknowledgment and processing of all NEO observations. Such a network would also recommend criteria and thresholds for notification of an emerging impact threat.

(b) The international asteroid warning (IAWN) network should interface with the relevant international organizations and programs to establish linkages with existing national and international disaster response agencies to study and plan response activities for potential NEO impact events as well as to recommend strategies using well-defined communication plans and procedures to assist governments in their response to predicted impact consequences. This does not limit the possibility of organizing, in this respect, additional international specialized advisory groups, if necessary.

(c) A space mission planning advisory group (SMPAG) should be established by Member States of the United Nations that have space agencies. The group should include representatives of spacefaring nations and other relevant entities. Its responsibilities should include laying out the framework, timeline and options for initiating and executing space mission response activities. The group should also promote opportunities for international collaboration on research and techniques for NEO deflection.

12. The groups recommended above should have their work facilitated by the United Nations on behalf of the international community.

13. The Working Group recommended that the Action Team on Near-Earth Objects should assist in the establishment of IAWN and SMPAG. The Action Team should inform the subcommittee of the progress in the establishment of both groups. Once established, IAWN and SMPAG should report on an annual basis on their work.

14. The Working Group agreed that all recommendations contained in the present report should be implemented with no cost to the regular budget of United Nations.

15. The Working Group encouraged the specialized agencies of the United Nations, member States and their institutions to follow near-Earth object developments on a regular basis, for example via the following web pages: http://neo.jpl.nasa.gov, www.jpl.nasa.gov/asteroidwatch, http://neo.ssa.esa.int and http://neoshield.net.

16. At its seventh meeting, on February 22, 2013, the Working Group adopted the present report.

After all that's been said in the previous chapters about what needs to be done, what isn't being done and what should be done if the crisis comes upon us, it has to be said that these are excellent developments. The Action Team's remit was Recommendation 14 at Unispace III, and it may have taken 14 years to get here, but it was very well worth doing. The year of debate that Jay Tate and Lembit Öpik thought would occur in a real impact alert may be significantly reduced by the work that has been done, and may be reduced still further if the international cooperation being recommended now comes to pass.

Already, we have a new European initiative that will be well worth watching. After the Space Situation Awareness program and NEOShield comes Stardust—The Asteroid and Space Debris Research and Training Network, "Pushing the Boundaries of Space Research to Save Our Future." In June 2013, Max Vasile and the Department of Mechanical & Aerospace Engineering at the Lord Hope building of the University of Strathclyde were among institutions seeking Early-Stage Researchers, in the first 4 years (of full-time equivalent research experience) of their research careers, to make a 3-year commitment. Here is a description of the program:

> Stardust is an EU-wide program, funded by the FP7 Marie Curie Initial Training Networks (ITN) scheme, which will provide Europe with a first generation of decision makers, engineers and scientists that have the knowledge and capabilities to address the asteroid and space debris issue now and in the future. The overriding goal of this network is to train researchers to develop and master techniques for asteroid and space debris monitoring, removal/deflection and exploitation such that they can be applied in a real scenario. It is our belief that the integration of all the disciplines involved in this network is a fundamental step towards the resolution of the asteroid and space debris issue and has beneficial consequences in all research areas.
>
> The scientific program focuses on a number of underpinning areas of research and development that are fundamental to any future and present initiative aiming at mitigating the threat posed by asteroids and space debris, and is divided into three major research areas: modeling and simulation, orbit and attitude estimation and prediction, and active removal/deflection of uncooperative targets.
>
> This project will bring together mathematicians, engineers, astronomers and mission analysts with a wide range of expertise. This is essential to ensure that all new concepts, and theoretical

and numerical techniques that will be developed, are driven by clearly identified requirements, and conversely that all new developments are applied immediately to practical problems.

Background and Motivations

Various technologies have been proposed for both the removal of space debris and the deflection of asteroids. Some techniques require direct contact with the target, while others are contactless; some are conceived to be installed on-board future spacecraft while others target the current debris population. For space debris, accelerated de-orbit through drag enhancement devices (e.g., sails, balloons, foam), tethers, ion beams, lasers or enhanced radiation pressure have been proposed. Sails, balloons, tethers require some attachment and are ideal to be attached on future spacecraft while ion beams, lasers are contactless and can remove existing debris. None of these techniques, however, are yet at the level of maturity to be implemented in a real mission. For large objects the control of the deorbiting and entry phase is still an open issue. Different techniques apply better to different types of targets in different orbits, and a complete picture, with a full comparative assessment is still missing. Even for asteroids, a plethora of concepts have been proposed in the last few decades. Many comparative assessments exist, but the high degree of uncertainty on the physical nature of NEOs makes some conclusions questionable. Some deflection techniques imply an instantaneous transfer of momentum (explosive techniques, impact techniques) along with a low level of control but a sizable effect, whereas others imply a controlled but fainter effect on the asteroid trajectory (gravity tractor, ion beam shepherd, laser ablation). Moreover, the problem of deflecting asteroids or satellites not originally designed for being captured, and/or tumbling in space with no attitude control capabilities, is an extremely difficult problem requiring extensive research to advance the enabling technologies, including robust guidance, navigation, and control in highly nonlinear and uncertain scenarios.

Be it noted, this is not a scenario for a science fiction movie, nor the abstract of a lone researcher's 'blue sky' speculative paper. When this book project began, it was frequently said that the total number of people working on near-Earth objects worldwide was less than the staff of a typical branch of MacDonald's. Now, the last four paragraphs are the preamble to the job description for just one university's participation in a pan-European program—and as Jay Tate said in 2001, the extraordinary thing is that nobody's laughing at them.

Conclusions

It's remarkable how far we have come and how generally positive the outcomes appear today as opposed to only a short time ago. In the late 1960s we were aware of the danger, but had no idea how imminent it was or how to deal with it. By the late 1970s we had a way to deal with it—industrialize Earth-grazing asteroids—but we were a long way from having existing capabilities. By the late 1980s we had a less demanding solution, with Gordon Ross's comet-chaser, but it had still to achieve publication. By the late 1990s that had happened, and the Jupiter impacts and some close Earth passes had at last stimulated debate, but even then there were huge gaps still to be closed.

Over 90% of the asteroids in the inner Solar System have now been mapped, as have almost all of the ones capable of extinction level events, without finding any immediate threats and with complete coverage attainable in a few more years. We have demonstrated that with existing space capabilities and the recreation of past ones, a 1-km rock asteroid could be deflected given 10 years' warning, and there are a number of ways in which to do it. If solar sail technology continues to develop, if the SLS development comes to fruition, and if NERVA or a more advanced system becomes available for lunar and planetary missions, by 2020 we should be able to say that the means to do it are in hand. And we have seen that even if an impactor couldn't be stopped, there are ways to improve dramatically on the loss of human life that currently would be unavoidable.

The same techniques could probably deal with short-period comets, with modifications still to be determined and beyond the scope of this book. Much new technology will be needed to counter comets from the outer Solar System, as the threat to Mars from Comet McNaught and the sheer size of Comet Swift-Tuttle both demonstrate. But for the moment it seems we can count ourselves lucky—perhaps luckier than we deserve.

Epilogue

"It has been a well-worn truism," said *The Times*, "that our human race are a feeble folk before the infinite latent forces which surround us. From the prophets of old and from the philosophers of our own time the same message and warning have reached us. But, like all oft-repeated truths, it has in time lost some of its urgency and cogency. A lesson, an actual experience, was needed to bring it home. It is from that salutary but terrible ordeal that we have just emerged, with minds which are still stunned by the suddenness of the blow, and with spirits which are still chastened by the realization of our own limitations and impotence. The world has paid a fearful price for its schooling. Hardly yet have we learned the full tale of disaster, but the destruction by fire of New York, of Orleans, and of Brighton constitutes in itself one of the greatest tragedies in the history of our race.

But the material damage, enormous as it is both in life and in property, is not the consideration which will be uppermost in our minds today. All this may in time be forgotten. But what will not be forgotten, and what will and should continue to obsess our imaginations, is this revelation of the possibilities of the universe, this destruction of our ignorant self-complacency, and this demonstration of how narrow is the path of our material existence, and what abysses may lie to either side of it. Solemnity and humility are at the base of our emotions today. May they be the foundations upon which a more earnest and reverent race may build a more worthy temple."

—Sir Arthur Conan Doyle, *The Poison Belt*, 1913

D. Lunan, *Incoming Asteroid!: What Could We Do About It?*,
Astronomers' Universe, DOI 10.1007/978-1-4614-8749-4,
© Springer Science+Business Media, LLC 2014

Appendix I

The Politics of Survival

> We have to minimize the after-effects of damage to ensure the survival of human and terrestrial life, which are unique in the Galaxy as far as we know.

> — Dr. Arthur Hodkin, SAAM Conference, 2003

The Politics of Survival, by Duncan Lunan

Updated March 2002 from "Notes towards a Politics of Survival," *Science & Public Policy*, February 1987.

Synopsis

Phrases such as "Limits to Growth" or "Alternatives to Growth" imply objections to continuing economic development, and there are valid objections in regard to overpopulation, destruction of irreplaceable natural resources and pollution of the environment. But any strategy we adopt to counter those dangers must also take account of the other threats to human survival: the manmade ones of weapons of mass destruction and of long-term genetic breakdown, and the external ones posed by giant meteor or comet impact, a change in the Sun or a nearby supernova, or by Contact with other intelligence. Big impacts, and probably supernova shockwaves, have brought about major changes on Earth in the past and will undoubtedly happen again in the future. To have a reasonable guarantee of surviving such catastrophes any future society will require extensive development of resources outside Earth. Technological and industrial development in space, and

D. Lunan, *Incoming Asteroid!: What Could We Do About It?*,
Astronomers' Universe, DOI 10.1007/978-1-4614-8749-4,
© Springer Science+Business Media, LLC 2014

self-supporting extraterrestrial settlements, must be regarded as essential to any policy for future development.

We therefore require a "Politics of Survival," a series of practical programs with the object of guaranteeing the survival of mankind against all foreseeable dangers: an objective which could be met within 300 years. Since continuing space development is required to meet the extraterrestrial dangers, it should be used to the full in solving terrestrial problems, and two practical programs are suggested: first, an international effort to free the world from hunger within the next 20 years; second, to remove all major raw materials gathering and industry from Earth's surface by the end of the twenty-first century. Ways are also suggested for space colonization to help, psychologically and sociologically, to meet the dangers of warfare, overpopulation and genetic breakdown.

The Politics of Survival

The philosophies of "Limits to Growth" [1] and "Alternatives to Growth" reflect a growing awareness of the dangers of population growth, uncontrolled industrial growth, and the uncritical use of the environment to supply raw materials and to absorb pollution. They were direct challenges to the assumptions made by "most decision makers on all levels... that past trends will continue and to rely on growth as a panacea" [2]. In the interests of survival such assumptions should be challenged, even if on examination they proved to be correct. However the initial challenges were followed by the widespread acceptance of philosophies which were little more attractive and very little less dangerous. It was argued that with the right social policies and alternative technologies a stable culture could be created, regulating the numbers of mankind, living in harmony with the other inhabitants of Earth's ecosystem, and neither adding to nor subtracting from the available pool of natural resources. Within the confines of the planet, obviously, such accommodations do indeed have to be reached.

But it also came to be said, or implied, and widely believed, that the technological level of such a culture would be lower, and the range of its exploratory, scientific and industrial activities would be less, than those of the present 'Developed World.' Such predictions have two major weaknesses:

1. Human nature being what it is, and natural phenomena being what *they* are, any stable culture—even a global one—will eventually be perturbed by social forces or natural changes drastic enough to upset the balance. The alternatives are then contraction, which if continued leads to extinction, or a new phrase of expansion leading back to the previous crisis level.
2. The findings of Earth sciences, space research and solar physics now strongly reinforce the conclusion—previously obvious from first principles, but little heeded—that any purely Earth-based civilization will be subject, sooner or later, to external natural forces strong enough to change its character utterly if not to destroy it altogether.

The objective of the "Politics of Survival" is to formulate a series of practical programs whose final aim, perhaps 300 years in the future, would be to guarantee the survival of mankind against any foreseeable catastrophe. To see clearly what has to be done, no qualification of the word 'any' can be permitted: the aim is to make sure that some self-supporting element of the human race will survive, in sufficient numbers and with the full resources of history at their disposal, so that what has been accomplished shall not be lost—even if Earth itself or even the Solar System did not remain able to support life.

To a great extent such a program, encompassing almost all human activities, would have to be proceed by persuasion rather than coercion. Its underlying principle would become 'that no individual interest, national, commercial, political, military or unscientific, should contribute to the potential extinction of mankind.' Such a "Politics of Survival" would become the international morality of the twenty-first and twenty-second centuries, and would be built up out of decisions on the applications of particular technologies—just the twentieth century had to arrive at a consensus on the uses of nuclear power, organ transplants and genetic engineering. Since we don't know what the controversial technologies of the next two centuries will be, we can't plan the route to 'guaranteed survival' in detail; but since it can be shown that space technology *has* to be incorporated, and since we already have a general idea of the resources available, hazards to be countered and techniques to be used, we can construct a

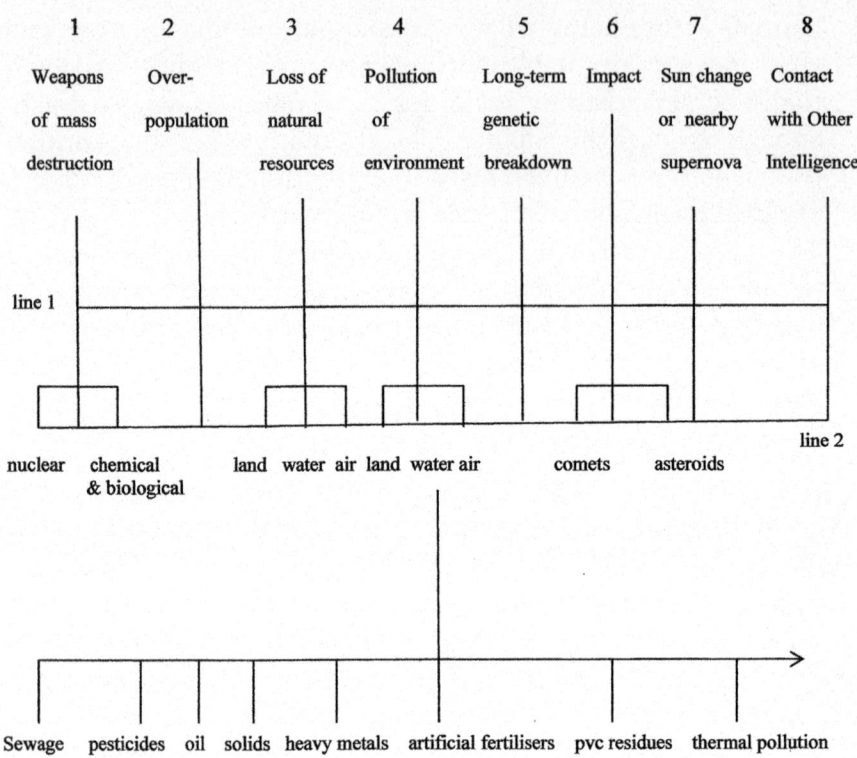

FIG. A.1 A classification scheme for natural and manmade hazards, partially completed (© Duncan Lunan, 2002)

generalized set of strategies which provide the Politics of Survival with a framework and a timescale.

Figure A.1 shows the beginning of a classification system for the dangers to be met and countered in accordance with such a philosophy. The eight headings on the top line represent categories of disaster that could, alone, bring about the annihilation of mankind. Those eight headings are the main subject matter of this essay. It should be remembered that they are closely linked and could occur in synergistic combinations—i.e., that accidents in two or more categories, though not drastic enough to wipe out the race, could combine to bring about that effect. For example, even a limited nuclear war, in a context of overpopulation, a polluted environment or dependence on advanced medical technology, could have far worse effects than the mathematics of yield and fallout alone would predict. The second line of Fig. A.1 shows

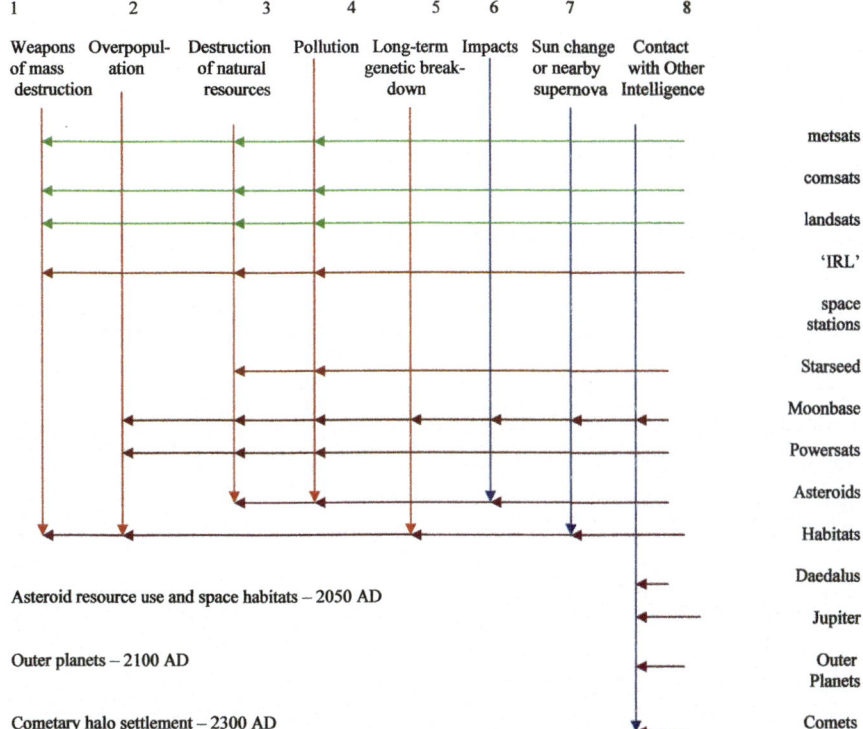

1	2	3	4	5	6	7	8
Weapons of mass destruction	Overpopulation	Destruction of natural resources	Pollution	Long-term genetic breakdown	Impacts	Sun change or nearby supernova	Contact with Other Intelligence

metsats

comsats

landsats

'IRL'

space stations

Starseed

Moonbase

Powersats

Asteroids

Habitats

Daedalus

Asteroid resource use and space habitats – 2050 AD

Jupiter

Outer planets – 2100 AD

Outer Planets

Cometary halo settlement – 2300 AD

Comets

Guaranteed survival of mankind

FIG. A.2 Interactions (*arrowheads*) of present space technologies (*green*), intermediate proposals (*light brown*), proposals under preliminary development (*dark brown*), and theoretical proposals (*purple*), with threats to human survival (*red* for manmade, *blue* for external). © Duncan Lunan, 2003

examples, a far from exhaustive list, of subheadings, two or more of which might act in concert to equal the annihilating outcome of a major disaster on line 1.

The eight headings of line 1 are (1) weapons of mass destruction, (2) overpopulation, (3) destruction of irreplaceable natural resources, (4) pollution of the environment, (5) long-term genetic breakdown, (6) large-scale impacts, (7) Sun change or nearby supernova, (8) Contact with other Intelligence (not necessarily with malevolent intent). The first five are dangers of our own making, the other three represent outside forces that could intervene drastically in human affairs. In Fig. A.2, the descending lines for the first five are coded in red and the external threats in blue.

Those three are normally overlooked in discussion of human survival, and if they *are* invoked, are usually dismissed as irrelevant or statistically remote by comparison with the more 'immediate' dangers posed under the other five headings. But another view of their immediacy and relevance can be taken. Unlike, say, pollution or overpopulation, which are cumulative over decades, they could strike without warning at any time, they threaten our extinction on a day to day basis and cannot be prevented at the present state of our technology. Sometime in the future, probably the near future relative to the timescale of biological evolution, Earth will be struck by an asteroid—or the Solar System by supernova shockwaves—with sufficiently destructive effects to remove the higher life forms, including ourselves, unless we acquire the means to prevent it.

Thus any long-term plan to guarantee the survival of the human race must include the cancellation of headings (6), (7) and (8), and solutions to headings (1)–(5) must be selected accordingly. In plain words, the human race dare not adopt 'alternatives to growth' that ignore or preclude what Dr. Krafft Ehricke has called 'the strategic approach to the Solar System' [3]—i.e., survey and exploitation of interplanetary resources for the practical benefit of mankind. Such a conclusion is a major departure from most publicly discussed 'alternative' strategies and obviously requires detailed justification. Headings (6), (7) and (8) are therefore discussed below in ascending order of immediacy, as defined above.

DIRECT CONTACT WITH OTHER INTELLIGENCE (heading 8) is the hazard that might *not* exist in reality. We simply do not know whether or not we are alone in the galaxy, and scientific opinion is divided. ASTRA's "Man and the Stars" discussion project discussed the possibilities [4], and although the technological background to that discussion is now out-of-date, the general rules established remain valid. It is not impossible that direct contact with some highly advanced spacefaring culture might disrupt our own, even accidentally, to a point where self-supporting survival became impossible; or even that some destructive group might deliberately attempt the extirpation of the human race. No matter how advanced *their* technology, however, if our culture is spread through the cometary halo of the Solar System in self-supporting habitats, hunting down all of its elements would seem to be out of the question [5]. So that is the level of advance at which the survival of the human race is guaranteed, and it's this heading, even if included only for the sake of completeness, which sets the bottom line of the Politics of Survival framework in Fig. A.2.

SUN CHANGE OR NEARBY SUPERNOVA (heading 7) groups together the dangers of which we have become aware through advances in stellar physics and related Earth-based studies. It is not yet clear just how much we are threatened by variations in our own Sun, but there is clear evidence—correlating tree ring studies, climatic studies, and visual observations since the invention of the telescope—that the 'mini-ice ages' of the last 1,000 years have coincided with periods of little or no sunspot activity, i.e., extended breaks in the regular 11-year sunspot cycle [6].

Another cause for concern was the continued failure of research groups to detect the neutrino flux from the core of the Sun predicted by nuclear physics. The determination of the mass of the neutrino may resolve the problem, but it may also be that conditions in the solar core may not match the standard model, and just how serious for us the differences may be remains to be seen. One suggestion was that fusion reactions in the core of the Sun might be intermittent and give rise to 'hiccups' of violent solar activity—perhaps responsible for the 'megadeaths' in Earth's history, when great numbers of species died out simultaneously [7]. A quite opposite idea is that the outer layers of the Sun have been enriched by passing through an interstellar dust cloud, and our model of the core may therefore be inaccurate. If so, the absorbing effect of dust between us and the Sun might have started the present ice age cycle [8], but there would seem to be less of a direct extinction threat (The indirect one comes under heading 6).

Supernova explosions are estimated to occur in the galaxy every 50 years or so on average, and at least three have been bright enough to be visible to the naked eye during the last 1,000 years. At its peak output a star exploding as a supernova can emit as much energy as all the other stars of the galaxy combined. Planets of any star for many light-years around would be subjected first to an intense radiation bombardment, later to magnetic disturbances and radioactive fallout from the shockwaves that once formed part of the mass of the exploding star. It was often suggested that some such event was responsible for the most famous 'megadeath,' the extinction of the dinosaurs, and one scenario attributed it to a rare and very violent 'Type 3' supernova more than 3,000 light-years away [9].

Although the dinosaur extinction is now known to have been an impact event (heading 6, again), there is some evidence that supernovae may have caused other megadeaths. At present it seems that only one of the rarer, more violent, events could harm us under heading 7, because there are no supernova candidates in the immediate stellar vicinity; but even so, we don't know just how much warning we would have. In the longer timescale of millions of years (the timescale of mankind on Earth thus far) only statistical chance determines when one of the more common, less violent supernovae will occur close enough to do serious damage.

A change in the Sun or a nearby supernova would force us to abandon the open surfaces of Earth and any other inhabited worlds 'for the duration of the emergency' in order to survive. The technology required to keep large numbers of people alive underground or on the seabed for years or decades would be very high indeed, and on Earth fighting around the available shelters might well be so fierce that no one would survive. Self-sustaining settlements in space, or on the Moon, Mars or the asteroids, will have to be fully shielded against the existing primary cosmic radiation and would therefore be virtually safe from the radiation flux of a supernova, though decontamination procedures would be needed for incoming personnel and materials during and after the supernova phase.

While the possible magnitude of a change in our Sun is uncertain, the supernova event is only a matter of time. Some future society will experience it, and the prospects for a low-technology one on Earth's open surface do not appeal. There is, however, a worse possibility, which is likely in the still shorter term.

GIANT METEOR OR COMET IMPACT (heading 6) is a danger that has become apparent over the last 30 years. In the early 1960s it could still be argued that the craters on the Moon were volcanic in origin, since Earth showed few signs of similar bombardment. The argument was weakened when the first successful Mars flyby showed a surface intermediate between ours and the Moon's— cratered, but showing extensive weathering. From later Mars missions, the Moon landings and Mercury flybys, the mapping of Venus, and the Voyager missions to Jupiter, Saturn, Uranus and

Neptune, we now know that the Solar System was subjected to an intense bombardment in the final stages of its formation, and many individual impacts have occurred since. By the mid-1970s, over 200 impact features had been identified on Earth's surface [10]; then came the evidence of a really big impact, probably one of a series, coinciding with the extinction of the dinosaurs and many other species.

The growing evidence for waves of impacts suggests that in many of the megadeaths comets rather than asteroids may have been responsible. A catastrophe of that kind would at least give decades or more of warning, as the skies filled with spectacular comets. But a single comet or asteroid approaching Earth at present might not be detected until scant minutes before impact—if at all.

The destructive effects of a big impact are proportional to the energy absorbed by the environment. On land, an impact equivalent to the formation of the Vredevoort Ring in Africa would sterilize the continent affected, but other parts of the world would suffer relatively little harm apart from widespread earthquakes and volcanic eruptions. The white-hot crater, initially penetrating through the crust of Earth to the magma, would radiate much of the impact energy back into space. But in the sea, which offers a much larger target, the effects are much worse. Tsunamis up to thousands of feet high would circle the planet, sweeping most populated areas virtually without resistance. Vast volumes of water would be converted into superheated steam before the crater walls and the rising magma were quenched. Half the world, at least, would be ravaged by the resulting storms, while earthquakes and eruptions reduced the chances of survival even above wave height in the other hemisphere. Lastly, enough matter might be lifted into the atmosphere to black out the surface for years and cause a temporary ice age.

Vredevoort events may not occur more than once in 200 million years. But it has been estimated that Earth suffers blows violent enough to reverse the magnetic field, by the effect of shockwaves passing through the core about every 170,000 years on average [11]; and the association with impacts is strengthened by simultaneous megadeaths in the seas as well as on land [12]. Less violent impacts are still more frequent. There were at least three in the twentieth century, two of them in the territory of the former Soviet Union, with results that would have been catastrophic in populated areas, and in 1972 a giant meteor flew through the atmosphere over the United States, miraculously not striking the ground [13]. Upper estimates of its mass are around 4,000 t, giving it enough kinetic energy to devastate an entire state. But at that time even a small impact, with an energy release comparable to a nuclear weapon, posed a terrible danger to the human race because it could have led to war between the major powers.

The danger cannot be met as popularly supposed, by intercepting the asteroid. A magnetic field reversal/megadeath event could be perpetrated by a body as small as 300 m across [11], virtually undetectable by present-day methods and especially so when coming straight towards Earth at tens of miles per second. Even if it was picked up on radar at, say, the distance of the Moon, there would not be time to program and aim a missile to intercept it. A direct hit would not significantly harm any but the smallest asteroids. Published scenarios involve explosions *beside* the asteroid, so that vaporizing material thrusts it off the colliding course [14], but the guidance requirements for close flybys are still more demanding and less likely to be met in time. Even fragmenting the asteroid provides little respite. Splitting a billion-ton mass into a million fragments would reduce some of the worst overall effects, but a million thousand-ton impacts would still devastate Earth's surface.

In short, the only way to protect Earth from big impacts is to maintain continuous radar mapping of the entire inner Solar System, to detect hazards in time to cancel them far from Earth. Merely shattering a big asteroid is still an inadequate answer, since it will add enormously to the number of smaller hazards to be traced, and is very much a risk if the deflection is to be achieved at long distance using nuclear weapons. Gordon Ross and the author suggest a gentler method using solar sails, which is particularly suitable for deflecting comets [15]. But in the long term, the most effective course would be to send an industrial task force to a threatening asteroid (or comet, if time allows) to process it entirely, launching the products on controlled trajectories to the various planets and leaving the residues (if any) to disperse as harmless dust.

In other words, to guarantee the survival of mankind against the impact threat we have to raise our technological level at least as far as the exploitation of the Asteroid Belt and the Earth-grazing asteroids. Since as already shown such an approach would also permit us to survive a change in the Sun or a supernova shockwave, that level of attainment must be an interim objective of the Politics of Survival.

It cannot be too strongly stressed that any alternative society that limits itself to Earth must face disaster or destruction, from the natural forces discussed, sooner or later. The chance of a giant impact in any given year may be statistically remote, but that will be scant consolation in the year that it happens. All viable models for future societies must therefore include an ongoing space effort. But to ensure that some self-supporting group of humans will survive anything we can foresee is only a minimum objective of the Politics of Survival. A still more worthy aim would be to use the same development in space to help solve the terrestrial threats to survival, headings 1–5 of Fig. A.1 and the ones coded in red in Fig. A.2.

The right-hand side of Fig. A.2 tabulates the technologies and systems, mainly space ones, which are needed to meet the Politics of Survival objectives. The first three are coded in green because they represent existing technologies—weather satellites, communications and Earth resources ones—which in more developed forms will contribute to solving the population, pollution and resources problems, the ones which "The Limits to Growth" considered to be collectively impossible to solve. Each horizontal arrowhead represents an interaction, and each descending line ends at one when the problem can no longer wipe out the entire human race. Horizontal lines colored light brown mark technologies and systems that have still to be developed and dark brown lines mark the ones that eventually will counter all five of the terrestrial threats and the first two of the external ones, still within the confines of the Earth-Moon system. Blue lines mark the ones that will take us to the planets and ultimately to the stars, putting us beyond the possibility of extinction by any currently known hazard.

As the author Michael Cobley pointed out, each arrowhead marks not only an interaction but a potential conflict; and those have to be foreseen and countered as best we can, starting with the discussion below.

Weapons of Mass Destruction (1)

There is of course no purely technological counter to a danger generated by human aggression—no defense is 100% effective. Indeed, when the first draft of this paper was written in 1969 and the second in 1987, further technological breakthroughs were to be feared, since if the balance of power was seriously disrupted the logic of deterrence drove both sides towards an attempt at a pre-emptive first strike. In 1977, it would have been impossible to believe that the leader of the West would deliberately drive us towards a resumption of that situation.

Nor can political, social or ideological answers be foreseen in detail at this stage; but in setting the objective of 'guaranteed survival' 300 years hence, we have to accept that the issues that divide the modern world will have changed, probably out of all recognition, in much less time than that. In a century from now we might hope that with intelligent planning, not merely deterrence but the institution of warfare itself may be a thing of the past [16].

In that time, furthering the Politics of Survival, there should be settlements on the Moon, in free space, probably on Earth-grazing asteroids, the moons of Mars and even Mars itself. Such settlements should be self-supporting. An article on that by the late Chris Boyce, written for *Asgard* in 1978, is not a whit less topical today [17]. As such, these will have much greater symbolic value than the research stations in Antarctica, to which they are so often compared at present [18].

It is an old argument that self-supporting extraterrestrial colonies would prevent humanity's total extinction in war on Earth. But even more important from the philosophical viewpoint is that such colonies will have to work in cooperation, not competition, because under existing international agreements the resources and territory of the heavenly bodies are 'the common heritage of mankind.' While disagreements are bound to arise, warfare between such settlements would be either impossible or suicidal. There have been societies on Earth in which warfare is unknown, but it has always been possible until now to deride them as primitive or irrational. When there are sophisticated offshoots of our own societies working together without the constant threat of war, they may provide models for less violent international relations on Earth.

Overpopulation (2)

Here, too, there is no technological panacea, and palliatives can only make the situation worse, by increasing the numbers who have to be convinced that they should not raise large families. The underlying social issues are often too complex to be settled by mere availability of birth control methods. But even here space technology has a contributory role to play, by making education in farming methods, hygiene, medicine and birth control available to large numbers through communications satellite television.

It's customary to reject outright the idea that interplanetary emigration is even a partial answer to population problems on Earth. However in the 1970s Gerard O'Neill outlined an answer that in theory at least it is feasible—the building of 'Island Three' space habitats, huge cylinders made of lunar materials and spun to generate the equivalent of Earth-surface gravity on their inner surfaces, which would become farmland and provide a reasonably close model of Earth's surface conditions. Each Island Three could house several million people, and in theory habitats could be built sufficiently rapidly to keep pace with Earth's present population increase [19].

O'Neill did not suggest that this is the best answer (what if nobody wants to go?), but he foresaw a less frenzied situation, where in a 100 years' time the total population of the Earth-Moon system might be 8,000 million, half of them living in space settlements. Earth's population would be lower than it is now, and stable, with most of the energy and raw materials for their needs coming from off-planet (see below).

The difficulty is to see how we get there from here. The combined effort of all the spacefaring nations falls far short of the program O'Neill envisages. However the gap can be bridged. One example is the Project Starseed concept, which would use nuclear waste disposal in space as its baseline. The fuel tanks of the boosters that deliver the waste to orbit would be used to build the Starseeds, which are mobile orbiting factories, and they in turn would build solar power satellites from lunar materials. Enough powersats could be built in 10 years to meet Earth's energy needs in the twenty-first century. Revenues from the powersats would pay for the waste disposal program, which would wind down after the first 10 years and virtually cease over the next 10–20 as the ground-based power plants went out of service. As a by-product, however, during the 10 years it would take to build enough Starseeds to build enough powersats to run the world, enough hull material could be produced to build 50 of the smaller 'Island One' space habitats, and enough 'volatiles' gathered to make six of them habitable [20].

In *Man and the Planets, The Resources of the Solar System* [5], it was argued that Island One habitats were the optimum size because they too could be mobile, and could spread through the Solar System to claim its resources for mankind. Eventually, when they reached the cometary halo, they would become sufficiently scattered to defeat even the most improbable aspect of heading 8.

Life in space settlements will be more immediately purposeful, and the need for restraint and mutual cooperation will be more apparent than in day-to-day life on Earth. Sagan has suggested that such settlements may provide new social models that will help to resolve tensions on Earth. At the other end of the social spectrum, Dyson has argued that 'city-states' in the outer Solar System will be needed as outlets for independent, innovative spirits, to escape the pressures towards conformity in a high-technology civilization [21].

The popularity of O'Neill's ideas showed that there is a real need for such outlets, and Andy Nimmo suggested in *Man and the Planets* that in a world subject to overall population control, not having space habitats, or having them, would be the difference between living in a box and living in a box with the lid off.

Exhaustion of Resources and Pollution (3 and 4)

Although they pose distinct threats to mankind, these two headings can be taken together here because the same space technologies are relevant to solving both. These technologies are Earth resources monitoring from orbit, and manufacturing in space (including energy generation). In both cases considerable work has already been done, but has tended towards small-scale, even exotic applications rather than global answers to problems.

The potential of orbital surveys for Earth resources management is enormous. There are applications in forestry, agriculture, mining and land use of all kinds, as well as in all branches of marine activity. Particular examples are in monitoring volcanic activity, earthquake threats, rainfall, snow, flooding, crop diseases, land spoilage and reclamation; in estimating fish stocks, plotting the availability of nutrients, reporting on ice build-up and on sea states in general; and in determining pollution of land, sea and air. It must be stressed that the breakthroughs were made on manned missions. Instruments can enhance but not replace the imaginative performance of the eye and brain. This trend began in the earliest manned missions and continues with reports from astronauts and scientists of phenomena observed but not recorded. For instance, speaking at the 1985 Space Development Conference in Washington, Charles Walker described observations of 25 haze layers in Earth's atmosphere where the instruments recorded only four. Such reports lead to the development of more sensitive instruments, but the present volume of manned and unmanned observation is far below what would be needed for global resource management.

Relatively little thought has been given so far to the use of orbital survey data on such a scale. The potential exists for world-wide synoptic assessment of all kinds, and detailed assessments of the damage (and repairs) happening in the environment. The concept of Spaceship Earth could become literally true, with all the world's cycles as closely monitored as those aboard a manned spacecraft. But such comprehensive coverage will be of little value if it merely calibrates a steady trend towards breakdown, or if it gives rise to one last gold-rush of resource grabbing that accelerates catastrophe.

Politics of Survival precepts will have to be compatible with the ideologies of the groups who are expected to put them into practice. With all the strains and conflicts of the modern world the earliest Politics of Survival objectives will therefore have to be simple ones, though the effects can be far-reaching nevertheless.

One of the first such programs, to tie together the data-gathering capability of orbital surveys with all related efforts at ground level into one overall effort, might be an international program with the target of removing hunger from the world within the next 20 years.

For simplicity of objective in relation to sheer scope it beats Kennedy's famous target (to put a man on the Moon within the decade) by several orders of magnitude, yet it is hard to see on what ideological grounds any group on Earth could object to it. In fact the program should generate a band-wagon effect, in which groups not conspicuously concerned about human suffering at present would cast around for ways to contribute to the program and lay claim to some of the credit. One minor example would be that merely from identifying and naming major sources of atmospheric and oceanic pollution, there would be an implication that the sources concerned were contributing to the possible failure of the program. By observation, the result of such publicity is often a voluntary cleanup accompanied by a loud affirmation of social responsibility. At any mention of such a program, the usual objection is that it takes no account of human nature. But a well-thought out program, whose steps can scarcely be opposed without apparently favoring starvation and poverty for others, can turn the most selfish of self-interests to its advantage. Self-interest is generally easier to work with than apathy.

Weather and crop monitoring on continental scales had already begun in the 1980s, in hopes of evaluating and getting ahead of the drought problem in Africa [22]. Such monitoring ties in directly with proposals for world food supply banks, equipment and medical supplies for early mobilization *before* famines take hold. That leaves a long way to go before reaching a level at which crises can be averted and the spread of the deserts reversed [23], and a common complaint is that the resources needed will not be diverted from military spending. The answer is that 'International Resources Liaison'—the imaginary international organization represented by the letters 'IRL' in Fig. A.2—could only operate by drawing on the military resources of governments around the world. No one else has the ships, aircraft, ground transport, communications and personnel for the 'ground truth' studies that would be needed and the practical action to follow. Absurd as such an idea might have seemed in the 1960s, the trend is now well established with the use of military ships and aircraft for conservationist purposes—as in fisheries protection—and humanitarian ones such as famine and disaster relief, and air-sea rescue. All these have been incorporated into the brief of the proposed European Rapid Reaction Force, and while many aspects of that are politically controversial, there's no argument that some military force shouldn't do them. There is a great deal more to be done along those lines [16].

The elimination of hunger program deals only with the organic resources of Earth, however—the food-producing capability of the land and oceans. Mineral and energy resources will likewise be delineated fully by the data-gathering net, but full exploitation of their potential would be disastrous—in immediate environmental effects and in the shortages to follow. The Politics of Survival should not merely prevent that, but also relieve the present industrial burden on Earth's resources and the environment. The objective, to be quantified in a second major program starting in the early twenty-first century, should be to remove all major raw materials gathering, energy generation and industrial processing from Earth's surface during the next 100 years.

Fantastic though such a statement may look, the groundwork for the transition has already been done, and detailed technical solutions have been proposed for many of the intermediate steps. The former USSR had an ongoing program of research into manufacturing in space, and the United States has worked on it intermittently; Europe and Japan have similar interest in it. The first products will be high-cost, low-mass supply items such as vaccines and electronic components, but once processing facilities exist in orbit, raw materials from the Moon become an attractive proposition because delivery can be made cheaper than from Earth, bringing costs down to the point where most industrial processes can be run profitably from orbit [24]. Once again, the need at present is for a program that bridges the gap between the exotic, limited applications and the large-scale operations that are needed; and once again, the Starseed proposal is one possible answer. The Starseeds would act as testbeds for the transfer of terrestrial industries into space; they and the lunar base would develop the delivery systems needed for raw materials, and the powersats can beam energy to orbiting factories even more easily than to Earth.

That point takes us straight into the political issues of developed world vs. developing, or (though the terminology does some violence to geography) the North–south conflict of interests [25]. Environmentalists believe that the rest of the world *cannot* aspire to the present per capita energy use of the United States without serious environmental effects [26]. Even a complete move away from burning fossil fuels and forests (which powersats would allow, even in the developing world, by tie-ins with appropriate ground-based technology [27]), while reducing the feared build-up of carbon dioxide and its greenhouse effect, would still leave serious disturbances due to waste heat. It would be folly to suggest that the developing world should cease its development and remain disadvantaged, in order to permit the present energy-extravagant lifestyle of the developed world, and as a result some have jumped to the conclusion that the collapse of industrial civilization is unavoidable. Once again, however, the choice is not simply between pollution and poverty, because development in space is an alternative. Roughly 60–70% of the United States' energy use is industrial; it should be moved systematically into space.

Why should the developing world opt for space development, and how can it be brought about in present circumstances? A partial answer lies in the vexed question of lunar and planetary resources, controversially described in U. N. treaties as 'the common heritage of mankind.' As the Starseed program (for example) aims to provide powersats for *every* nation on Earth, the developing world should receive its full share of those benefits. But how is their stake in the program to be acquired? One interesting suggestion by John Braithwaite (unpublished until the first version of this paper came out) was that the World Bank should advance credit to developing nations, based on the resources revealed by orbital surveys that they allow to remain *unused*. It is then easy to imagine the same system being used to secure the developing world's stake in space manufacturing and raw material supply from the Moon and the asteroids.

The detailed requirements for space industrialization—lunar oxygen plants, solar-powered electromagnetic launchers, geosynchronous orbit power stations, processing facilities for lunar and asteroidal materials—have been studied not in government programs but in private projects such as funded by the Space Studies Institute in Princeton. The Lunar Polar Orbiter, to complete the lunar surveys begun in the Apollo program, has now (2002) been using two spacecraft of relatively limited capacity, and they have turned in intriguing but controversial results. In his closing speech at the 1985 Space Manufacturing Conference in Princeton, O'Neill called for just such a program, to include a probe in solar orbit to search for Earth-grazing asteroids. That, too, is being improvised, under the auspices of the international Project Spaceguard (see below), but at a far less thorough level than O'Neill envisaged.

He also called for a small, high-performance re-entry vehicle for personnel transport to and from space stations. Research into such a vehicle was begun when the British space program had begun in the 1950s and was canceled soon after 1960. It is ASTRA's flagship, the Waverider, designed by Terence Nonweiler of the Royal Aircraft Establishment, later Cranwell, Queen's University Belfast, Glasgow University and lastly the University of Wellington in New Zealand. In the 1990s Waverider studies have been conducted in the United States, Europe, Japan, China, the former USSR and elsewhere, but much of this renewed interest has been due to the efforts of ASTRA's Waverider Aerodynamic Studies Project (WASP), headed by Gordon Ross.

From the Politics of Survival viewpoint, larger Waveriders (probably unmanned) could be crucial to space industrialization, in political terms, because raw materials or finished products could be landed from Moon-Earth transfer orbit, from the asteroids or from Earth-orbiting factories, in any latitude and on conventional runways (see below) [28].

Industrialization of the Moon will lead on to the moons of Mars and to Earth-grazing asteroids, in search of their known content of volatile elements that are needed for life support but on the Moon are found only in solar wind deposits trapped by the soil. For more plentiful supplies of carbon, nitrogen, hydrogen and helium a spacefaring habitat civilization would inevitably be drawn to the giant planets, and so we move into a full 'strategic' development of the Solar System, countering all foreseeable threats to human survival—only one of which has still to be discussed.

Long-Term Genetic Breakdown (5)

Heading no. 5 is a controversial subject, and we cannot be drawn here into fine distinctions about desirable characteristics. The principal fear is that inherited conditions such as hemophilia might become so prevalent, in a high-technology civilization, that the effect of any disaster might be magnified, and too few unaffected individuals might survive to continue the race.

For example the dentists' expertise has virtually removed the pressures of natural selection against soft teeth. Space colonization would help to counter the trend, partly by reducing the survival threat from any one disaster, but mostly because such disabilities would be more keenly felt in space, and new attitudes of social responsibility would arise. These in turn would have an effect on Earth. While every effort must be made to avoid Earth becoming 'a slum for unwanted genes,' it's only dimly apparent at present what resources for improvement may become available through genetic engineering, and what the moral issues will be. It is however clear that expansion into space makes genetic breakdown no longer a threat to overall human survival.

Conclusion

The object of the Politics of Survival is to implement a strategic approach to the Solar System and provide a broad enough base of space settlements, raw material and energy supplies to guarantee human survival against all foreseeable hazards. It is not intended that population and industrial growth should continue unchecked. For example, Freeman Dyson has foreseen a future in which, for purely Malthusian reasons, humanity tears up the planets and builds a shell around the Sun to maximize living space and available energy, while emigrant ships spread our intelligence "like a technological cancer" through the galaxy. In the ASTRA discussions that led to *Man and the Planets*, the view emerged strongly that conservationist approaches would dominate long before such extremes were reached [5]. It is certainly to be hoped that by the time its survival was assured, by the Politics of Survival route, the human race would have acquired new perspectives and more sophisticated objectives than those Dyson fears.

The first change in our perspectives to come from space research was the awareness of our own vulnerability, from the Apollo images of the fragile Earth seen from the distance of the Moon. The correct response to that realization is a Politics of Survival program that does not remove that vulnerability but takes away its potential finality. Another major shift in perspective may come (removing all fear from heading 8) through peaceful contact with other intelligences. At the very least, it would prove that high-technology cultures are not doomed to collapse and that space development *is* a route to survival. Arthur C. Clarke suggested that expansion into space may be as significant as life's evolution from the sea to dry land; when we can set our affairs in an interstellar perspective, we may begin to see the further choices between the lines of development that lie ahead of us.

Appendix 2

Contingency Stores for the Aftermath of Asteroid Impact

By Dr. Arthur Hodkin, Sphere Consultants Ltd.

"No one needs to starve, with prudent reserves."

The policy is analogous to Swiss fall-out (nuclear) shelter provisions.

Assuming we have some years' warning:

1. *Set up a Ministry of Food Security.*
2. *Stockpile in warehouses one extra year's supply of key foodstuffs.*

   (a) Wheat as grain
   (b) Rice as brown grain/maize
   (c) Dried potatoes
   (d) Unskimmed dried milk
   (e) Dried soy (tofu)
   (f) Canned meat and fish
   (g) Dried vegetables
   (h) Canned fruit
   (i) Multivitamins

   All this stock to be rotated annually.

3. *Energy*
   Build up 3 years' stockpile of coal and hydrocarbons, plus harnessing of previously untapped reserves to provide 50% spare capacity for electricity generation.
4. *Household emergency stock*
   Six months' supply:

   (a) Flour
   (b) Polished rice

(c) Dried potatoes
(d) Unskimmed dried milk
(e) Canned meat and fish
(f) Canned fruit
(g) Multivitamins

Goods to be issued at cost against ration book, returned every 6 months to exchange for fresh supplies.

Returns to be used up in institutions—hospitals, schools, prisons, care homes, work canteens and budget (civic) restaurants/ soup kitchens to minimize waste of food.

References

Preface

1. Lunan, D. *Man and the Stars*. Souvenir Press, London (1974).
2. Lunan, D. *New Worlds for Old*. David & Charles, Newton Abbot (1978).
3. Lunan, D. *Man and the Planets*. Ashgrove Press, Bath (1983).
4. Lunan, D. "Notes towards a Politics for Survival" (sic). *Science & Public Policy*, 14, 1, 43–50 (Feb. 1987).
5. Tate, J. "Spaceguard UK – the Inside Story", "Spaceguard UK – Executive Summary." *Asgard*, ASTRA, 4, 2, 22–28, 29–41 (Mar. 2002).
6. Lunan, D., and Ross, G. "Keep Watching the Skies!" *Ibid.*, 41–49.

Chapter One: Comets

1. Wells, H. G. "The Star" (1897). In: Wells, H.G., *Tales of Wonder*. Collins, London & Glasgow (1953).
2. Wells, H. G. *In the Days of the Comet* (1906). Collins, London & Glasgow (1954).
3. Conan Doyle, A. "The Poison Belt" (1913). In: *The Professor Challenger Stories*. John Murray, London (1952).
4. Rambosson, J., *Rambosson's Astronomy*. Chatto & Windus, (1897).
5. An Astronomical Lecture by J. Braithwaite, Gent., illustrated by lantern slides including the late Comet of 1862'. ASTRA, Hamilton, Lanarkshire (17 Apr. 1982).
6. Lockyer, J. N., Proctor, R. A. (eds.), Guillemin, A. *The Heavens*, 4th edition. Richard Bentley, London (1871).
7. Chambers, F. *A Handbook of Descriptive and Practical Astronomy*, 4th edition. John Murray, London (1889).
8. Proctor, R. A. *Myths & Marvels of Astronomy*. Longmans, Green & Co., London (1896).
9. Roth, E. (trans.), Verne, J. *To the Sun? Off on a Comet!* (*Hector Servadac*, Parts 1 & 2). Dover, New York (1960).
10. Clarke, A. C. "Into the Comet." In: Clarke, A.C. *Tales of Ten Worlds*. Harcourt, Brace & World, New York (1962).
11. Whipple, F. L. *Earth, Moon and Planets*, Third Edition. Pelican Books, London (1971).
12. Asimov, I. "Comets," introducing Lunan, D. "The Comet, the Cairn and the Capsule" (1972). In: Asimov, I., Greenberg, M. H. & Waugh, C. G. (eds.). *The Science Fictional Solar System*, Harper & Row, New York (1979).

D. Lunan, *Incoming Asteroid!: What Could We Do About It?*, 359
Astronomers' Universe, DOI 10.1007/978-1-4614-8749-4,
© Springer Science+Business Media, LLC 2014

13. NASA. *Astronautics & Aeronautics, 1970. Chronology on Science, Technology and Policy*. NASA SP-4015, US Government Printing Office, Washington (1972).
14. Niven, L., Pournelle, J. *Lucifer's Hammer*. Futura, London (1978).
15. Berry, A. "Comet May Be on a Disaster Course." *Daily Telegraph*, 24 Mar. 1982, Gibson, R. "Comet of Doom Is a Long Shot." *Daily Star*, 25 Mar. 1982.
16. Bortle, J. "Comet Digest." *Sky & Telescope*, 62, 1, 29 (July 1981), Bortle, J. "Comet Digest." *Sky & Telescope*, 63, 7, 102 (July 1982).
17. Flammarion, C. *Les Étoiles*. Paris. (1882).
18. Proctor, R. A. *The Orbs Around Us*. Longmans, Green & Co., London (1899).
19. Marsden, B. "Comet Swift-Tuttle: Does It Threaten Earth?" *Sky & Telescope*, 85, 1, 16–19 (Jan. 1993).
20. Tucci, L. "Funds to Track Comets Get Big Boost, but Killer Comet Threat Dismissed." *Space News*, 3, 42, 8 (Nov. 9–15, 1992).
21. Whipple, F. L. "The Spin of Comets." *Scientific American*, March 1980. In: Brandt, J.C. (ed.). *Comets*. W. H. Freeman, San Francisco (1981).
22. Anon. "News Update: Halley's Outburst." *Astronomy Now*, 5, 8, 9 (August 1991).
23. Cribb, J. "Scientists Allay Fears of Comet Collision." *The Weekend Australian*, 12–13 Dec 1992.
24. David, L. "Defense Experts Duck Asteroid Threat Hearing." *Space News*, 4, 13, 6 (March 29-April 1, 1993).
25. Jet Propulsion Laboratory. JPL Small-Body Database Browser: 109P/Swift–Tuttle, accessed 2013.
26. Burghardt, D. "Scientists Still at Odds on Tunguska after 100 Years." Novosti Press Agency, Moscow, online (24 July 2009).
27. Rincon, P. "Fire in the sky: Tunguska at 100." BBC News, Science and Environment, online, 30 June 2008.
28. Atkinson, N. "Possible Meteorite Fragments from 1908 Tunguska Explosion Found." Universe Today, 2 May 2013.
29. Major, J. "Claims of Tunguska Meteorite Fragments 'Ridiculous', Scientist Says." Universe Today, 7 May 2013.
30. Herschel, J. *Outlines of Astronomy*, (11th ed.). Longmans, Green & Co., London (1871).
31. Miller, W. *The Heavenly Bodies*. Hodder & Stoughton, London (1883).
32. Anon. "Amateur Astronomers." *Sky & Telescope*, 56, 4, 312 (Oct 1978).
33. Sagan, C., Druyan, A. *Comet*. Michael Joseph, London (1985).
34. Pournelle, J. "Halfway to Anywhere." *Galaxy*, 34, 7, 94–101 (Apr. 1974). In: Pournelle, J. *A Step Further Out*. W.H. Allen, London (1980), Drexler, E. "Laser Propulsion to Geosynchronous Orbit." *L5 News*, 3, 7, 8–10 (July 1978).
35. Hindley, K. "Fireball Networks – a Mixed Blessing." *New Scientist*, 72, 1032, 695–698 (23–30 Dec. 1976).
36. Ridpath, I. *Messages from the Stars*. Fontana/Collins, London (1978).
37. Cooper, K. "The Sky Is Falling." *Astronomy Now*, 27, 2, 7–74 (Feb. 2013).
38. Howell, E. "We've Found 10,000 Near-Earth Objects. How To Step Up The Search?" Universe Today, 27 June 2013.
39. O'Leary, B. "To Catch a Falling Star." In: Pournelle, J. (ed.). *The Endless Frontier*. Ace Books (1979).

40. Davies, J. K. "Is 3200 Phaeton a Dead Comet?" *Sky & Telescope*, 70, 4, 317–318 (Oct 1985).

41. Steel, D. *Target Earth*. Time-Life Books, London (2000).

42. Steel, D., Snow, P. "The Tapanui Region of New Zealand: Site of a "Tunguska" Around 800 Years Ago?" In: Harris, A., Bowell, E. (eds.). *Asteroids, Comets, Meteors 1991*. Lunar & Planetary Institute, Houston, Texas (1992).

43. Anon. "Did a Passing Dwarf Star Form the Ice Giant Planets?" *Astronomy Now*, 19, 12 Dec. 2005.

44. "Astronomy Notes." *Journal of the British Interplanetary Society*, 25, 12, 467 (Dec. 1972).

45. Vsekhsvyatskiy, S. K. *The Nature and Origin of Comets and Meteors*. 'Prosveshcheniye' Press, Moscow (1967), NASA Technical Translation TT F-608, U. S. Government Printing Office, Washington (April 1970).

46. Haines, T., and Riley, C. *Space Odyssey*. BBC Books, London (2004).

47. Clube, V., and Napier, B. *The Cosmic Serpent*. Faber, London (1982).

48. Brownlee, D. Lecture, Fifth Annual Space Development Conference, Seattle, May 1986, Mason, J., and Moore, P. Lecture, Theatre Royal, Glasgow, 16 May 1986.

49. Anon. "Comet Theories Called into Question." *L5 News*, 2, 1, 2 (16 Jan. 1987).

50. Atkinson, N. "Fizzy Comet Hartley 2 Is Throwing Snowballs." Universe Today, 21 Nov. 2010.

51. Cowen, R. "Comets Take Pole Position as Water Bearers, matching chemical signatures indicate that Kuiper comets brought water to Earth." *Nature*, Online News, 5 Oct. 2011.

52. Hoagland, R.C. "Rendezvous in 1985." *Analog* XCV, 9, 59–76 (Sept. 1975).

53. Asimov, I. "The Rocks of Damocles." *Fantasy & Science Fiction* (Mar. 1966). In: *Asimov on Astronomy*, Macdonald & Jane's, London (1974).

54. Hall, R. A. "Secondary Meteorites." *Analog*, LXXII, 5, 8–16 & 81–85 (Jan. 1964).

55. Enever, J. E. "Giant Meteor Impact." *Analog* LXXVII, 1, 61–84 (Mar. 1966).

56. Asimov, I. "Updating the Asteroids." *Fantasy & Science Fiction* (Aug. 1974). In: Asimov, I. *Of Matters Great and Small*. Ace Books, New York (1976).

57. Steel, D. *Rogue Asteroids and Doomsday Comets*. Wiley, New York (1995).

58. Anon. "Australia's Henbury Craters." *Sky & Telescope* 49, 5, 287–290 (May 1975).

59. Baillie, M. *Exodus to Arthur, Catastrophic Encounters with Comets*, revised edition. Batsford, London (2000).

60. Mooney, D. Secrets of the Irish Landscape, Part 2. Raidió Teilifís Éireann, Cork, 12 May 2013.

61. Arago, F. *Popular Astronomy*. Longman, Brown, Green, Longman & Roberts, London (1858).

62. Keys, D. "Comet May Have Caused Catastrophe on Earth." *The Independent*, 25 July 1994.

63. Keys, D. Catastrophe. Channel 4, UK, May 2000.

64. Man on Earth, Channel 4, UK, 28 Dec. 2008.

65. Asimov, I. *A Choice of Catastrophes, The Disasters that Threaten Our World*. Hutchinson, London (1980).

66. Duplaix, N. "Fleas, the Lethal Leapers." *National Geographic Magazine*, 173, 5, 672–694 (May 1988).

67. Baillie, M. "The Camelot Comet." *Daily Telegraph*, 26 Aug. 1998.
68. Withington, J. *A Disastrous History of the World, Chronicles of War, Earthquake, Plague and Flood*. Piatkus, London (2008).
69. Kiple, K.F. *Plague, Pox and Pestilence, Disease in History*. Weidenfeld & Nicolson, London (1997).
70. Ashe, G. *King Arthur's Avalon: the Story of Glastonbury*. Collins, London (1957).
71. Fowler, J. T. (ed.). *Adamnani Vita S. Columbae*. Clarendon Press, Oxford (1920).
72. Gore, J. E. *The Worlds of Space*. Innes, London (1894).
73. Ryan, T. "Moon Shadow, Eclipses in the Ancient Annals, Part 2." *Astronomy & Space*, July 1999.
74. Hamilton, N. E. S. A. (ed.). *William of Malmesbury, Gesta Pontificorum Anglorum*. Rolls Series 52, Longman & Trübner, London (1870).
75. Shaffer, C. H. "If a Tree Falls." *Analog*, CXXIII, 12, 50–58 (Dec. 2003).
76. Luard, H. R. (ed.). *Roger de Wendover, Flores Historiarum*. Rolls Series No.95, Eyre & Spottiswoode, London (1890).
77. Matthews, R. "Did St. Patrick Get Help from a Falling Star?" *Sunday Telegraph*, 17 Mar. 1996.
78. Richards, H. (ed.), *Chronica Majora of Roger de Wendover & Matthew Paris*, (7 vols.). Rolls Series No.57, Longmans & Co., London (1883).
79. Graves, R. *The White Goddess*. Faber, London (1948), Tolstoy, N. *The Quest for Merlin*. Hamish Hamilton, London (1985).
80. Stubbs, W. (ed.). *Gervase of Canterbury, The Chronicle of the Reigns of Stephen, Henry II and Richard I*. Rolls Series No.73, Longman & Co., London (1879).
81. Sagan, C. *Cosmos*. Macdonald Futura, London (1980).
82. Mulholland, D. Calame, O. "Lunar Crater Giordano Bruno." *Science* 199, 875–877 (24 Feb. 1978).
83. Lunan, D. *Children from the Sky, A Speculative Interpretation of a Mediaeval Mystery – The Green Children of Woolpit*. Mutus Liber, Edinburgh (2012).
84. Withers, P. "Meteor Storm Evidence against the Recent Formation of Lunar Crater Giordano Bruno." *Meteoritics & Planetary Science*, 36, 525–529 (2001).
85. Nockolds, P. "Reply to Paul Withers", "The Date of Gervase's Event of June 1178." Cambridge Conference Net (20 Dec. 2001), Waddington, G. letters, "The Date of Gervase's 'Wonderful Sign' in 1178", "The Date of Gervase's Event of June 1178", Cambridge Conference Net, online (28 Mar. 2001).
86. Schenk, P. M. "Comet to Hit Jupiter." *Lunar and Planetary Information Bulletin*, 69, 3–5 (November 1993), Hanlon, H. *The Worlds of Galileo*. Constable, London (2001).
87. Masursky, H., Colton, G.W., El-Baz, F. (eds.). *Apollo Over the Moon, A View from Orbit*. NASA SP-362, US Govt. Printing Office, Washington (1978).
88. Musgrove, R. G. (ed.). *Lunar Photographs from Apollos 8, 10 and 11*. NASA SP-246, US Govt. Printing Office, 1971.
89. Moore, P. "What We Know About the Moon" (with tentative Farside map). In: Carter, L. J. (ed.). *Realities of Space Travel*. Putnam, London (1957).
90. Oberg, J. E. *New Earths*. Stackpole Books, Harrisburg, Pa. (1981).

91. Safronov, V. S., Rushkol, Y.L. *History of the Lunar Atmosphere and the Possibility of Ice and Organic Compounds Existing on the Moon.* Vosprosy Kosmogonii, 9 (1963), NASA Technical Translation TT F-232, U.S. Government Printing Office, Washington D.C. (Sep. 1964).

92. Gillis, J. "Inside the Batcave: the Clementine 1 Mission." *Lunar and Planetary Information Bulletin*, 71, 2–5 (May 1994), (Anon). "Clementine's Lunar Mapping Mission: an Overview of the Science Results." *Ibid*, 72, 16–17 (Aug. 1994).

93. Bond, P. "Prospector Finds Ten Times More Ice." *Astronomy Now*, 12, 10, 51–52 (Oct. 1998), Berger, B. "Scientists: Moon Contains More Ice." *Space News*, 9, 34, 10 (7–13 Sep. 1998).

94. Morning, F. "Images Appear to Show Water Ice on Moon." *Aviation Week*, 28 July 2010, Atkinson, N. "Radar Images Reveal Tons of Water Likely at the Lunar Poles." Universe Today, 2 Aug. 2010.

95. Firsoff, V. A. Letter, "Could Mercury Have Ice-Caps?" *The Observatory*, 91, 85–87 (April 1971).

96. Moore, P. *New Guide to the Planets.* Sidgwick & Jackson, London (1993).

97. Stephenson, J. (ed.). *Ralph of Coggeshall, Chronicon Anglicanum.* Rolls Series No.66, Longman & Co., London, 1875.

98. Uhlig, R. "Chunk of Moon Rock Seen Orbiting the Sun." *The Daily Telegraph*, 25 Feb. 1999.

99. (Anon). "Nebraska Crater Only a Pup." *Astronomy Now*, 8, 2, 13 (Feb. 1994).

100. Spedicato, E. "Evidence of Tunguska-type Impacts over the Pacific Basin around the Year 1178 A.D.", SIS Conference. Natural Catastrophes during Bronze Age Civilisations (11–13 July 1997).

101. Mason, J. A. *The Ancient Civilizations of Peru.* Penguin Books, London (1969).

102. Sagan, S. *The Cosmic Connection.* Doubleday, New York (1973).

103. Steel, D. Lecture, Third Charterhouse Conference on Britain's Achievements in Space, British Rocketry Oral History Programme, Charterhouse, Day 2, 'Near-Earth Objects' seminar (12 Apr. 2001).

104. Clarke, A. C. *The Hammer of God.* Gollancz, London (1993).

Chapter Two: Asteroids

1. Herschel, W. "Observations on Two Newly Discovered Bodies (Vesta and Ceres)." Quoted in Hoyt, W. G. *Planets and Pluto.* University of Arizona Press, Tucson (1980).

2. Sir John Herschel quoted in Crosswell, K. *Planet Quest.* Oxford University Press, Oxford (1997).

3. Anon. "Hubble Images of Asteroids Help Astronomers Prepare for Spacecraft Visit." Hubble News Release STScI-2007-27, 20 June 2007.

4. Cain, F. "Three Trojans Found in Neptune's Orbit." *Universe Today*, 15 June, 2006.

5. Howell, E. "Uranus Is Being Chased by Asteroids!" *Universe Today*, 18 June, 2013.

6. Coffey, J. "Asteroid Cruithne." *Universe Today*, 11 Aug. 2009.

7. Lunan, D. *New Worlds for Old*. David & Charles, Newton Abbot, UK (1979).

8. Cole, D. M., Cox, D. W. *Islands in Space, the Challenge of the Planetoids*. Chilton, Philadelphia (1964).

9. Lunan, D. *Man and the Planets*. Ashgrove Press, Bath, UK (1983).

10. O'Leary, B. "To Catch a Falling Star," op cit.

11. The Planetary Society, news release. "That Asteroid Has a Name: Bennu!" 1 May 2013.

12. James, P. D., quoted in Jarski, R. (ed.). *Great British Wit*. Ebury Press, London, 2005.

13. Hecht, J. "Closest Asteroid Yet Flies Past Earth." *New Scientist*, online (2 Oct. 2003).

14. Atkinson, N. "Airburst Explained. NASA Addresses the Russian Meteor Explosion." *Universe Today*, 15 Feb. 2013, Cooper, C. "Focus: Target Earth – What If it Had Hit?" *Astronomy Now*, 27, 4, 58–67 (Apr. 2013).

15. Atkinson, N. "Astronomers Calculate Orbit and Origins of Russian Fireball." *Universe Today*, 25 Feb. 2013.

16. Asimov, I. "The Rocks of Damocles," op cit.

17. Atkinson, N. "Mars Gets Bombarded by 200 Small Asteroids and Comets Every Year." Universe Today, 15 May 2013.

18. Jacchia, L. G. "A Meteorite that Missed Earth." *Sky & Telescope*, 48, 1, 4–9 (July 1974).

19. McInnes, C. "The Nuclear Trigger." *Asgard*, 5, 1, 44–46 (Nov. 2003).

20. Wooley, C. L. *Ur of the Chaldees*. Ernest Benn (1929).

21. Pritchard, J. B. (ed.). *Ancient Near Eastern Texts Relating to the Old Testament*, 2nd edition, revised and enlarged. Princeton University Press, Princeton (1955).

22. Wilkins, W. J. *Hindu Mythology*. Thacker, Spink & Co., Calcutta (1882), Maspero, H. "Legends Mythologiques dans le Chou King." *Journal Asiatique*, 24, 1–100 (1924).

23. Vyse, R. W. H. H. *Operations Carried on at the Pyramids of Gizeh in 1837*, Vol. 2, Appendix. Jas. Fraser, London (1840).

24. Herbeger, C. F. "Samson Strides the Skies." *Griffith Observer*, 51, 3, 2–13, Mar. 1987.

25. Baillie, M. *Exodus to Arthur, Catastrophic Encounters with Comets*, revised edition, op cit.

26. Arago, F. *Popular Astronomy*, op cit.

27. Hodges, J. "Somewhere Out in Space There Is an Asteroid.." *The Scotsman*, 17 Nov. 1998.

28. Bond, A., and Hempsell, M. *A Sumerian Observation of the Köfels' Impact Event*. Alcuin Academics, York (2008).

29. Ley, W., and Bonestell, C. *The Conquest of Space*. Sidgwick & Jackson, London (1951).

30. (Anon). "Nebraska Crater Only a Pup", op cit.

31. Edmonds, J. *North West Panorama*. Periodicals Division, West Australian Newspapers Limited, Perth, Western Australia (1980s, undated).

32. (Anon). "The remains of a gigantic, three-billion-year-old meteorite impact discovered in Greenland." GUES News, Geological Survey of Denmark and Greenland, Copenhagen (28 June 2012).

33. Enever, J. E. "Giant Meteor Impact", op cit.

34. Baxter, J. *The Hermes Fall*. Panther, London (1976).

35. Cramer, John G. "The Alternate View: Dinosaur Breath." *Analog*, CVIII, 7, 140–143 (July 1988).

36. Gray, R. "Gardeners' Debt to the Fires that Once Swept Earth." *Sunday Telegraph*, 19 Dec. 2010.

37. Anon. "Chicxulub: Site of the K/T Impactor?" *Lunar & Planetary Information Bulletin*, 60, 3–6 (Aug. 1981).

38. Anon. "Continuing Work on Chicxulub Establishes New Links between Impact Crater in the Yucatán and Massive Extinction of Life." *Lunar and Planetary Information Bulletin*, 65, 4 & 18 (November 1992), Ross, J. "Asteroid 'Finished Off Dinosaurs.'" *The Australian*, 8 Feb. 2013.

39. Clark, S. "The Lunar and Planetary Science Conference: Was It a Comet that Killed the Dinosaurs?" *Astronomy Now*, 27, 5, 20 (May 2013).

40. Mainzer, A., et al. "Characterizing Subpopulations within the Near Earth Objects with NEOWISE: Preliminary Results." *Astrophysical Journal*, 752, 2, article id. 110, 16 pp. (2012).

41. Steel, D. *Target Earth*, op cit, *Rogue Asteroids and Doomsday Comets*, op cit.

42. Kramer, J.G. "Killer Asteroids and You." *Analog*, CXII, 1 & 2, 208–213 (Jan. 1992).

43. Steel, D. Lecture, Third Charterhouse Conference on Britain's Achievements in Space, op cit.

44. Tate, J. "Spaceguard UK – The Inside Story", op cit.

45. Peattie, C., and Taylor, R. "Alex." *The Daily Telegraph*, 2 Feb. 2013.

46. Tate, J. Lectures, ASTRA, Airdrie Arts Centre, 20 Mar. 1998, ASTRA, Central Hotel, Glasgow, 21 Mar. 1998, BROHP Charterhouse Conference, 12 Apr. 2001.

47. Wright, L. Lecture, 'Why Don't We Ask the Dinosaurs?' BROHP Charterhouse Conference, 12 Apr. 2001.

Chapter Three: A Designer Hazard

1. Maxwell, G. Wolfpit. In: Maxwell, G. *Plays One: The Lifeblood, Wolfpit, The Only Girl in the World*. Oberon Modern Playwrights, London (2006).

2. Worden, Brig-Gen. S. P. "NEOs, Planetary Defence and Government – a View from the Pentagon." CCNet-ESSAY, Cambridge-Conference Net, 7 Feb. 2000.

3. Atkinson, N. "NASA Scientists Discuss Potential Comet Impact on Mars." Universe Today, 28 Mar. 2013, "New Calculations Effectively Rule Out Comet Impacting Mars in 2014." Universe Today, 12 Apr. 2013.

4. Whipple, F. L. "The Spin of Comets," op cit, Ball, Sir R. S. *Text-Books of Science: Elements of Astronomy*. Longmans, Green & Co., London (1900).

5. Clube, V., and Napier, B. *The Cosmic Serpent*, op cit.

6. Hancock, G., Bauval, R., Grigsby, J. *The Mars Mystery*. Michael Joseph, London (1998), citing Clube, V. and Napier, W. *The Cosmic Winter*. Blackwell, Oxford, 1990.

7. Bellenden, J. (trans.). *The History and Chronicles of Scotland. written in*

Latin by Hector Boece, canon of Aberdeen, Vol. II. W. & C. Tait, Edinburgh (1821), Stapledon, T. (ed.). *Chronicon Petroburgense*. Camden Society, Vol. XLVII (1849).

8. Lunan, D. "The Comet, the Cairn and the Capsule", op cit.
9. Paine, M. "Defending Earth: Fact v Fiction." Cambridge-Conference Net Essay, 16 Feb. 2000.
10. Zaitzev, A.V. "A Russian View of the Impact Hazard and Planetary Defence." Cambridge-Conference Net Essay, 21 Dec. 2000.
11. Steel, D. *Target Earth*, op cit.
12. Steel, D. *Rogue Asteroids and Doomsday Comets*, op cit.
13. Grondine, E. P. "Workshop for Mitigating the Effects of Public Concern on the NASA Bureaucracy: Two Days in Washington, 2002." Cambridge-Conference Net Correspondence, 26 June 2003.
14. Atkinson, N. "Solving the Asteroid-Meteorite Puzzle." *Universe Today*, 13 Aug. 2008.
15. Rudaux, L., de Vaucouleurs, G. *Larousse Encyclopedia of Astronomy*. Trans. Guest, M., Sidgwick, J. B., revised Kopal, Z., Batchworth, London (1959).
16. Yoshikawa, M., Asami, A., Asher, D., Fuse, T., Hashimoto, N., and Isobe, S. "Current Status of Asteroid Observations in Bisei Spaceguard Center." *Memorie della Società' Astronomica Italiana*, 73, 3, 772 (2002).
17. Atkinson, N. "Will Asteroid 2011 AG5 Hit Earth in 2040?" *Universe Today*, 1 Mar. 2012.
18. Anon. "In Brief: Earth Safe from Asteroid Impact." *Astronomy Now*, 27, 2, 10 (Feb. 2013).
19. Atkinson, N. "Recent Earth-Passing Asteroid is Much Bigger Than Originally Estimated." Universe Today, 22 June 2012, SPACE.com Staff. "Surprise! Big Asteroid That Flew By Earth Larger Than Thought." SPACE.com, online, 22 June 2012.
20. Atkinson, N. "Apophis' Odds of Earth Impact Downgraded." *Universe Today*, 7 Oct. 2009.
21. Planetary Society News Release. "Tag an Asteroid, Win a Prize." *Universe Today*, 14 Dec. 2006.
22. Atkinson, N. "Asteroid Apophis: Bigger, Darker But Not a Threat in 2036." *Universe Today*, 11 Jan. 2013.
23. Hancock, G., Bauval, R., Grigsby, J. *The Mars Mystery*, op cit, citing Scott, W. (ed.). *Hermetica*. Shambhala, Boston (1993).
24. Christie, M. "Coming One Day Near You – a Mega-Tsunami." Reuters, 26 Feb. 2002, citing Bryant, E. *Tsunamis – The Underrated Hazard*. Springer Praxis, Chichester, UK (2008).
25. Cain, F. "Did a Killer Asteroid Drive the Planet Into An Ice Age?" Universe Today, 20 Sep. 2012, Goff, J., et al. "The Eltanin asteroid impact; possible South Pacific palaeomegatsunami footprint and potential implications for the Pliocene–Pleistocene transition." *Journal of Quaternary Science*, 27, 7, 660–670 (3 Sep. 2012).
26. Dickinson, D. "Giant Ancient Impact Crater Confirmed in Iowa." *Universe Today*, 6 Mar. 2013.
27. Lunan, D., and Dick, G. "Flight in Non-terrestrial Atmospheres, or the Hang-Glider's Guide to the Galaxy." *Analog*, CXIII, 1 & 2, 56–77 (January 1993).
28. Peiser, B. J. Lecture. Third Charterhouse Conference on British Achievements in Space, op cit.

Chapter Four: Detection and Reaction

1. Sinnott, R. W. "An Asteroid Whizzes Past Earth." *Sky & Telescope*, 78, 2, 30 (July 1989).
2. Anon. "Frontier." *Space-faring Gazette*, 6, 3, 4, National Space Society Golden Gate Chapter (Mar. 1990).
3. Steel, D. *Target Earth, op cit*, Clarke, A.C. *Rendezvous with Rama*. Gollancz, London (1973).
4. Beeston, R. "'Spacewatch' Plan to Attack Meteors with H-bombs." *The Daily Telegraph*, 13 Feb. 1981.
5. Fitzsimmons, A. "Target Earth." *Astronomy Now*, 7, 2, 38–40 (Feb. 1993).
6. Tucci, L. "Funds to Track Comets Get Big Boost, but Killer Comet Threat Dismissed", op cit.
7. Berry, A. "Film Plot Could Save World from Disaster." *Daily Telegraph*, 8 May 1990.
8. Reuters. "U. S. Air Force says Space Fence program safe for now." *News Daily*, 18 Apr. 2013.
9. Matthews, R. "Missiles to Zap Meteor Menace." *The Sunday Telegraph*, Feb. 1992.
10. Smith, F. "A Collision over Collisions: a Tale of Astronomy and Politics." *San José Mercury News*, 22 Mar. 1992, reprinted *Mercury*, May-June 1992, 97–102 & 110.
11. Clarke, A. C. *The Hammer of God*, op cit, Berry, A. "Writer's Explosive Theory." *The Daily Telegraph*, 31 Mar. 1993.
12. Steel, D. Lecture, Charterhouse 2001, op cit.
13. Rincon, P. "Cuts Cast Doubts on Asteroid Plan." BBC News, online, 25 Mar. 2010.
14. Kremer, K. "NASA's KaBOOM Experimental Asteroid Radar Aims to Thwart Earth's Kaboom." Universe Today, 12 Mar. 2013.
15. Moore, P., and North, C. *The Sky at Night, Answers to Questions from Across the Universe*. BBC Books, London (2012).
16. European Space Agency. "Lost, potentially hazardous asteroid rediscovered." *Phys.Org. News*, 15 Oct. 2012.
17. Richardson, R. S. "Astronomical Observations from the Moon." In: Richardson, R. S. *Exploring Mars*. McGraw-Hill, New York (1954), Richardson, R.S. (ed.). *Man and the Moon*. World Publishing Company, New York (1961); Shrunk, D., Sharpe, S., Cooper, B. and Thangalevu, M. *The Moon, Resources, Future Development and Colonization*. Wiley-Praxis, New York (1999).
18. Takahashi, Y. D. *New Astronomy from the Moon. A Lunar Based Very Low Frequency Radio Array*. Thesis submitted to the University of Glasgow for the degree of Master of Science, Department of Physics and Astronomy, University of Glasgow (July 2003).
19. Cooper, K. "Focus. the End of the World - The Sky Is Falling." *Astronomy Now*, 27, 2, 70–74 (Feb. 2013).
20. Peiser, B. J. Lecture. Third Charterhouse Conference on British Achievements in Space, op cit.
21. Howell, E. We've Found 10,000 Near-Earth Objects. How To Step Up The Search? op cit, Anon. NASA Announces Asteroid Grand Challenge. NASA News, online, 21 June 2013.

22. Lovett, R. A. "The Golden Age Comes to Seattle: Is Asteroid Mining Really Part of Our Future?" *Analog*, 22–28 (May 2013).
23. David, L. "Experts Push for a NASA Asteroid-Hunting Spacecraft." Space.com, 21 Dec. 2010.
24. Schweickart, R., quoting Yeomans, D. 2013 Planetary Defense Conference, International Academy of Astronautics, Flagstaff, Arizona, 15–19 Apr., 2013.
25. Cooper, K. "Focus: Target Earth – Why Didn't We See It Coming?" *Astronomy Now*, 27, 4, 60–61 (Feb. 2013).
26. Anon. "Russian Radar to Help Save Earth?" Russia Today, online, 19 June 2008.
27. Baxter, J. *The Hermes Fall*, op cit.
28. Vasile, M. Lecture. Tunguska Centenary Seminar, ASTRA, Ogilvie Centre, Glasgow (30 June 2008), Sanchez, J. P., Colombo, C., Vasile, M. and Radice, G. "Multi-criteria Comparison among Several Mitigation Strategies for Dangerous Near Earth Objects." *Journal of Guidance, Control and Dynamics*, 32, 1, 121–142 (Jan. 2009).
29. McInnes, C. Lecture, 'NEOs, Hazards and Opportunities'. Ayr Astronomical Society, Ayr Academy, 23 Jan. 2011, repeated ASTRA, Ogilvie Centre, Glasgow, 21 Mar. 2011.
30. Lunan, D. "How to Blow Up an Asteroid." *If*, 22, 2, 37–58 (Nov.-Dec. 1973). This story was heavily rewritten by 'person or persons unknown' before publication and the author denies responsibility for male chauvinism, overwriting and logical absurdities that were added.
31. Melosh, J., et al. "Averting Armageddon." ABC News, 14 Aug. 2003.
32. Niven, L., and Pournelle, J. *The Moat Around Murchison's Eye*. HarperCollins, 1993.
33. Vasile, M. Lecture, NEO Impacts. ASTRA, Ogilvie Centre, Glasgow (21 Feb. 2011).
34. Ahrens, T. J., Harris, A. W. "Deflection and fragmentation of near-Earth asteroids." *Nature*, 360, 429–433 (3 Dec. 1992).
35. Ahrens, T. J., Harris, A. W. "Deflection and fragmentation of near-Earth asteroids." In: Gehrels, T. (ed). *Hazards Due to Comets and Asteroids*. University of Arizona Press, Tucson (1994).
36. Holloway, N. Personal communication, 9 Mar. 2004.
37. Lunan, D. *Man and the Planets*, op cit.
38. Lunan, D. *New Worlds for Old*, op cit.
39. Dick, G. Lecture, 'Light-sail Propulsion in Space'. ASTRA Ayrshire Branch, Largs Community Centre, 16 July 1986.
40. Melosh, J. "Non-nuclear Strategies for Deflecting Comets and Asteroids." In: Gehrels, T. *Hazards Due to Comets and Asteroids*, op cit.
41. Anon. "Solar Sail Moon Race, Basic Reference Document." Union pour la Promotion de la Propulsion Photonique, Venerque, France (Second Printing, Oct. 1982), Bird, J. "Columbus's Heirs Set Out to Sail to Mars." *The Sunday Times*, 10 Dec. 1989, Lloyd, C. "Sun Rises on Race of the Century." *Sunday Times*, 30 Jan. 1990, Berry, A. "Mars Race Is Far from Plain Sailing." *Daily Telegraph*, 23 Mar. 1990.
42. Bye, N. Personal communication, for The Millennium Commission, 7 Oct. 1998, Berry, A. "Millennium Space Race for Pennies from Heaven." *Daily Telegraph*, 12 Apr. 1995, Nuttall, N. "Sailing through Space." *The Times*, 17 Apr. 1995.

43. Simmons, J. Lecture, 'Solar Sailing'. ASTRA, Central Hotel, Glasgow, 6 June 1987, Simmons, J.F.L., McDonald, A.J.C., Brown, J.C. "The Restricted 3-Body Problem with Radiation Pressure." *Celestial Mechanics*, 5, 145–182 (1985), McDonald, D. "Scot's Satellite to Set Sail for the Sun." *The Express*, 29 Nov. 1998, anon. "Solar Sailing to Mercury Could See "Dream" Demonstrate New Space Propulsion." *Glasgow University News Review*, 17–18 (June 2002).

44. McInnes, C. Personal communication, 30 Sept. 2003.

45. McInnes, C. "Deflection of Near-Earth Asteroids by Kinetic Energy Impacts from Retrograde Orbits." *Planetary and Space Science*, 52, 7, 587–590 (June 2004).

46. Nerlich, N. "Astronomy Without A Telescope – Impact Mitigation." Universe Today, 13 Aug. 2011.

47. Deimos Space. "Near Earth Objects Space Mission Preparation: Don Quixote Mission Executive Summary." Astrium, 14 Mar. 2003.

48. Takanaoi, S., et al. "Small Carry-on Impactor of Hayabusa 2 Mission." *Acta Astronautica*, 84, 227–236 (Mar.-Apr. 2013).

49. Kremer, K. "European Asteroid Smasher Could Bolster Planetary Defense." *Universe Today*, 24 Feb. 2013.

50. Lu, E.T., and Love, S. G. "A Gravitational Tractor for Towing Asteroids." NASA Johnson Space Center, undated, Atkinson, N. "How to Deflect an Asteroid with Today's Technology." Universe Today, 15 Oct. 2010.

51. Bombardelli, C., et al. "The ion beam shepherd: a new concept for asteroid deflection." *Acta Astronautica*, in press, online 7 Nov. 2012.

52. Bombardelli, C., and Peláez, J. "Ion Beam Shepherd for Asteroid Deflection." *Journal of Guidance, Control, and Dynamics*, 34, 4, 1270–1272 (July-Aug. 2011).

53. Atkinson, N. "Solving the Asteroid – Meteorite Puzzle." *Universe Today*, 13 Aug. 2008.

54. Atkinson, N. "Deflecting Incoming Asteroids with Paintballs." *Universe Today*, 26 Oct. 2012.

55. Atkinson, N. "How to Save the World from Asteroid Impact: Plastic Wrap." *Universe Today*, 25 Aug. 2008.

56. Ehricke, K. A. "A Strategic Approach to Interplanetary Flight." In: Roadman, Strughold and Mitchell (eds.). *Fourth International Symposium on Bioastronautics and the Exploration of Space*. Aerospace Medical Division (AFSC), Brooks Air Force Base, Texas, 1968.

57. Hutchinson, L. "New F-1B rocket engine upgrades Apollo-era design with 1.8M lbs of thrust." Ars Technica, online, 15 Apr. 2013.

58. Perrett, B. "China's Long March 5 will not launch until 2015." *Aviation Week & Space Technology*, online, 25 Mar. 2013.

59. Bond, A. Lecture, 'Travelling at the Edge of Space – Reaction Engines and Skylon in the Next 20 Years'. Senate Court Suite, Collins Building, University of Strathclyde, 10 Mar. 2010.

60. Lunan, D., and Ramsay, W. "Crossroads in Space." *Asgard*, 2, 3, 3–25, (Nov. 1991), abridged, *Space Policy*, 8, 1, 3–8 (Feb. 1992).

61. Cole, D. M., and Cox, D. W. *Islands in Space, the Challenge of the Planetoids, op cit.*

62. Law, R. Lecture, 'NASA's Near Earth Object Mission study'. ASTRA, Glasgow Council for the Voluntary Sector, 13 Aug. 2007.
63. Couvault, C. "Dual Orion Capsules Studied for Manned Asteroid Missions." Spaceflight Now, online, 17 Aug. 2009.
64. Svitak, A. "Ariane V Launches Europe's Fourth ATV to Space Station." *Aviation Week & Space Technology*, online, 5 June 2013, Atkinson, N. "Official Confirms NASA Plan to Capture an Asteroid." Universe Today, 6 Apr. 2013.
65. Morning, F. "Robotic Probes Scouting Human Gravity Pathways." *Aviation Week & Space Technology*, online, 15 Oct. 2012.
66. Wright, L. Lecture, 'Why Don't We Ask the Dinosaurs?' *op cit.*
67. Sagan, S. *Pale Blue Dot*. Headline, London (1995).
68. Oberg, J. "Terraforming Earth." Futures Focus Day Symposium, 'Planetary Climate Modification and the U. S. Space Command – As-Yet Unrecognized Missions in the Post-2025 Time Frame,' sponsored by the Commander-in-Chief, US Space Command, Colorado Springs, Colorado (23 July 1998).
69. Holloway, N. Lecture, 'Roid Wars. 'Celebrating British Achievements in Space', BROHP Conference, 2001, op cit, Asher, D. Lecture. ASTRA, Glasgow Council for the Voluntary Sector, 25 May 2001.
70. Bradley, M. *The Cronos Complex I, a personal enquiry into the temporal origins of human culture and psychology*. Nelson, Foster and Scott, Toronto (1973).
71. Nield, T. *Incoming! Or, Why We Should Stop Worrying and Learn to Love the Asteroid*. Granta, London (2011).
72. Zigel, F. Yu. *The Minor Planets*. NASA TT F-700, US Govt. Printing Office, Washington, D.C. (1972).
73. Allaby, M., and Lovelock, J. *The Great Extinction*. Secker & Warburg, London (1983).
74. Enever, J. E. "Giant Meteor Impact", op cit.
75. Nield, T. Lecture. ASTRA, Ogilvie Centre, Glasgow, 10 Mar. 2011.
76. Dixon, D. *After Man, A Zoology of the Future*. Granada, London (1981).
77. Milligan, C. *Muses with Milligan*. Adapted from the BBC-2 TV series, produced by John Furness, Decca LK 4701, 1965.
78. Benford, G., and Rotsler, W. *Shiva Descending*. Avon Books, New York (1980).
79. Moran, S. *The Secret World of Cults, from Ancient Druids to Heaven's Gate*. Bramley Books, Godalming, Surrey (1999).
80. Dickinson, D. "Debunking Comet ISON Conspiracy Theories (No, ISON is not Nibiru)." *Universe Today*, 29 Apr. 2013.
81. Sagan, S. *Broca's Brain, The Romance of Science*. Hodder & Stoughton, London (1979).
82. Lovell, J. Lecture. 'A View from Earth 1984' Conference, Venture Sciences Association, Big Bear Lake, CA (30 June – 1 July 1984).
83. Hoyle, F., and G. "The Monster of Loch Ness." In: Hoyle, F. and G. *The Molecule Men*. Harper & Row, New York (1972).
84. Horizon – Asteroids. BBC-4 TV, 2 Feb. 2012.

Chapter Five: The First Scenario: Deflection

1. Lunan, D. *New Worlds for Old, op cit, Man and the Planets*, op cit.
2. Waddell, P. "Stretchable Concave Mirrors." *Spectrum, British Science News*, 198, 5–8 (1986).
3. Lunan, D. "The Wonderful Mirror." *Free Space News*, reprinted *Infinity* (both Feb. 1986).
4. Shimuzu, H. "Ultralightweight Reflector for Lidar Applications." *Applied Optics*, 25, 9, 1467–1469 and 1475 (May 1986).
5. Miller, S. "First Light at the Keck." *Astronomy Now*, 5, 4, 21–23 (April 1991), "Last Pieces of Mountain Mosaic." *Astronomy Now*, 6, 1, 10–11 (Jan. 1992).
6. Anon. "REOSC. a Sharp Eye on Space Optics." *News from Prospace*, Prospace, Paris, 33, 32–41 (Dec. 1991).
7. Zeeman, C. Lecture, 'Introduction to Catastrophe Theory, with applications in science and medicine'. IBM 'Science and the Unexpected' conference, Excelsior Hotel, London Airport, Heathrow (18–19 Feb. 1982).
8. Farquhar, R. W. *The Control and Use of Libration-Point Satellites*. NASA TR R-346, US Government Printing Office, Washington D.C. (Sep. 1970).
9. Lunan, D. Letters, "Need we protect Earth from space objects and if so, how?" *Space Policy*, 8, 1, 90–91 (Feb. 1992), "NASA considers asteroid threat." *Space Policy*, 8, 4, 366 (Nov. 1992).
10. Lunan, D., and Ross, G. "Keep Watching the Skies." *Analog*, XCIV, 12, 70–85, October 1994, extended version, *Asgard*, 3, 1, 3–23 (May 1995), updated and reprinted *Asgard*, 4, 2, 44–52 (Mar. 2002), and Cambridge Conference Net, 2002.
11. Melosh, J. "Non-nuclear Strategies for Deflecting Comets and Asteroids", op cit.
12. Horizon. Averting Armageddon. BBC-2, 23 Jan. 2003.
13. Oberg, J. *Terraforming Earth*, op cit.
14. Lunan, D. "Waverider Entry Spacecraft, a History." In: Sales, I. (ed.). *Rocket Science*. Mutation Press, Edinburgh (2012).
15. Norris, G. "X-51A Waverider Achieves Hypersonic Goal on Final Flight." *Aviation Week & Space Technology*, online, 2 May 2013.
16. Vessot, R. F. C. "Rockets, Clocks and Gravity." In: Cornell, J. and Lightman, A.P. (eds.). *Revealing the Universe, Prediction and Proof in Astronomy*, Chapter 2. Einstein's Perception of Space and Time. MIT Press, Cambridge, Mass. (1982).
17. Amos, J. "Gravity Probe B Confirms Einstein Effects." BBC News, Science and Environment, 4 May 2011.
18. Bernasconi, M.C. and Zurbuchen, T. "Lobed Solar Sails for a Small Mission to the Asteroids." *Acta Astronautica*, 35, Supplementary Issue, 645–655 (1995).
19. Ross, G. "Archimedes Asteroid Burner, Description and Schematic." Personal communication, 6 Oct. 2003.
20. Vasile, M. Lecture. ASTRA, Glasgow Council for the Voluntary Sector, 10 Oct. 2005.
21. The Times. "On Reflection, Mirrors Best Asteroid Defence." *The Australian*, 6 Oct. 2007.
22. Vasile, M. Lecture, Tunguska Centenary Seminar, op cit.

23. Sanchez, J. P., Colombo, C., Vasile, M., and Radice, G. "Multi-criteria Comparison among Several Mitigation Strategies for Dangerous Near Earth Objects", op cit.

24. Vasile, M. "A Multi-Mirror Solution for the Deflection of Dangerous NEOs." Personal communication, 2008. *Communications in Nonlinear Science and Numerical Simulation*, 14, 12, 4139–4152 (2008).

25. Vasile, M. lecture, 'NEO Impacts.' ASTRA, Ogilvie Centre, Glasgow, 21 Feb. 2011.

26. Vasile, M., Maddock, C. A., Radice, G., and McInnes, C. "NEO Deflection through a Multi-mirror System." Call for Ideas. NEO Encounter 2029, Mid-Term Review Meeting, Department of Aerospace Engineering, University of Glasgow. Final Report, Ariadna ID 08/4301, ESA/ESTEC, 12 May 2009.

27. Kahle, R., et al. "Physical Limits of Solar Collectors in Deflecting Earth-Threatening Asteroids." *Aerospace Science and Technology*, 10, 3, 256–263 (2006).

28. Shen-Ping Gong, Jun-Feng Li and Yun-Feng Gao. "Dynamics and control of a solar collector system for near Earth object deflection." *Research in Astron. Astrophys.*, 11, 2, 205–224 (2011).

29. Oberg, J. *UFOs & Outer Space Mysteries, a sympathetic skeptic's report.* Donning, Norfolk, VA (1982).

30. Gibbons, A., Vasile, M. et al. "Laser Bees, A Concept for Asteroid Deflection and Hazard Mitigation." University of Strathclyde Advanced Space Concepts Laboratory, Final Report and Presentation, The Planetary Society, Pasadena, USA (2012).

31. Lunan, D. *The Stones and the Stars, Building Scotland's Newest Megalith.* Springer, New York (2012); Bradley, J. "Scot Brings New Dimension to 3-D TV." *The Scotsman*, 20 June 2010.

32. Staff writers. "City Tech Research Team Casts Light on Asteroid Deflection." The City University of New York, online, 31 Jan. 2011, Matloff, G.L. Deflecting Asteroids. *IEEE Spectrum*, online, 28 Mar. 2012.

Chapter Six: The Second Scenario: Manned Mission

1. Stewart, R.J. and Williamson R. *Celtic Bards, Celtic Druids.* Blandford, Cassell, London (1999).

2. As quoted in *Joe Miller's Complete Jest Book*, Anecdote Number 1528, Page 494, Scott, Webster & Geary, London (1840).

3. Clarke, A. C. *The Hammer of God*, op cit.

4. O'Neill, G. K. *The High Frontier, Human Colonies in Space.* Jonathan Cape, London (1977).

5. White, J. "Deadly Litter." In: White, J. *Deadly Litter.* Corgi, London (1968).

6. Lunan, D. *Man and the Planets*, op cit.

7. Olds, J., A. Charania, A.C., and Schaffer, M.G. "Multiple mass drivers as an option for asteroid deflection missions." In: American Institute for Astronautics and Aeronautics. 2007 Planetary Defense Conference, Washington, DC, pp. S3-7 (2007).

8. Irwin, J. B. and Emerson, W. A., Jr. *To Rule the Night, The Discovery Voyage of Astronaut Jim Irwin.* Hodder & Stoughton, London (1973).

9. Scott, D., and Leonov, A. *Two Sides of the Moon, Our Story of the Cold War Space Race.* Simon & Schuster, London (2004).

10. Nonweiler, T. R. F. Lecture, 'The Apollo 13 Disaster'. ASTRA, Hamilton, Lanarkshire (29 Apr. 1970).

11. Fister, J. L. "Payload Assist Module: Failure Investigation, Background Information." News from McDonnell Douglas (Nov. 1984).

12. Slayton, D., Lecture. 'A View from Earth 1984' Conference, op cit.

13. Hutchinson, L. "New F-1B rocket engine upgrades Apollo-era design with 1.8M lbs of thrust", op cit.

14. Feynman, R. P. *What Do You Care What Other People Think?* Unwin Hyman, London (1988).

15. Clarke, A. C. *The Promise of Space.* Hodder & Stoughton, London (1968), citing Gray, E.Z. and Dixon, F.P. "Manned Expeditions to Mars and Venus." Fifth Goddard Memorial Symposium, Washington D.C. (Mar. 1967).

16. Turner, M. J. L. *Expedition Mars.* Springer-Praxis, Chichester, UK (2004).

17. von Braun, W., Ley, W., and Bonestell, C. *The Exploration of Mars.* Sidgwick & Jackson, London (1956).

18. Richardson, R., and Bonestell, C. *Mars.* Allen & Unwin, London (1965).

19. Ross, M. "Now for Mars. US Plans Voyage to Planet by 1986." *Daily Express*, 24 Sep. 1969.

20. Parkinson, R. C. *Citizens of the Sky.* 2100 Ltd., Stotfold, Herts., UK (1987).

21. *America at the Threshold, Report of the Synthesis Group on America's Space Exploration Initiative.* U. S. Govt. Printing Office, Washington D. C. (1991).

22. Singer, S. F. "The PhD Project in Perspective." In: McKay, C.P. (ed.). *The Case for Mars II.* American Astronautical Society, Science and Technology Series, 62, 221–1224 (1985).

23. O'Leary, B. "Phobos and Deimos as Resource and Exploration Centres." *Ibid*, 225–244.

24. Lunan, D. Lecture, 'Are the Russians Going to Mars?' ASTRA, Strathclyde University Union, 10 Nov. 1979, "To Mars – Will There Be Life on Mars? Future Missions." *Spacereport*, ASTRA, 11, 3, 28–36, (Jan. 1997), Hansson, A. lecture, 'Soviet Mars Exploration'. ASTRA, Airdrie, 9 Mar. 1990.

25. "MARSEMI Final Presentation", Aerospatiale, TF/E No. 106710, Paris (2 Oct. 1992).

26. Atkinson, N. "Will Russia's Next Rocket Be Nuclear?" Universe Today, 28 Oct., 2009.

27. Clarke, A. C. *The Lost Worlds of 2001.* Sidgwick & Jackson, London (1972).

28. Dyson, G. *Project Orion, the Atomic Spaceship 1957–1965.* Allen Lane, The Penguin Press, London (2002).

29. Martin, A. R. (ed.). *Project Daedalus. Journal of the British Interplanetary Society* Supplement, 1978.

30. Anon. "Fusion Speeds Mars Flight." *Flight International*, 18 June 1988, anon. "Trip to Mars May Use Nuclear Fusion." *The Australian*, 31 May, 1988, anon. "Mars Trip Fuels Speculation." *Flight International*, 3 July 1988.

31. Zubrin, R. "The Magnetic Sail." *Analog*, May 1992, 58–75, in: Schmidt, S. and Zubrin, R. (eds.). *Islands in the Sky: Bold New Ideas for Colonising Space.* Wiley, New York (1996), Aron, J. "New Satellite Sail Is Propelled by Solar

Protons." *New Scientist*, 7 May 2013, Wang, B. "Electric Sail Components and Technical Details." nextbigthing.com, online, 1 June, 2013.

32. Lunan, D., and Ramsay, W. "Crossroads in Space", op cit.

33. Bücker, H., et al. "Biostack Experiment." In: Lyndon B. Johnson Space Center. *Apollo 17 Preliminary Science Report*. NASA SP-330, U.S. Government Printing Office, 1973.

34. Linenger, J. M. *Off the Planet, Surviving Five Perilous Months aboard the Space Station Mir.* McGraw-Hill, New York (2000.)

35. Foale, C. *Waystation to the Stars, The Story of Mir, Michael and Me.* Headline, London (1999).

36. Tolins, E. "Mannequins Warn of the Perils in Manned Flights to Mars." *Daily Telegraph*, 26 Apr. 2007, Major, J. "Plastic Protection against Cosmic Rays." *Universe Today*, 12 June 2013.

37. Carpenter, S. Lecture, 'A View from Earth 1984' conference, op cit.

38. Isaacs, N. Lecture, 'Protein Structures, Hormones and Sunlight'. ASTRA, Airdrie Arts Centre, 31 May 1996, repeated, Glasgow Council for the Voluntary Sector, 26 Oct. 1996.

39. Johnson, R. D., and Holbrow, C. (eds.). *Space Settlements, a design study.* NASA SP-413, U. S. Government Printing Office, Washington D. C. (1977).

40. Iannotta, B. "Fantastic Voyager." *New Scientist*, 2202, 26–29 (4 Sep.1999).

41. Zeitlin, C., et al. "Measurements of Energetic Particle Radiation in Transit to Mars on the Mars Science Laboratory." *Science*, 340 (6136). 1080–1084. The key data are summarized with different emphases in Grossman, L. "Return trips to Mars pose unacceptable radiation risk." *New Scientist*, online, (30 May 2013), Anon. "Radiation Assessment Detector." Wikipedia, online, accessed 1 June 2013, Anon. "Round trip to Mars would push radiation safety limits." Ars Technica/Science, online, 1 June 2013, Chang, K. "Data Point to Radiation Risk for Travelers to Mars." *New York Times*, 30 May 2013, which includes the quotes from Zubrin and MacCallum.

42. Lovell, J. Lecture, 'A View from Earth 1984', op cit.

43. Conrad, P. Lecture, 'A View from Earth 1984', op cit.

44. Jones, C. *So Far from Home, A Story of Life and Death in Space.* Vintage, London (2008). See also Burroughs, B. *Dragonfly, NASA and the Crisis aboard Mir.* Fourth Estate, London (1999).

45. Lavin, P. Lectures, UK Students for the Exploration and Development of Space 'Moon on a Stick' conference, University of Newcastle, 10–12 Nov., 2001, 'Mars Direct Plus', ASTRA, Airdrie Arts Centre, 8 Mar. 2002, 'Manned versus Unmanned Spaceflight', ASTRA/Mars Society UK, Glasgow Council for the Voluntary Sector, 9 Mar. 2002.

46. Atkinson, N. "Hanging Out with Astronauts." Universe Today, 24 May 2013.

47. Palmer, J. "Gravity suit mimics Earth's pull for astronauts." BBC News, Science & Environment, 4 November 2010, citing James M. Waldie, J. M., and Newman, D. J. "A Gravity Loading Countermeasure Skinsuit." *Acta Astronautica*, 68, 722–730 (2011), online 16 Sep. 2010.

48. Haines, T., and Riley, C. *Space Odyssey*, op cit.

49. Ley, W., and Bonestell, C. *Beyond the Solar System.* Sidgwick & Jackson, London (1965).

50. Broad, W. J. "2 Soviet Astronauts Escape Death after Being Trapped Outside Mir." *New York Times*, 19 July, 1980.
51. Crossfield, S. *Always Another Dawn*. Hodder & Stoughton, London (1960).
52. Schirra, W. Lecture, A View from Earth 1984, op cit.
53. Clarke, A. C. "Which Way Is Up?" In: *The Challenge of the Spaceship*, Ballantine Books, New York (1961), "A Breath of Fresh Vacuum" (1966). In: *The View from Serendip*. Gollancz, London (1978).
54. Cook, R. "The Long Stern Chase: a Speculative Exercise." *Analog*, CVI, 7, 32–43 (July 1986).
55. Trafton, A. "One giant leap for space fashion: MIT team designs sleek, skintight spacesuit." MIT News, online, 16 July 2007.
56. Hsu, J. "Future Spacesuit to Imitate Gravity on Long NASA Missions." InnovationNewsDaily, TechNewsDaily, online, 30 Aug. 2011.
57. Atkinson, N. "Lunar Dust Transport Still a Mystery." *Universe Today*, 15 Dec. 2010.
58. Anon. "Transient Lunar Phenomena." *Spaceflight*, 14, 9, 353 (Sep. 1972).
59. Howell, E. "'Avalanche' Risk Higher Than Thought For Asteroid Landings: Study." Universe Today, 5 July 2013.
60. Atkinson, N. "Underwater Asteroid Mission Ends Early." *Universe Today*, 26 Oct. 2011.
61. Anon. "Aquanauts' Complete Mock Asteroid Mission on Ocean Floor." World News International, online, 22 June 2012.
62. Turnill, R. *The Moonlandings, an Eyewitness Account*. Cambridge University Press, Cambridge, 2003.
63. Jordan, S. *Jeff Hawke*, 'The Bees on Daedalus'. *Daily Express*, 19 July 1971–7 Oct. 1971. In: *Jeff Hawke's Cosmos*, 6, 3, 20–34 (Apr. 2011); Lunan, D. "Hawke's Notes, 'The Bees on Daedalus'." *Ibid.*, 34–37.
64. Atkinson, N. "NASA to BEAM Up Inflatable Space Station Module." Universe Today, 16 Jan. 2013.
65. Collins, M. *Carrying the Fire, an Astronaut's Autobiography*. W.H. Allen, London (1975).
66. Smith, A. *Moon Dust, In Search of the Men Who Fell to Earth*. Bloomsbury, 2005.
67. Meier, R. L. "Solar Energy, Space Technology, and Community Ecology, Second Draft." College of Environmental Design, University of California, Berkeley, Mar. 1980.
68. Atkinson, N. "Chris Hadfield Teams Up With Tested.com to Try Food and Games in Space." *Universe Today*, 22 May 2013.
69. Clarke, A. C. *The Sands of Mars*. Sidgwick & Jackson, London (1951).
70. Jordan, S. *Jeff Hawke*, 'The Helping Hand', story by Willie Patterson. *Daily Express*, 12 December 1964 to 3 March 1965. In: *Jeff Hawke's Cosmos*, 4, 2, 3–17 (Dec. 2007).
71. Wheeler, M. "Football Scores Big in Zero-Gravity Experiment." *The West Australian*, 31 May 2013.
72. Illustrated in Whitfield, S. E., and Roddenberry, G. *The Making of Star Trek*. Ballantine Books, New York (1968).

Chapter Seven: Final Options

1. Clancy, T. *Submarine*. HarperCollins, London (1993).
2. Lunan, D. *Man and the Planets*, op cit.
3. Holloway, N. Personal communication, op cit.
4. Clarke, A. C. *The Hammer of God*, op cit.
5. Baxter, J. *The Hermes Fall*, op cit.
6. Lunan, D. "Project Starseed: an integrated program for nuclear waste disposal and space solar energy." *Journal of the British Interplanetary Society*, 36, 426–432 (Sept. 1983).
7. News Release. "California Scientists Propose System to Vaporize Asteroids that Threaten Earth." University of California at Santa Barbara, Santa Barbara, CA, 14 Feb. 2013.
8. Niven, L., and Pournelle, J. *Lucifer's Hammer*, op cit.
9. Turner, M. J. L. *Expedition Mars*, op cit.
10. Ball, Sir R. *The Earth's Beginning*. Second edition, Cassell, London (1909).
11. Anon. "AVOID ash detection technology enters final test stage." *Advance*, ADS, online, 13 May 2013.
12. Hoyle, F. *Ice*. Hutchinson, London (1981).
13. McBride, N., and Gilmour, I. (eds). *An Introduction to the Solar System*. The Open University, Cambridge University Press, Cambridge (2007).
14. See for example Ryan, P., and Pesek, L. *Solar System*. Penguin Books, London (1978), Moore, P. and Hunt, G. *The Atlas of the Solar System*. Royal Astronomical Society, Mitchell Beazley, London (1983).
15. Hartmann, W. K. "The Smaller Bodies of the Solar System." *Scientific American*, 233, 3, 142–159 (Sep. 1975), Hindley, K. "The Debris." In: Stubbs, P. (ed.). *New Science in the Solar System, a New Scientist special review*. IPC Magazines, London (1975).
16. Shoemaker, E. "The Collision of Solid Bodies." In: Beatty, J.K., O'Leary, B. and Chaikin, C. (eds.). *The New Solar System*. Cambridge University Press, Cambridge (1981).
17. Clarke, A. C. "The Forgotten Enemy." In: Clarke, A.C. *Reach for Tomorrow*. Ballantine, New York (1956).
18. Muller, R. *Nemesis, the Death Star*. Weidenfeld & Nicolson, New York (1988).
19. *The Eruption of Krakatoa and Subsequent Phenomena*, Report of the Krakatoa Committee of the Royal Society, London (1888).
20. Steel, D. *Rogue Asteroids and Doomsday Comets*, op cit.
21. Allaby, M., and Lovelock, J. *The Great Extinction, op cit*; Atkinson, N.: Satellite Watches Dust from Chelyabinsk Meteor Spread around the Northern Hemisphere. Universe Today, 14 Aug. 2013.
22. Enever, J.E. "Giant Meteor Impact," op cit.
23. European Space Agency. "Herschel Links Jupiter Water to Comet Impact." ESA, Herschel, online news release, 23 Apr. 2013.
24. Laurie, P. *Beneath the City Streets, A Private Inquiry into the Nuclear Preoccupations of Government*. Second edition, Penguin, London (1972).
25. Jago, L. *The Northern Lights*. Hamish Hamilton, London (2001).

Chapter Eight: The Starseeds Grow

1. McAllister, A. 'The Ownership of Extraterrestrial Resources'. ASTRA discussion meeting, Glasgow Council for the Voluntary Sector (11 Dec. 2006).
2. Shrunk, D., Sharpe, S., Cooper, B., and Thangalevu, M. The Moon, Resources, Future Development and Colonization, op cit.
3. Pournelle, J. "Those Pesky Belters and their Torchships." In: Pournelle, J. A Step Father Out, op cit.
4. Whitehouse, D. The Moon. a Biography. Headline, London (2001).
5. Palmer, J. "Moon's Interior Water Casts Doubt on Formation Theory." BBC News, Science and Environment, 26 May 2011.
6. Major, J. "Earth-Passing Asteroid is 'An Entirely New Beast'." Universe Today, 17 June 2013.
7. Bilderdijk, W. A Short Account of a Remarkable Aerial Voyage and Discovery of a New Planet. (Trans.) Wilfion Books/UNESCO, Paisley, Scotland (1987).
8. Verne, J. From Earth to the Moon, All Around the Moon. (Trans.) Dover, New York (1960s, undated).
9. Moore, P. On the Moon. Cassell, London (2001).
10. Bracewell, R. N. "Communications from Superior Galactic Communities." Nature, 186, 670 (1960). In: Cameron, A.G.W. (ed.). Interstellar Communication. Benjamin, New York (1963).
11. Strong, J. Flight to the Stars. Temple Press, London (1965).
12. Sassoon, G. "A Correlation of Long-Delay Radio Echoes and the Moon's Orbit." Spaceflight, 16, 7, 258–265 (July 1974).
13. Lunan, D. "Space Probe from Epsilon Boötis." Spaceflight, 15, 4, 122–131 (Apr. 1973), Man and the Stars, op cit.
14. Lunan, D. "Long-Delayed Echoes and the Extraterrestrial Hypothesis." Journal of the Society of Electronic and Radio Technicians, 10, 8, 180–182 (Sep. 1976).
15. Lunan, D. Children from the Sky, op cit.
16. Lawton, A. T., Newton, S.J. "Long Delayed Echoes: the Search for a Solution." Spaceflight, 16, 5, 181–187 (May 1974).
17. Freitas, R. A., Valdes, F. "A Search for Natural or Artificial Objects Located at Earth-Moon Libration Points." Icarus, 42, 442–447 (1980), "A Search for Objects Near Earth-Moon Libration Points." Icarus, 53, 453–457 (1983).
18. Southall, I. Woomera. Angus & Robertson, Sydney (1962), Hill, C.N. A Vertical Empire, The History of the UK Rocket and Space Program, 1950–1971. Imperial College Press, London, 1971, Black Knight, Britain's First Ballistic Rocket. British Rocketry Oral History Program, 2007.
19. The Times. "How US Searched for Film that Fell to Earth." The Australian, 27 Dec. 2012.
20. Taylor, J. W. R., and Allward, M. Satellites and Space Travel. Ian Allan ABC, Hampton Court, Surrey, UK (1960), Gatland, K. Spacecraft and Boosters. Iliffe, London (1964).
21. Anon. "Putting the Planets in Their Place." ASTRA Far Future archive, source unknown.
22. Shepherd, L. R. "Interstellar Flight." In: Carter, L.J. (ed.). Realities of Space Travel, op cit; Yunge-Bateman, Cmdr. J., RN. "Across the Frontier of Space:

Landing on the Asteroid", "Design for an Asteroid." *Rocket, The First Space-Age Weekly*, 21 July & 28 July, 1956.

23. Chilton, C. *The World in Peril*. Herbert Jenkins, London (1960).

24. McInnes, C. Lecture, 'NEOs, Hazards and Opportunities,' op cit.

25. von Braun, W., and Ley, W. *The Exploration of Mars*, op cit.

26. Phillips, T. "Mystery Object Orbits Earth." Science@NASA, online, 20 Sep. 2002.

27. Anon. "The Fate of Satellites of the Moon." *Journal of the British Interplanetary Society*, 27, 5, 385 (May 1974).

28. McInnes, C. Personal communication, 10 Oct. 2012.

29. Yarnoz, D. G., Cuartielles, J-P. S., and McInnes, C. "Easily Retrievable Objects among the NEO Population." *Celestial Mechanics and Dynamical Astronomy*, August 2013.

30. Steel, D. "SETA and 1991 VG." *The Observatory*, 115, 78–83 (Apr. 1995), "Of Asteroids and Aliens." *The Skeptic*, 15, 1, 9–10 (1995).

31. Uhlig, R. "Chunk of Moon Rock Seen Orbiting the Sun", op cit.

32. Brin, D. "Just How Dangerous Is the Galaxy?" *Analog*, CV, 7, 80–94 (1985).

33. Lunan, D. *Man and the Planets*, op cit.

34. Cutler, A., et al. Panel discussion, 'Space Resources'. Fourth Annual Space Development Conference, The Shoreham Hotel, Washington D.C. (27 Apr. 1985); Kremer, K.: NASA Alters 1st Orion/SLS Flight – Bold Upgrade to Deep Space Asteroid Harbinger Planned. Universe Today, 9 July 2013; Atkinson, N.: A New Look at NASA's Capture Plan. Universe Today, 23 August, 2013.

35. Lunan, D. "Nuclear Waste Disposal in Space." *Journal of the British Interplanetary Society*, 36, 147–152 (Apr. 1983), "Project Starseed: an integrated program for nuclear waste disposal and space solar energy", *op cit*.

36. Lunan, D. "Project Starseed, or, Nuclear Waste Saves the World." *Analog*, CV, 2, 54–73 (Feb. 1985).

37. Lunan, D. "Project Starseed," revised version. *Settlers Sentinel*, Space Settlers Society (1987), "Project Starseed", (fourth version). *Asgard* 2, 3, 3–25, (Nov. 1991), "Project Starseed." *Asgard*, 5, 1, 47–53 (Dec. 2003).

38. O'Neill, G. K. *The High Frontier, Human Colonies in Space*, op cit.

39. Lee, Capt. C.M. Two-part presentation, 'STS Operations,' 'Space Transportation Systems Overview.' ASTRA/Third Eye Centre seminar, 'Nuclear Waste Disposal in Space,' Glasgow Film Theatre, 26 Sep. 1979.

40. Johnson, R. D., Holbrow, C. *Space Settlements, a design study*, op cit.

41. Anon. "Plan for a Sea City." *New Scientist*, 37, 586, 477 (29 Feb. 1968).

42. Sheppard, D. J. "Concrete Space Colonies." *Spaceflight*, 21, 1, 3–8 (Jan. 1979).

43. Lunan, D. "Nuclear Waste – Into Space?" *The Journal of Practical Applications in Space*, 1, 3, 47–51 (Spring 1990).

44. Benford, G. *Deep Time*. Harper-Collins, New York (2000).

45. Roberts, L. E. J. "Radioactive Waste – Policy and Perspectives", condensed version. United Kingdom Atomic Energy Authority, Apr. 1979.

46. Anon. "Products from Basalt." *Spaceflight*, 17, 2, 63 (Feb. 1975).

47. O'Leary, B. talk, 'The Case for Living Modules.' 'Human Communities in Space' program track, Fourth Annual Space Development Conference (see above).

48. Diamandis, P. Lecture, 'Algae-Based Closed Life-Support Systems,' Meinel, C., reply. Space Manufacturing Conference, Session III – Space Stations and Habitats, Princeton University, 8 May 1985.

49. Meier, R. L. "Solar Energy, Space Technology, and Community Ecology," op cit.

50. Martin, A. R. (ed.). *Project Daedalus, a design study*, op cit.

51. Anon. "Fusion Speeds Mars Flight," op cit, anon. "Trip to Mars May Use Nuclear Fusion," *op cit*, anon. "Mars Trip Fuels Speculation," op cit.

52. Anon. "Lunar Initiative Update." *Lunar & Planetary Information Bulletin*, 44, 11 (1986).

53. Lunan, D. "The Sky Above You: A Return to the Moon, Please." Various newspapers and magazines, Dec. 1986.

54. *America at the Threshold, Report of the Synthesis Group on America's Space Exploration Initiative*, op cit.

55. Anon. "Noble gases hitch a ride on hydrous minerals." Brown University Space & Earth/Earth Sciences News Release, 16 June, 2013, citing Jackson, C.R.M. et al. Noble gas transport into the mantle facilitated by high solubility in amphibole. *Nature Geoscience*, online, 16 June 2013.

56. Chu, J. "Battered Asteroid May Have Warm Core." MIT News Office, 28 Oct. 2011, Plotner, T. "Asteroid Lutetia May Have a Molten Core." *Universe Today*, 29 Oct. 2011.

57. Matloff, G. L. Lecture, 'Utilisation of O'Neill's Model 1 Lagrange Point Colony as an Interstellar Ark'. British Interplanetary Society Interstellar Conference, Imperial College, London, 31 Mar. 1975. In: *Journal of the British Interplanetary Society*, 29, 775–785 (Dec. 1976).

58. Lunan, D., and Dick, G. "Flight in Non-terrestrial Atmospheres, or the Hang-Glider's Guide to the Galaxy." *Analog*, CXIII, 1 & 2, 56–77 (Jan. 1993), Lunan, D. "Hawke's Notes: Chalk Circle." *Jeff Hawke's Cosmos*, 8, 1, 26–29 (July 2013).

59. Haines, T., and Riley, C. *Space Odyssey*, op cit.

60. Clarke, A. C. *Interplanetary Flight*. Temple Press, London (1950).

61. Ehricke, K. A. "A Strategic Approach to Interplanetary Flight," op cit.

62. Asimov, I. "Steppingstones to the Stars." In: *Asimov on Astronomy*, op cit.

63. Lunan, D. *Man and the Stars*, op cit. It can now be revealed that the anonymous astronomer in Chapter 1 of that book was Archie Roy, nervous of being associated with speculation on interstellar travel, before he 'came out' as a thriller-writer and founder of the Scottish Society for Psychical Research.

64. Finney, B. R., and Jones, E. M. (eds.). *Interstellar Migration and the Human Experience*. University of California Press, Berkeley and Los Angeles (1985).

65. Fogg, M. J. "Temporal Aspects of the Interaction among First Galactic Civilizations: the 'Interdict Hypothesis'." *Icarus*, 69, (1987).

66. Hewes, J. J. *Redwoods, the World's Largest Trees*. Hamlyn, London (1981).

67. Benford, G., and Brin, D. *Heart of the Comet*. Bantam, New York (1986).

68. Asimov, I. "The Universe and the Future." In Asimov, I. *Is Anyone There?* Ace, New York (1967).

69. Dyson, F. *Disturbing the Universe*. Harper & Row, New York (1979).

70. Ridpath, I. *Worlds Beyond*. Wildwood House, London (1975).

71. Jordan, S. "I Talk to the Trees." *Daily Record*, Glasgow, 5 Dec. 1983–24 Feb. 1984, coda 16 Mar. 1984. Story by Duncan Lunan.

72. Greig-Smith, P. "The Trees Bite Back." *New Scientist*, 7 May 1986.
73. Crichton, M. *Jurassic Park 2, The Lost World*, 'Fourth Configuration: The Red Queen'. Arrow, London (1997).
74. Highfield, R. "Biomechanics: If Bees Can Fly, Why Can't a Vine Dance?" *Daily Telegraph*, 2 Oct. 1991.
75. Anon. "Sense Is Common in Plants." *Daily Telegraph*, 9 Sep. 1994.
76. Bamford, M. "The Seed of a Difference." *The West Australian*, 8 July 2002.
77. Attenborough, D. The Secret Life of Plants. BBC-2, UK (1995).
78. Sands, N. "Tasmanian Plant May Be World's Oldest." *The West Australian*, 9 July 1997.
79. Monmaney, T. "Nature's Antennas", *Science 85*, 1985.
80. Clarke, A. C. *Imperial Earth*. Gollancz, London (1976).
81. Highfield, R. "Fame of the Giant Fungus Is Spreading." *Daily Telegraph*, 2 Apr. 1992.
82. Bailey, E. "The Thing that Ate Washington." *Daily Telegraph*, 4 July 1992.
83. 'R. H.' "A Fungi to Talk To." *Daily Telegraph*, 10 July 2002.
84. Moore, P. The Sky at Night. BBC-1, UK (6 Feb. 2006.)
85. Dyson, F. J. Letter, *Scientific American*, 210, 4 (Apr. 1964).
86. Kardashev, N. S. "Transmission of Information by Extraterrestrial Civilisations." In: Tovmasyan, G.M. (ed.). *Extraterrestrial Civilisations, Proceedings of the First All-Union Conference on Extraterrestrial Civilisations and Interstellar Communication, Byurakan, 20–23 May 1964*. Israel Program for Scientific Translations, Jerusalem (1967).
87. Dyson, F. J. "Search for Artificial Stellar Sources of Infra-red Radiation." In: Cameron, A.G.W. (ed.). *Interstellar Communication*, op cit.
88. Dyson, F. J. "The Search for Extraterrestrial Technology." In: Marshak, R.E. (ed.). *Perspectives in Modern Physics*. Interscience, New York (1966).
89. Lunan, D. "The Moon of the Thin Reality." *Galaxy*, 30, 3, 43–55, 152–154 (June 1970).
90. Clarke, A.C. *The City and the Stars*. Muller, London (1956).

Chapter Nine: Keep Watching the Skies

1. Hicks, C. Lecture. Third Charterhouse Conference on British Achievements in Space, op cit.
2. Crowther, R. Personal communication, 12 Oct. 2012.
3. Interim Report of the Action Team on Near-Earth Objects. United Nations General Assembly, Committee on the Peaceful Uses of Outer Space, Scientific and Technical Subcommittee, Forty-ninth session, Vienna (6–17 Feb. 2012).
4. Draft Recommendations of the Action Team on Near-Earth Objects for an international response to the near-Earth object impact threat. *Ibid,,* 15 Dec. 2011.
5. Amos, J. "Asteroids: When the Time Comes to Duck." BBC News, Spaceman Blog, 29 Oct. 2010.
6. Cowing, K. "International Agreement on Asteroid Threats Reached." United Nations Office for Outer Space Affairs, online (25 Feb. 2013).

Appendix 1: The Politics of Survival

1. Meadows, D. R., Meadows, D.L., Randers, J. and Behrens, W.W. III. *The Limits to Growth*. New American Library, New York (1972).
2. Anon. "Mitchell Prize Contest Guidelines, 'Alternatives to Growth.'" *New Internationalist*, 1972.
3. Ehricke, K. A. "A Strategic Approach to Interplanetary Flight," op cit.
4. Lunan, D. *Man and the Stars*, op cit.
5. Lunan, D. Man and the Planets, op cit.
6. Anon. "Monitor." *New Scientist*, 70, 996, 129 (15 Apr. 1976).
7. Caughlan, G. R. and Fowler, W.A. Mean Lifetimes and Equilibrium Abundances in the Fast CN Cycle. *Nature*, 238, 80, 22–24 (1972).
8. Anon. "Solar Neutrino Problem May Be a Remnant of the Ice Age." *New Scientist*, 71, 1015, 436 (1976).
9. Hughes, V. A., and Routledge, D. "An Expanding Ring of Interstellar Gas with Centre Close to the Sun." *Astronomical Journal*, 77, 210 (1973).
10. Anon. "Another Giant Meteor Crater Identified." *Sky & Telescope*, 50, 3, 156 (Sep. 1975).
11. Anon. "Earthquakes and Earth's Magnetism." *Journal of the British Interplanetary Society*, 27, 2, 150 (Feb. 1974).
12. Sagan, C. *The Cosmic Connection*, op cit.
13. Jacchia, L. G. "A Meteorite that Missed Earth." op cit.
14. Brownlee, S. "Cycles of Extinction." *Discover*, 5, 5, 22–32 (May 1984).
15. Lunan, D., and Ross, G. "Keep Watching the Skies!" op cit.
16. Lunan, D. "A Politics of Survival Option: an End to Warfare." *Asgard*, 4, 3, 8–12 (Aug. 2002).
17. Boyce, C. "Space Colonies and Independence." *Asgard*, 1, 2, 2–8 (Oct. 1977), reprinted *Asgard*, 4, 4, 6–11 (Nov. 2002).
18. Lunan, D. "Shelfspace: *The Moon*, by Bonnie Cooper et al." *Asgard*, 4, 1, 14–17 (July 2001), (abridged), *Space Policy*, 17, 4, 303–305 (Nov. 2001).
19. O'Neill, G. K. *The High Frontier*, op cit.
20. Lunan, D. "Project Starseed. an integrated program for nuclear waste disposal and space solar energy", op cit, "Project Starseed, or, Nuclear Waste Saves the World", op cit, "Project Starseed, revised version", op cit, "Project Starseed, (fourth version)", op cit, "Project Starseed", op cit.
21. Dyson, F. J. "Human Consequences of the Exploration of Space." In: Rabinowitch, E. and Lewis, R.S. (eds.). *Men in Space*. Medical & Technical Publishing Co., Aylesbury, UK (1970).
22. Tierney, J. "Drought in Africa, the Bigger Picture." *Science*, 85, 6, 3, 14 (Apr. 1985).
23. Lunan, D. "Terraforming Earth, the Sahara Spearhead." *Asgard*, 4, 3, 21–27 (June 2002).
24. von Puttkamer, J. "The Next 25 Years; Industrialisation of Space: Rationale for Planning." *L5 News*, 15, 1–7 (Nov. 1976).
25. Sampson, A. (ed.). *North–South: a Program for Survival*. Pan Books, London (1980).

26. Dooling, D. "Outlook for Space." *Spaceflight*, 18, 422–425 (1976).
27. Vajk, J. P. *Doomsday Has Been Cancelled*. Peace Press, Culver City, CA (1978).
28. Lunan, D. "The Rôle of Nonweiler Waverider Spacecraft in Exploring and Developing the Solar System." L5 Society (Western Europe) Conference, 1977, (abridged) *Asgard*, 1, 2, 14–22 (1978), (in full) *Spacereport*, 3, 2, 1–19 (Dec. 1980), (abridged) *Journal of the British Interplanetary Society*, 35, 1, 45–47 (Jan. 1982), (in full, updated), *ISTRA Journal*, 1982. Many subsequent articles on Waverider, most recently. "Waverider Entry Spacecraft, a History." In: Sales, I. (ed.). *Rocket Science, op cit*.

Index

D. Lunan, *Incoming Asteroid!: What Could We Do About It?*,
Astronomers' Universe, DOI 10.1007/978-1-4614-8749-4,
© Springer Science+Business Media, LLC 2014

The manufacturer's authorised representative in the EU is Springer
Nature Customer Service Centre GmbH, Europaplatz 3, 69115 Heidelberg,
Germany. If you have any concerns regarding our products, please
contact ProductSafety@springernature.com

Printed and bound by CPI Group (UK) Ltd, Croydon, CR0 4YY
19/05/2026
02113605-0003